Assembling the Dinosaur

Assembling the Dinosaur

FOSSIL HUNTERS, TYCOONS, AND THE MAKING OF A SPECTACLE

LUKAS RIEPPEL

HARVARD UNIVERSITY PRESS

Cambridge, Massachusetts, and London, England

2019

Printed in the United States of America

First printing

Library of Congress Cataloging-in-Publication Data

Names: Rieppel, Lukas, author.

Title: Assembling the dinosaur : fossil hunters, tycoons, and the making of a spectacle / Lukas Rieppel.

Description: Cambridge, Massachusetts : Harvard University Press, 2019. | Includes bibliographical references and index.

Identifiers: LCCN 2018053315 | ISBN 9780674737587 (alk. paper)

Subjects: LCSH: Carnegie, Andrew, 1835–1919. | Fossils—Collection and preservation—United States—History. | Dinosaurs in popular culture—United States—History. | Science museums—Public relations—United States—History.

Classification: LCC QE718 .R54 2019 | DDC 560.75—dc23

LC record available at https://lccn.loc.gov/2018053315

For Christine

Contents

Assembling the Dinosaur

Introduction

The dinosaur is a chimera. Some parts of this complex assemblage are the result of biological evolution. But others are products of human ingenuity, constructed by artists, scientists, and technicians in a laborious process that stretches from the dig site to the naturalist's study and the museum's preparation lab. The mounted skeletons that have become such a staple of natural history museums most closely resemble mixed media sculptures, having been cobbled together from a large number of disparate elements that include plaster, steel, and paint, in addition to fossilized bone. For example, the adult *Barosaurus* that stands in the Theodore Roosevelt Rotunda at the American Museum of Natural History in New York City is entirely made out of casts. The specimen is shown rearing up on its hind legs to protect its young from the ferocious predator *Allosaurus,* ominously circling nearby. Original fossils would have been far too heavy, fragile, and valuable to make such a lively and vivid display. Another example is the famous *T. rex* skeleton named Sue, which stands in Chicago's Field Museum of Natural History. Although this exhibit is primarily composed of real fossil bones, the articulation of Sue's skeleton was informed by recent and still controversial theories about the anatomy and behavior of a long-extinct creature that no human being has seen in the flesh. At a minimum, all mounted specimens require an armature of some kind to hold the pieces in place, and they usually incorporate a great many other sculptural elements, too. When standing before one of these towering creatures, it is surprisingly difficult to distinguish which features are ancient and which ones are modern, where prehistory ends and imagination begins.[1]

If dinosaurs in museums are chimeras, their prehistoric antecedents are unobservable entities. In this respect, dinosaurs resemble subatomic particles like electrons, neutrons, and positrons. Both are inaccessible to direct observation, but for different reasons. Whereas subatomic particles are too small to be seen, dinosaurs are too old. And in both cases, scientists gain access to their objects of study by interpreting the effects they produce: electrons leave characteristic marks on a photographic emulsion as they pass through a cloud chamber, and dinosaurs supply us with clues to their former existence in the form of fossilized bones. But dinosaurs are unlike electrons in a number of important ways. For one thing, dinosaurs cannot be experimented upon. Instead, scientists have to interpret the fossil record, which is spotty at best. The first dinosaur discoveries consisted of only a few bones and a handful of teeth. Before long, more complete skeletons began to be found, but the individual pieces were usually scattered about in a jumbled mess of material. Often, they had also been crushed and distorted by the immense pressures at work during and after the process of fossilization. For that reason, paleontologists had to work hard to assemble dinosaurs into something that resembled real, live animals. In doing so, they relied not only on the available evidence, but also on inference, judgment, and their imagination.[2]

Because dinosaurs are in part creatures of the imagination, they reveal a great deal about the time and place in which they were found, studied, and put on display. Often, paleontologists tasked with reconstructing the fragmentary remains of these animals have been guided in their pursuits by analogies to more familiar objects and circumstances. In the mid-nineteenth century, the British anatomist Richard Owen modeled dinosaurs on pachyderms such as the elephant, whereas early American paleontologists looked to the kangaroo as an anatomical guide. It was not until the turn of the twentieth century that dinosaurs came to be seen as massive, hulking, and lumbering behemoths of prehistory. More recently, many museums have completely overhauled their aging dinosaur displays yet again, to better reflect contemporary views of these creatures as bird-like, active, and fast-moving, with complex social structures. Dinosaurs simultaneously occupy two widely divergent temporal regimes: they hail from a world in which humans did not exist, yet they are also a product of human history.

Dinosaurs tell us a great deal about ourselves. Their immense size and outlandish appearance all but ensured that dinosaurs would become a mass public spectacle. But the scarcity of their fragmentary remains and the vast temporal chasm that separates their world from ours meant that it was difficult to know much about these creatures with certainty. The mystery of what life might have been like during the depths of time allowed people to project their fears and anxieties, as well as their hopes and fantasies, onto these alien creatures. Taken together, these features helped to make dinosaurs into a favorite target for the philanthropic largesse of wealthy elites, which ensured that there would be plenty of resources devoted to the science of vertebrate paleontology. For all these reasons, dinosaurs allow the historian to assemble a remarkably vivid and rich snapshot of American culture during the late nineteenth and early twentieth centuries, the period in which they rose to immense prominence among scientists and the broader public alike.

Although no shortage of ink has been spilled on dinosaurs, they remain a fresh source of historical insight.[3] Indeed, the wealth of existing scholarship makes dinosaurs an especially rewarding topic of study, in that it invites a broader discussion that integrates their status as an object of technical knowledge with social, cultural, and economic history. That is the aim of this book. Specifically, I bring thematic and methodological insights from recent work on the history of capitalism to bear on the history of science, and vice versa.[4] To that end, this book follows dinosaurs as they circulated across social, geographic, and institutional space, in hopes of revealing some of the ways in which science and capitalism were mutually entangled in the late nineteenth- and early twentieth-century United States.[5]

The time period I focus on covers roughly five decades, from the end of Reconstruction to the start of the Great Depression, a period I call the Long Gilded Age. This was a tumultuous chapter in US history, and there is no doubt that a great many aspects of American culture and society were transformed in these decades. But there was a coherence to the period as well. During this time, financial elites like J. P. Morgan and industrialists like Andrew Carnegie rose to enormous power and influence, and they oversaw the transition of the country's political economy from an unruly and highly competitive form of proprietary capitalism to a more managed economy dominated by large corporate firms. This was precisely when dinosaurs from the

American West became an icon of science, and the transition to corporate capitalism affected the practice of vertebrate paleontology in surprisingly concrete and far-reaching ways. Not only did dinosaurs reflect the obsession with all things big and powerful that prevailed at the time, but the science of paleontology itself was profoundly influenced by the creation of large, corporately organized, and bureaucratically managed museums of natural history.[6]

At first blush, dinosaurs may appear far removed from the political economy of modern capitalism. Unlike physics or chemistry, vertebrate paleontology did not yield practical insights that easily translated into lucrative investments or technological innovations. This is true despite the fact that paleontology was and remains closely tied to the mineral industry, which relied upon so-called index fossils to identify geological strata containing sought-after resources like coal. But dinosaurs were too rare to serve as a reliable guide for those hoping to strike it rich in the mines. While dinosaurs were strongly associated with mining, then, they did not play a direct role in America's booming extractive economy. It was precisely this complex interplay between distance and proximity—between the useful and the ornamental—that helped to make dinosaurs so popular among the period's financial elites, many of whom funded expeditions to discover fossil vertebrates and built museums to house them. While the science of paleontology did not offer a reliable source of technological innovation for industries to exploit, it was embedded within a more rarified but no less important prestige economy. Dinosaurs were prized for their ability to confer social legitimacy and cultural status, not to generate profits.

To untangle these connections, I follow dinosaurs from the point of discovery in the field to the museum, where their fossil remains were subjected to scientific investigation and assembled into imposing displays. From there, I escort them into a broader context by examining how they encountered a much larger and more popular audience in urban centers such as New York. This book therefore offers neither a traditional microhistory of a single object, event, or person, nor a broad, macrohistorical survey of the links between science and capitalism. Rather, I combine elements of both approaches, weaving stories about particular objects together with larger claims about the communities and institutions they encountered to provide insights about the practice of science in a deeply commercial culture.[7]

Focusing on the dinosaur's movement from the field to the museum and thence into more widespread circulation, I highlight circuits of exchange between science, commerce, and popular culture. In so doing, I join other historians who refuse to maintain a clear demarcation between the creation and the dissemination of knowledge—between scientific research and its popularization—to show that, far from being mere afterthoughts, communication and circulation are constitutive elements of the scientific enterprise. At the same time, this book also emphasizes acts of exclusion, the policing of boundaries, and the erection of hierarchies in science and commerce alike. Dinosaurs did not always travel easily, and museums patronized by wealthy philanthropists worked hard to control the precise contexts in which they circulated. Doing so was essential to shore up the dinosaur's status as a priceless object of reliable knowledge. Just as art galleries during this period erected a hierarchy between fine art and commercial decoration, natural history museums were keen to uphold the distinction between science and showmanship that modern historians have worked so hard to break down.[8]

The first dinosaur fossils were uncovered in England during the 1820s and 1830s, and they acquired the name Dinosauria from the British anatomist Sir Richard Owen in 1841. During the decades that followed, many additional fossils came to light, including an especially rich quarry found in a Belgian coal mine that contained dozens upon dozens of *Iguanodon* specimens. Nonetheless, the earliest dinosaurs did not stand out among all of the other large, impressive, and strange-looking creatures from prehistory that were being unearthed, which included extinct mammals, such as the *Megatherium*, and marine reptiles, such as ichthyosaurs and plesiosaurs. This suddenly changed during the last third of the nineteenth century, with a series of new discoveries in the American West that elicited enormous excitement. American dinosaurs were a scientific and popular sensation, especially once their fossil remains were mounted as free-standing skeletons in urban museums at the turn of the twentieth century. In part, this was due to the fossils themselves. American dinosaurs struck many observers as bigger and more imposing than their European counterparts. But the United States also proved an especially receptive environment for these creatures, a fertile niche that promoted their development into the towering behemoths that continue to wow museum visitors.

At precisely the same time that dinosaur bones became a public sensation, the United States was transforming into an industrial powerhouse of global proportions. Between the end of the Civil War and the beginning of the First World War, the country's economic output grew to exceed that of England, France, and Germany combined.[9] This was due, in no small part, to the development of a robust extractive economy.[10] As a result, the Rocky Mountain region, where many of the continent's rich mineral resources were concentrated, came to be seen as a land of almost unlimited possibility, and white settlers looking to profit from its abundant resources quickly colonized the region. Simultaneously, more and more people were moving to cities like New York, Philadelphia, Pittsburgh, and Chicago. This included a growing class of wealthy merchants, bankers, and entrepreneurs who bankrolled the process of industrialization. The railroad linked these two worlds together, tying the city and the countryside into an increasingly dense network of supply and demand. Resources flowed in one direction and capital in the other, with a good many people siphoning off a sizeable profit along the way.[11]

Because they were so prodigious in size, dinosaurs came to stand in for the power and fecundity of the United States. In a striking coincidence, three major dinosaur quarries were simultaneously discovered in the American West during a single field season in the summer of 1877. They contained some of the most recognizable fossils, including *Stegosaurus*, *Brontosaurus*, and *Allosaurus*, a close relative of *T. rex*.[12] Subsequent decades brought more discoveries to light, catapulting the United States into position as a world center for vertebrate paleontology. This was still a relatively new science at the time, but the wealth of amazing specimens being unearthed quickly became a scientific and popular sensation. For a people still emerging from the shadow of a bloody civil war, this was a welcome development, and the industrial elite were quick to embrace dinosaurs as their nation's most iconic extinct creatures. As a result, dinosaurs came to symbolize the country's economic might and power, offering material proof of its exceptional history and outstanding promise.

With the best specimens hailing from the country's interior, dinosaurs became associated its celebrated western frontier. Their discovery was deeply embedded within the extractive economy that dominated the region at this time. In part because the exploitation of mineral resources in precisely this part of the country was so instrumental in propelling the United States into the position of an economic superpower, dinosaurs from the American West were elevated into a symbol for the entire political economy. Widely heralded as having been larger, fiercer, and more abundant than prehistoric animals from Europe, they meshed well with a conventional narrative that celebrated American exceptionalism. Their origin in the deep past also ensured that dinosaurs would be associated with evolutionary theory, which was often invoked to explain social, cultural, and economic developments. But dinosaurs did not function as a straightforward image of progress. The mass extinction event that killed them off at the end of the Cretaceous period mirrored the era's widespread anxieties about degeneration and decline, and dinosaurs were often inserted into a cyclical narrative that characterized evolutionary development as a predictable series of fits and starts. The same evolutionary process was understood, in turn, to result in a familiar pattern of boom and bust that mirrored the emerging conception of what came to be called the business cycle.

The link between dinosaurs and American capitalism was material as well as symbolic. The rapid process of industrialization created riches that had been almost unimaginable just decades before. But the period's wealth and prosperity was not equally shared among all parts of society. During the late nineteenth century, a small group of financial and industrial capitalists coalesced into an elite social class that supplanted an older generation of merchant families. Because wealthy industrialists often hailed from fairly humble, artisan backgrounds, they signaled their newfound class status to themselves and each other using traditional markers of high social standing. In addition to wearing expensive clothes and adopting erudite modes of speech, they invested considerable resources in amassing impressive collections of artworks and natural history specimens. Whereas artworks largely functioned as a display of refined aesthetic sensibilities, natural history represented another form of social distinction, one that combined epistemic virtues like objectivity with notions of good stewardship and civic munificence. While industrial capitalists primarily coveted artworks from Europe, however, they generally agreed that Wyoming, Colorado, and Utah harbored the biggest, most complete, and most spectacular dinosaur fossils. Not coincidentally, this region was precisely where the raw materials consumed by their factories could also be found. Industrialists who had grown rich by exploiting America's natural resources thus turned to dinosaur paleontology as an efficient means of cultural resource extraction.[13]

Although the economy was booming, American capitalism was in a state of crisis during this period. The industrial juggernaut was responsible for unprecedented levels of economic growth, but it also produced frequent financial panics and economic depressions. Working people were especially hard hit during these downturns, and inequality rose sharply. This led to a widespread backlash against a system of economic production that seemed to yield almost equal measures of growth and precarity, of gratification and misery. The most visible response came in the form of frequent strikes and labor disputes, which could be remarkably violent and bloody. The Federal Bureau of Labor Statistics estimated that there were more than 36,000 strikes between 1881 and 1905, averaging four a day.[14] More telling still is that US National Guard soldiers were mobilized over three hundred times between 1877 and 1903 to take care of "labor troubles."[15] A sense of revolutionary uprising was in the air, leading to widespread moral panic among the social and financial elite, who feared that radical immigrants and incendiary labor leaders were spreading an anarchist message that could bring the industrial economy to its knees. Some even worried that a new civil war might be brewing over the issue of wage rather than slave labor. In response, the well-to-do literally armed themselves, forming militias and building ostentatious fortresses that doubled as clubhouses in cities across the United States.[16] At the same time, they also became avid philanthropists, founding organizations designed to uplift, edify, and educate working people by exposing them to the highest achievements of modern civilization. In the process, they created the nonprofit corporation. These institutions were designed to demonstrate that capitalism could be altruistic as well as competitive—that it worked for the good of all in society, not just the wealthy few. As Andrew Carnegie put it, philanthropy was "the true antidote for the temporary unequal distribution of wealth, the reconciliation of the rich and the poor," showing that "ties of brotherhood may still bind together the rich and poor in harmonious relationship."[17]

In addition to establishing universities, libraries, symphonies, and art galleries, wealthy capitalists such as Carnegie also founded natural history museums. Natural history doubled as both a popular leisure pursuit and a pious devotional exercise at the time, which made it an especially effective means for showing off one's generosity among a broad and socially diverse but respectable audience. Of all the branches of natural history, dinosaur paleontology offered a particularly attractive target for philanthropic investment. Dinosaurs lent themselves to the building of spectacular displays that attracted throngs of visitors to the museum, which was crucial to cement the argument that industrial capitalism could produce genuine public goods in addition to profits. Imposing dinosaur displays thus helped philanthropists such as Carnegie make the claim that because industrial capitalism concentrated wealth in the hands of the few, it unlocked the power for truly awesome achievements.

Philanthropists were also drawn to dinosaurs as a powerful tool to help naturalize the evolution of American capitalism. Before the Civil War, the business landscape of the United States was dominated by small, family-owned firms that specialized in a single product or service. But that changed dramatically during the last third of the nineteenth century, as sole proprietorships were increasingly replaced by large, capital-intensive, and often vertically integrated corporate firms. As these corporate behemoths gobbled up their competitors in a wave of mergers and acquisitions, some grew so large that they threatened to monopolize an entire industrial sector. During the height of its powers in 1905, for example, Rockefeller's Standard Oil Company controlled over 85 percent of the market for kerosene, while U.S. Steel had a market share of 66 percent at the time of its founding in 1901.[18] This restructuring of America's political economy elicited enormous controversy, especially among rural populations who found themselves on the receiving end of the bureaucratic machine. Complaints about the anticompetitive and corrupt business tactics of financial and industrial conglomerates grew so vociferous that, in 1903, President Theodore Roosevelt created a federal Bureau of Corporations to investigate industrial malfeasance, and the US Congress called upon "robber barons" like Andrew Carnegie, J. P. Morgan, and J. D. Rockefeller to defend themselves before a skeptical public.[19]

Wealthy elites responded by framing the transition to a political economy ruled by vast corporations as an example of evolutionary progress, celebrating the capacity of rational administration and organized planning to replace what they characterized as wasteful and "ruinous" competition between small, independent firms. Dinosaurs offered an especially powerful means to make that claim convincing. Paleontologists consistently portrayed these animals as vicious and solitary predators whose terrible reign had come to a sudden and ignominious end at the close of the Cretaceous period. But their mass extinction opened up the ecological space for a kinder and gentler world to emerge. According to this evolutionary narrative, the cutthroat competition of the deep past gave way to a more enlightened modernity, as intelligent mammals—including early hominids—put the struggle for existence behind them and began to cooperate for the greater good. The exhibition of dinosaurs in philanthropic museums thus helped to bolster the argument that the evolution of modern capitalism did not depend upon social conflict or lead to class warfare. On the contrary, it could be framed as a means to promote enlightened administration and organized teamwork over ruthless self-interest and incessant competition.

The history of dinosaur paleontology offers an instructive contrast to how the relationship between science and capitalism is often framed. Science has traditionally been portrayed as a higher calling, one that is insulated from the demands of the marketplace. This led early- to mid-twentieth-century historians and sociologists to stress the autonomy of science, emphasizing the extraordinary measures that researchers take to police the bounds of acceptable conduct, guard against misinformation, and prevent fraud. On this view, membership in the scientific community is governed by a set of normative expectations like objectivity and value-neutrality, as well as a commitment to share the results of one's work free of charge.[20] However, more recent developments have made these ideas seem hopelessly naive. In today's world of patented gene sequences, technology transfer offices, and Silicon Valley startups, it has become increasingly difficult to sustain the fiction that science is fundamentally divorced from the marketplace.[21] Instead, newer accounts are more likely to stress the extent to which powerful actors and institutions leverage their access to capital to shape the research priorities of the scientific community.[22] Rather than stress the autonomy of science, many historians now tend to examine how the boundary between science and capitalism has become blurred.[23] Nevertheless, our normative expectations have remained surprisingly stable, and many are troubled to learn that an important medical trial was bankrolled by the pharmaceutical industry, or that a study on climate change was funded by energy companies.[24]

Dinosaurs offer a very different perspective. It is precisely because dinosaurs were so deeply entangled with both science and capitalism that vertebrate paleontologists were especially careful to distance themselves from the world of commercial affairs. Dinosaurs rose to international prominence during a period when the economic elite in the United States had plenty of money but suffered from a deficit in legitimacy. In contrast, vertebrate paleontologists were engaged in a prestigious but expensive undertaking. Dinosaurs were not just immensely popular, they were also exceedingly difficult to find and collect. Access to a steady source of funds was therefore essential for anyone wishing to work on these remarkable creatures. As a result, paleontologists might be expected to have touted all the ways they could help further the interests of the wealthy donors who underwrote their endeavors. But in fact, exactly the opposite happened. Rather than broadcast their willingness to enter a quid pro quo agreement with philanthropists, paleontologists instead chose to guard the institutional autonomy of their discipline by insisting that funding be offered with no obvious strings attached. Ironically, this only made paleontologists more attractive to philanthropists, who were eager to distance themselves from their commercial and industrial roots. The two communities thus forged a strategic alliance with mutual benefits. Paleontologists acquired a steady stream of funding, while wealthy capitalists could claim to be engaged in a genuinely altruistic endeavor. And what better way to support this claim than by investing in a lost world that had entirely disappeared before human beings and the industrial economy had even come into existence?

Dinosaurs thus help to reveal the subtle complexity of American capitalism, which functioned as far more than a regime of profit maximization designed to satisfy narrow self-interests. Historians typically view the political economy of modern capitalism as one in which all manner of social interactions have been reframed as market transactions. The emergence of modern capitalism, economic historians often argue, involved the creation of a distinct economic sphere wherein the price mechanism rules supreme, coordinating the allocation of goods and services by mediating between supply and demand.

According to this "market revolution" narrative, the United States became a capitalist country once everything—land, labor, and even time—began to be exchanged through the cash nexus. But even during the Long Gilded Age, a capitalist society if there ever was one, not all values were expressed in the terse language of dollars and cents. Indeed, it was wealthy capitalists themselves who were perhaps most anxious to maintain clear boundaries between the generation of wealth and the circulation of more noble cultural goods, including scientific knowledge. Just as paleontologists worked hard to maintain their autonomy and protect the purity of the sciences, then, so too did philanthropists feel that it was necessary to cordon off some sectors of American culture from the realm of commercial exchange. In both cases, upholding rather than blurring the boundary between science and capitalism was deemed an essential means to make the entanglements between these cultural institutions productive for all parties involved.[25]

Finally, dinosaurs also lay bare the deep time horizons of modern capitalism, confounding a widespread belief that capitalist economies are especially future-oriented. Because investors and entrepreneurs train their attention on what lies ahead, a number of scholars have argued that a forward-looking cognitive disposition was required for the development of modern capitalism. In turn, this helped to create a new temporal order, one that is open-ended, directional, and linear rather than predetermined, static, or cyclical.[26] Of course, human beings have always wondered what tomorrow may bring. But it was only during the past several centuries that technical means to assign precise, quantitative values to the future were developed in tandem with new economic institutions like modern contract law that made such forecasts possible in the first place.[27] In a striking convergence, the very same time period also witnessed the invention of "deep time," and, with it, the emergence of new scientific fields such as archeology and paleontology that offered a way to use material traces like fossils or ancient monuments to begin writing a history of the earth itself.[28] Just as people were developing formal techniques to project themselves forward in time, they were beginning to use remnants of prehistory to project themselves back into the deep past.

Whereas the future emerged as a privileged space in which to locate pecuniary value, the past served as a lucrative site for the extraction of another value: legitimacy. In a sense, this too was not new. People have always looked to the past to explain, justify, and motivate their actions in the present. But once it was accepted that earth history resembled human history in being episodic and linear, it became possible to project the legitimating power of tradition beyond our species' temporal confines and into the furthest recesses of deep time. Nowhere was this more clear than with the excavation, study, and exhibition of fossil vertebrates. As dinosaurs were assembled into an icon of science during the Long Gilded Age, the political economy of American capitalism revealed that it valued the deep past as much as, if not more than, the next earnings report.

1

Prospecting for Dinosaurs

William Harlow Reed was walking home from a successful antelope hunt during the summer of 1877 when he stumbled across several large fossil bones weathering out of a hillside near Como Station in southeastern Wyoming. At the time, Reed was employed as a section foreman by the Union Pacific Railroad, which meant he was charged with maintaining about ten miles of track (Figure 1.1). Although the proud frontiersman would later deny it, he did not at first recognize the gigantic bones he had found to be those of a dinosaur. He did, however, realize they might be important. And before long, he got an inkling they might also be worth a considerable sum of money. Just how much money would turn out to be a matter of contention, however.[1]

A few weeks later, Reed shared his discovery with the stationmaster at Como, William E. Carlin, and the two men decided to sell the bones. They contacted the noted American paleontologist Othniel Charles Marsh, who had visited Como Station several years earlier to collect specimens of a curious salamander that locals referred to as the "fish with legs."[2] In their initial letter to Marsh, Carlin and Reed made it clear they were principally after money. "We are desirous of disposing of what fossils we have and also the secret of the others," they wrote, explaining, "We are working men and not able to present them as a gift, and if we can sell the secret of the fossil bed, and procure work in excavating others we would like to do so." Just in case that was not sufficiently clear, the two men reiterated the point toward the end of their letter, stating, "We would be pleased to hear from you, as you are well known as an

Figure 1.1. Portrait of William Harlow Reed as a young man, 1879.

enthusiastic geologist, and a man of means, both of which we are desirous of finding, more especially the latter." Tellingly, Carlin and Reed were deliberately circumspect about the precise location of their "secret" discovery, saying only that they had found "a large number of fossils" in an area "not far from this place." Divulging too much information was risky because Marsh might decide to avoid paying them by excavating the quarry himself. The cagey tone of their missive thus stemmed from an anxiety to protect their intellectual property and guard against others profiting from their discovery. To that end, Carlin and Reed used only their middle names—Harlow and Edwards—in their first correspondence, which they posted from Laramie rather than Como Station in hopes of further concealing the precise location of their valuable find.[3]

Carlin and Reed did not know it, but theirs was actually one of three separate dinosaur quarries independently discovered in the American West during the summer of 1877, none of which was found by a trained paleontologist. Rather, they were all located by people who were driven, in large part, by pecuniary motives. Besides Carlin and Reed, there was a clergyman and mining engineer from Morrison, Colorado, named Arthur Lakes. The third was a schoolteacher, Oramel Lucas, who found a rich deposit of fossil bones in Canon City, Colorado. Together, these three quarries would yield tons of valuable fossils belonging to dozens of new species, including some of the most famous dinosaurs of all time: *Brontosaurus* (now sometimes called *Apatosaurus*), *Stegosaurus,* and *Allosaurus,* a close relative of *T. rex.* The year 1877 revealed the American West to be a paleontologist's El Dorado, bringing it to the attention of the international scientific community and establishing the United States as a leader in dinosaur research, publication, and display.

In much the same way that the remarkable rise of the United States as an industrial powerhouse was partly a consequence of its rich storehouse of natural resources, the success of its paleontologists largely derived from their access to abundant fossil resources.[4] In the wake of the 1877 discoveries, it was quickly agreed that the largest, most spectacular, and most complete dinosaurs hailed from the United States. But the good fortune of paleontologists such as Marsh was neither an act of God nor a product of nature. Instead, the almost mythical status of the American West as a land of geological plenty grew out of the development of a booming extractive economy. Directly or indirectly, it was the mineral industry that brought people like Reed to the Rocky Mountains, and it was the hope of striking it rich that supplied an incentive for them to scour the landscape in search of a promising prospect. The rapid development of American paleontology and the remarkable expansion of its extractive economy thus went hand in hand, mutually informing, promoting, and reinforcing each other.

The emergence of the United States as a world center for dinosaur research was materially bound up with its bourgeoning mineral industry. But that does not mean transactions such as the one between Marsh and Reed always went smoothly. On the contrary, the exploitation of the region's abundant resources (fossil and otherwise) often brought with it a great deal of social friction. In the pages that follow, we will encounter one source of tension in particular. Because buyers and sellers tended to be separated by thousands of miles, and because they sought to exchange something that was buried deep underground, the information required to determine a discovery's value was extremely hard to come by. Before a specimen was disinterred, paleontologists could not know how complete it was, nor whether its bones were of high quality. This made them suspicious of prospectors like Reed, who had a financial incentive to misrepresent a discovery in hopes of inflating its value. To make matters worse, buyers and sellers usually did not know one another personally, which made it difficult for them to form relationships based upon trust. In the face of these difficulties, fossil collectors and paleontologists often looked to transactional practices from the mineral industry for cues on how to behave. Notably, these included negotiation tactics designed to exploit rather than overcome the deficit of trust between the two parties. The booming extractive economy of the American West therefore not only shaped the material context in which dinosaurs were discovered. It conditioned the social networks in which they circulated as well.

While the 1877 discoveries were an epoch-making event in the history of paleontology, these were not the first dinosaurs ever found. In fact, the first such discoveries took place in England some fifty years earlier. At that time, however, dinosaurs were seen to be far less noteworthy than other prehistoric creatures, especially marine reptiles such as the ichthyosaur and the plesiosaur. Nor was 1877 the first time that fossils were traded for cash. The purchase and sale of natural history specimens was a fairly routine practice going back to medieval crusades and Renaissance voyages of discovery, if not before. To illustrate why fossils from America's late nineteenth-century mining frontier were so distinctive, this chapter begins with a detour backward in time and across the Atlantic. From there, I argue that what made dinosaurs from the American West special was not only their immense size, teeming abundance, and relative completeness, although those features certainly mattered a great deal. It was also the manner in which they came to circulate through the scientific community. In particular, while long-distance circuits of exchange had long brought collectors into contact with naturalists, the traditional codes of civility that usually structured both commercial and epistemic exchanges in the scientific community did not obtain in the proverbial Wild West. Thus, negotiations about the 1877 discoveries were about far more than simply establishing a mutually satisfactory price. They also indexed the creation of a new market, one that was embedded within a specific cultural context and functioned according to a distinct economic logic.

On the mining frontier, dinosaurs were often treated like any other scarce resource one could dig out of the ground, such as gold, silver, or coal. This meant that in place of the shared codes of civility that traditionally structured scientific exchanges, the fossil trade was instead shaped by business practices that prevailed in the mineral industry.[5] In particular, whereas trust had long served as a preferred means to promote the free flow of information among members of the scientific community, the mineral industry was notoriously plagued by deception and fraud.[6] While trust was not entirely absent on the mining frontier, it functioned as far more than a condition of possibility for fruitful exchanges to take place. Rather, it came to be seen as a scarce resource in its own right, and it was treated as such by self-interested agents who sought to exploit its absence for strategic gain. In a world where trust was crucial but also in short supply, success came to hinge on the ability to overcome a competitor's privileged access to information while leveraging one's own. This proved to be just as true in vertebrate paleontology as it was in the mineral industry.[7]

Mary Anning's Fossil Depot

Paleontology has long been connected to mining, but that does not mean it was concerned with practical matters alone. As the hands-on knowledge required for success in the mines became a subject of learned contemplation during the seventeenth century, scientists grew more and more curious about the history of the earth. The questions they began asking, and the answers they elicited, represented a radically new way of thinking on at least two counts. First, it became clear the earth is immensely old, far older than the four thousand or so years that biblical scholars had often supposed. And second, perhaps even more significant was an increasingly widespread agreement that the planet's history was episodic, contingent, and linear, much like humankind's. In fact, the historian Martin Rudwick contends, the new sciences of geology and paleontology came into being by transposing the basic narrative structure of human history onto the deep past, or, as he puts it, "from culture into nature."[8]

In much the same way that archeologists began to use coins, artwork, and monuments to supplement the textual record of human history, early geologists learned to use fossils to create a kind of natural archive.[9] In so doing, they reconstructed a lost world that was far stranger and more alien than had previously been imagined. Most surprising, perhaps, was the veritable menagerie of new plants and animals being unearthed, some of which seemed to resemble the mythical creatures of ancient fables. These developments took place at a time of European colonial expansion, and it is no accident that many of the most spectacular fossils were disinterred in the New World. For example, a giant pachyderm that resembled the Siberian mammoth was initially uncovered by a French military expedition into the Ohio River valley in 1739. Not long thereafter, in 1787, an extraordinarily large and strange-looking creature—the *Megatherium*—was dug up near Buenos Aires, which caused an international sensation when Juan Bautista Bru mounted it as a freestanding display in the Royal Cabinet of Natural History in Madrid. The rapid and often violent expansion of European colonial states overseas thus fed directly into the development of vertebrate paleontology.[10]

Paleontological theories were bound up with the period's immense social and political upheavals as well. Perhaps chief among these was the idea that some organisms might have disappeared from the earth long ago, which was most ardently championed by the savant Georges Cuvier. Cuvier was not the first to invoke the concept of extinction to explain disparities between living and fossil forms. But previously, it had always seemed possible that the strange-looking creatures preserved as fossils might still exist in some hitherto unexplored parts of the globe. When Thomas Jefferson enlisted Meriwether Lewis and William Clark to explore a large territory that he purchased from France in 1803, for example, Jefferson explicitly asked them to keep an eye out for live mastodon roaming the continent's interior. Almost immediately after taking up a position at the prestigious *Muséum d'histoire naturelle* in 1795, however, Cuvier used the vast collection of extant and fossil specimens now at his fingertips—some of which had been recently plundered by Napoleon's troops—to dismiss these concerns out of hand, convincing his peers that animals like the mammoth and mastodon really had gone extinct. To back up this bold pronouncement, Cuvier teamed up with Alexandre Brongniart, a mineralogist who directed a porcelain factory just outside Paris, to assemble a remarkably complete collection of fossils from a single locality. By carefully mapping the occurrence of various specimens in different layers of rock, Cuvier and Brongniart ascertained that the deeper one dug down into a geological formation, the more unrecognizable its fossils became. This allowed Cuvier to determine the relative sequence in which new kinds of organisms appeared during the earth's history.[11]

Cuvier's discoveries were of a piece with a broader transformation in the way many eighteenth-century Europeans understood their place in the long arc of history. As is often remarked, the development of modern capitalism was made possible by the creation of a new temporal regime, one that conceived of the future as open-ended and full of almost limitless possibility: a lucrative space wherein profits are found. In much same way that so-called projectors imagined themselves into the future, geologists such as Cuvier extended the temporal horizon of history in the other direction, casting themselves into the deep past. In so doing, they raised the question of what could have happened to this "primitive earth," whose life-forms were so different from our own. Drawing an implicit connection to the political uprising he had witnessed in France, Cuvier deployed highly charged language to explain the remarkable disparities between living and extinct organisms. Nature "also had its civil wars," he remarked, and "the surface of the globe has been upset by successive revolutions and various catastrophes." Cuvier imagined these "revolutions" as massive geological upheavals, powerful enough to overturn the world's flora and fauna completely. Because such geological revolutions were rare, the fossil record tended to exhibit long periods of stasis and continuity, followed by sudden breaks in which most plants and animals disappeared completely, only to be replaced by new ones. This led Cuvier to divide the earth's history into several distinct stages, each dominated by a different group of organisms. Because it was ruled over by especially strange-looking creatures, the Age of Reptiles (now called the Mesozoic era) would prove to be particularly captivating.[12]

Self-consciously styling himself a "new species of antiquarian" who could "decipher and restore" the creatures that once populated the globe, Cuvier began reconstructing the extinct life-forms that formerly inhabited our planet in vivid and lifelike detail. To do so, he articulated a sweeping conception of the animal body as an "organized" entity, "a unique and closed system, in which all the parts correspond mutually." In much the same way that financial speculators began to elaborate detailed plans of their future endeavors to convince potential investors their imagined projects were creditworthy, Cuvier boasted that a knowledgeable naturalist could "reconstruct the whole animal" from a single tooth, claw, or shoulder blade. Sometimes, he even went so far as to sketch fully articulated skeletons complete with muscles and skin. Such feats of the imagination inspired a whole generation of paleontologists, giving them tools to "burst through the limits of time," as Cuvier described it, and enter another world. As a result, paleontology quickly became a runaway public success, attracting a wide and diverse audience beyond the small coterie of learned naturalists. This was especially true in Great Britain, whose contributions to paleontology soon grew to rival those from the Continent.[13]

As the new science of paleontology captured the public's attention, the market for specimens also grew more robust. In England, a group of enterprising collectors began to unearth a series of ancient reptiles that could command a remarkably high price. The most prolific among these fossilists, as they were often called, was Mary Anning, who possessed an uncanny ability to discover beautifully preserved specimens near her home in Lyme Regis, on England's southwestern coast. Born to a cabinetmaker in 1799, Anning began supplementing her family's modest income by scouring the exposed cliffs running alongside the English coastline for fossils, which she initially sold to wealthy tourists who flocked to the seaside during the summer months. When Anning's father died in 1810, her family expanded its natural history business to make ends meet, and she quickly distinguished herself as having an especially keen eye. Over the next several decades, Anning's many discoveries would earn her an international reputation, and by the age of twenty-seven she was able to open her own shop, which she named Anning's Fossil Depot.[14]

A particularly significant discovery Anning made with her brother Joseph between 1811 and 1812 became a sensation after having been purchased by

William Bullock, who operated a commercial museum in London. While naturalists at first could not make sense of the fossil, Charles König, a curator (or "keeper") from the British Museum, eventually identified it as a marine reptile, which he christened *Ichthyosaurus* (fish-lizard). The taxonomic confusion surrounding this creature only added to the specimen's appeal, and in 1819 the British Museum purchased it for forty-seven pounds and five shillings at auction, which was a considerable sum of money at the time.[15] In part due to the excitement over the ichthyosaur, the British naturalists William Conybeare and Henry De la Beche embarked on a quest to thoroughly reexamine the abundant fossils preserved in the rocks near Anning's home. In the course of their work, they came across fragmentary evidence of a particularly striking creature the two men suspected might constitute a missing link between ichthyosaurs and crocodiles. That suspicion was borne out in 1823, when Anning found a nearly complete specimen of the creature, which Conybeare and De la Beche christened *Plesiosaurus* (almost-lizard) and which was eventually purchased for one hundred pounds by the Duke of Buckingham.[16]

News of such prices spread quickly, inducing more people to try their luck digging for fossils. In some ways, this was a boon for paleontology, as it exposed many new specimens to light. But the growing influx of cash also brought with it an increased threat of deception and fraud. Because the price of a specimen depended on its novelty, its completeness, and how visually arresting it was, fossil hunters had an economic incentive to augment their discoveries in various ways. On rare occasions, they even combined different specimens into a larger and more impressive assemblage. Sometimes, they also used plaster to beautify what they had found. This posed a significant challenge to the viability and authority of paleontology, a relatively new science that was often accused of being overly speculative and more reliant on the scientific imagination than on material evidence.[17]

During the mid-1830s, the British Museum in London purchased an extensive collection of marine reptiles from Thomas Hawkins, an eccentric collector who would go on to write *The Book of the Great Sea-Dragons.*[18] To vouch for the quality of Hawkins's specimens, the respected paleontologists William Buckland and Gideon Mantell personally brokered the sale. After a careful examination, they valued the entire collection at £1,250 and assured the museum it was worth every penny. But when König unpacked Hawkins's fossils, he grew suspicious that several had been doctored by their original owner. Upon further inspection, König's worst fears were confirmed, especially with regard to a seemingly complete and astonishingly well-preserved ichthyosaur. Writing to inform the museum's trustees of this unsettling realization, he lamented that some fossils were "of much less value" than Buckland and Mantell had thought "on account of their being made up, and all over restored with plaster of Paris, and altogether unfit to be exhibited to the public, without derogation from the character of the British Museum." Eager to avoid controversy, the museum's trustees decided to let the matter drop, but not without first having the offending plaster painted a different color so that visitors could distinguish between real and fake bone. Writing to the American geologist Benjamin Silliman during the height of the scandal, Mantell accused Hawkins of outright "deception" for having "added many parts to his specimen" that gave an erroneous impression about its true quality.[19]

For the most part, however, such accusations of fraud were remarkably rare. This was largely due to the social structure of the natural history community at the time. In nineteenth-century Europe, a strict division of labor distinguished the contributions of learned naturalists from those of collectors. Whereas geologists and paleontologists were nearly all gentlemen who belonged to the social and economic elite, collectors usually hailed from a more humble, working-class background. In part for that reason, the latter were only expected to offer new specimens and interesting observations for more expert examination, whereas the former had the authority to place new discoveries within a broader philosophical framework.[20] But these distinctions did not prevent the two communities from forging a tight-knit social network built on mutual bonds of trust. In fact, the relationship between artisan collectors and gentlemen naturalists could be remarkably intimate. Many collectors treasured the reciprocity they could achieve through the exchange of valuable specimens with gentlemen naturalists, and they coveted the social legitimacy that came with acknowledged participation in a shared intellectual enterprise.[21] Here too, Anning serves as an excellent case in point. Not only did she receive international renown for her remarkable skill and intelligence—Mantell, for example, described her as a "Geological Lioness"—but the Prussian explorer Ludwig Leichhardt even flattered her as "the Princess of paleontology."[22]

The temptation of tampering with a fossil to augment its value was therefore checked by feelings of mutual respect, even admiration, as well as the desire to maintain a reputation for providing high-quality specimens, which was crucial for long-term commercial success. As we shall see, these social conventions did not transfer to North America's late nineteenth-century mining frontier, where deception and fraud would pose a more dire threat.

Although the most spectacular fossils in early nineteenth-century England were marine reptiles such as the ichthyosaur, Europe was not entirely devoid of dinosaur fossils. In fact, soon after the first plesiosaur was discovered by Anning, British paleontologists began to unearth a number of terrestrial reptiles that would later be designated as dinosaurs. The first was an extinct carnivore—*Megalosaurus*—whose fragmentary remains were described by Buckland in 1824. Not long thereafter, Mantell compared several large teeth he had previously found in the Tilgate Forest of northern Sussex to those of an iguana. Surmising that they must have belonged to an extinct relative of that animal, he christened it *Iguanodon.* Roughly a decade later, Mantell found the bones of a third terrestrial reptile, which he named *Hylaeosaurus.* These three specimens would eventually be judged sufficiently similar to warrant the creation of a new biological category, and in 1841 the noted anatomist Richard Owen coined the word Dinosauria to designate this new "tribe, or sub-order of Saurian Reptiles."[23] While the first dinosaurs were discovered by English naturalists during the 1820s, it therefore took over a decade for them to acquire that name.

The discovery of numerous fossils ranging from the ichthyosaur to dinosaurs lent the Age of Reptiles a tactile reality, causing it to become a mainstay of Victorian popular culture. When a group of investors decided to build a permanent version of the Crystal Palace in the London suburb of Sydenham during the mid-1850s, for example, they commissioned the sculptor Benjamin Waterhouse Hawkins to produce life-size reproductions of these and other "antediluvian" monsters. Designed to instruct as well as delight, Hawkins's three-dimensional figures created a huge public sensation and were seen by over a million people per year. However, despite the active participation of Owen, who had been hired to advise Hawkins on the finer points of comparative anatomy, some of these sculptures soon came in for intense criticism. Naturalists especially singled out Hawkins's dinosaurs as woefully inaccurate, and Owen himself admitted that some of their features were "more than doubtful."[24] Much of the trouble stemmed from the scant fossil evidence on which Hawkins was forced to rely. Whereas plesiosaurs and ichthyosaurs were sufficiently abundant that collectors like Anning could unearth a steady stream of exquisitely preserved and remarkably complete fossils, the same was not true for dinosaurs, whose remains at the time consisted of only a few scattered teeth and a small number of bones. But all of that was about to change, thanks to a remarkable turn of events across the Atlantic.

In 1857 an American lawyer and avocational naturalist named William Parker Foulke discovered a dinosaur fossil in his neighbor's marl pit just outside Haddonfield, New Jersey. After digging it up, Foulke donated it to the Academy of Natural Sciences in Philadelphia, where it was closely examined by Joseph Leidy, who christened it *Hadrosaurus* to commemorate its place of discovery. Unlike the fragmentary specimens on which previous paleontologists had to rely, well over half of this skeleton was available for scientific scrutiny.[25] Thus, despite being unearthed several decades after *Megalosaurus* and *Iguanodon, Hadrosaurus* offered the first relatively complete glimpse of what dinosaurs might have looked like. And it turned out to be very different from what Owen and Hawkins had thought.

Since the time of Cuvier, the reconstruction of extinct creatures had constituted an act of informed speculation involving a great deal of inference and outright guesswork in addition to the careful examination of material evidence. But even in cases such as that of *Iguanodon,* where naturalists had only a few scattered bones to work with, they did not simply make things up out of thin air. Rather, paleontologists often turned to living animals as a model to guide their imagination. Owen, for example, used modern pachyderms in this capacity, which is why Hawkins's dinosaur sculptures resembled an elephant or rhinoceros (Figure 1.2). Owen's decision to model dinosaurs on modern pachyderms reflected his staunch opposition to the contested theory of Lamarckian evolution, which held that living things progressively evolved to become more complex over geological time. By characterizing dinosaurs as mammalian creatures complete with an advanced circulatory system, Owen could argue that, if anything, reptiles had actually degenerated into the creeping, crawling, and slow-moving lizards that were familiar to a Victorian audience.[26]

Figure 1.2. A pair of *Iguanodon* dinosaurs, sculpted by Benjamin Waterhouse Hawkins for the Crystal Palace in 1854.

Leidy used the relative completeness of *Hadrosaurus* to argue in favor of a different anatomical model. Given the pronounced disparity between the skeleton's diminutive forelimbs and its large, powerful hind legs, Leidy drew an explicit comparison to the kangaroo, reasoning that *Hadrosaurus* must have stood upright and used its tail like a tripod for added stability.[27] This image was further reinforced when Leidy's younger colleague Edward Drinker Cope described another terrestrial reptile from the New Jersey marl, which he named *Laelaps aquilunguis.* As the playful etching Cope made to illustrate his ideas in the *American Naturalist* indicates, *Laelaps* had even tinier arms and a stronger tail than *Hadrosaurus,* making the analogy all the more powerful (Figure 1.3).[28]

The discovery of dinosaurs in the United States had a transformative impact on vertebrate paleontology, completely revising the way these creatures were conventionally understood. But *Laelaps* and *Hadrosaurus* were just the beginning—things really took off as paleontologists expanded the frontiers of their research to include the American West. As these scientists began

Figure 1.3. *Laelaps* confronting *Elasmosaurus* with *Hadrosaurus* foraging in the background. From E. D. Cope, "The Fossil Reptiles of New Jersey," *American Naturalist,* 1869.

digging up new specimens such as *Stegosaurus, Allosaurus,* and *Triceratops,* North American dinosaurs came to be regarded not only as more complete and better preserved than their European counterparts but also as especially large, commanding, and powerful. Another illustration, this one showing Marsh's reconstruction of a *Brontosaurus* skeleton that was found in Reed's quarry at Como Bluff in 1879, makes the point well. First published in 1883 and subsequently reprinted in Marsh's monumental *Dinosaurs of North America,* this illustration depicts what arguably remains the most well-known inhabitant of the Age of Reptiles in a style that would soon be ubiquitous (Figure 1.4). Not only was *Brontosaurus* massive—Marsh himself estimated that, when alive, it must have weighed an astonishing twenty tons—the skeleton was sufficiently complete that one could gain a vivid sense of the animal's likeness from its fossil bones alone.[29] A few decades later, a second *Brontosaurus* fossil was found near Reed's quarry and mounted into an impressive display at the New York Natural History Museum. Featuring a pose nearly identical to that in Marsh's illustration, the New York exhibit further cemented the creature's status as the period's most iconic dinosaur. Thus, while the

Figure 1.4. A skeletal restoration of *Brontosaurus* from Como Bluff by Othniel C. Marsh, from *Dinosaurs of North America,* 1896.

earliest dinosaur fossils were found in England, the modern conception of these creatures as lumbering behemoths of the prehistoric era only emerged once paleontologists began scouring the deep canyons, windswept plains, and desert highlands of the American West.

The Expanding Frontiers of American Paleontology

There is a strong temptation to treat the 1877 discoveries as a function of the region's physical geography, which is particularly well suited to fossil hunting. While the eastern United States is mostly covered in forest and grass, the Rocky Mountain West is arid and dry, with a great many exposed rock faces dating back to the Age of Reptiles. This all but ensured dinosaur bones would be well known to local people inhabiting the region before 1877. Indeed, several Native American origin stories indicate that indigenous tribes had a keen interest in the fossil bones of prehistoric life that littered their ancestral homelands.[30] For example, the Lakota storyteller James LaPointe explains that during the "first sunrise of time" all the "land was covered with a seething mass of animals," whose fierce combat left their skeletal remains strewn across the badlands of Nebraska and the Dakotas.[31] Because the bones of these "thunder beings" were often found after a hard rain exposed them to light, they were widely associated with extreme weather events. In one particularly memorable telling, the Lakota holy man Lame Deer recalled being trapped in a severe thunderstorm perched on a ridge, only to awaken at dawn to find that "I was straddling a long row of petrified bones, the biggest I had ever seen," from which he concluded that he had spent the night "along the spine of the great *Unktehi*," a serpentine creature that plays an important role in Sioux creation narratives.[32]

Paleontologists frequently relied upon the local knowledge of Native American guides, but their contributions to science have been largely forgotten. A notable exception is Marsh, who paid homage to Sioux origin stories when he named an especially large species of extinct ungulates *Brontotherium* (or "Thunder-Beast"), whereas he christened an even bigger and more familiar plant-eating dinosaur *Brontosaurus* (or "Thunder-Lizard"). A more typical example is Cope, who was guided by stories about the bones of "evil monsters" near the Grand River that had been killed "by the Great Spirit" when he traveled to South Dakota in search of new fossils during the early 1890s. In a letter to his wife, Cope explained that the Sioux "would not touch the bones for fear that a like fate would befall them," adding that, as a result, a large number of specimens "were fortunately preserved for the more intelligent white man who is not troubled by such superstitions."[33] Cope's easy dismissal of Sioux creation narratives as "superstitions" suggests that, in addition to their land and their culture, white settlers also expropriated the deep past from indigenous tribes. As paleontologists moved into the region during the late nineteenth century, Sioux thunder-beings were refigured as dinosaurs. Doing so not only involved excavating their fossil remains, removing them from the rock matrix so they could be transported East and accumulated in museums and university collections. It also involved removing these creatures from their epistemic matrix and inserting them into a scientific narrative about the evolution of life on earth.[34]

The extent to which Native American knowledge about vertebrate fossils has been written out of the history of paleontology helps to reveal the deeply social basis of scientific discovery. In effect, to make a scientific discovery is to be recognized for one's contributions to a specific knowledge community by the members of that community. To say this is not to disparage or belittle the knowledge of Native Americans but rather to argue that discoveries functioned as part of the reward system of science, an honorific the community bestowed on its members to celebrate and commemorate their achievements.[35] This points to a broader lesson about the way knowledge was constituted in nineteenth-century natural history. For a claim about prehistoric nature to become knowledge, the community of learned naturalists had to accept it as such: to know something was not just to have reasons for believing it to be true but also to have the community sanction those reasons as adequate and persuasive. Far from being a mere consequence or an afterthought of scientific research, the circulation of specimens, ideas, and publications was therefore a constitutive feature of the knowledge-making enterprise.[36]

While the paleontological literature and the public imagination alike tend to locate the place of scientific discovery out in the field, a fossil's movement through space was therefore an essential feature of its identity as a scientific specimen or an object of knowledge. This makes it significant that frontier collectors like Reed told naturalists such as Marsh about the rich diggings at places like Como in hopes of a financial reward, as it implies that the commercial specimen trade was indispensable to the discovery of North American dinosaurs. It was the creation of a market for vertebrate fossils that brought about the simultaneous discovery of dinosaurs in the American West, not the other way around.

In much the same way that the active construction of a robust infrastructure was required to support the region's extractive economy, social and technological factors helped spur the emergence of a commercial market for dinosaur bones.[37] For one thing, the rags-to-riches stories that had characterized western lore since the California Gold Rush of 1848 drew hordes of white settlers to colonize the area and dispossess its native populations.[38] Moreover, these settlers had been primed to view the land as a storehouse of valuable resources and the West more generally as a place of almost limitless economic opportunity.[39] Those who chose not to prospect for precious metals and minerals raised livestock, built homesteads, or set up shop as lawyers, adjudicating the disputes that invariably arose between miners working the same lode.[40] The federal government was an active participant too. In addition to passing numerous laws encouraging the exploitation of the region's "unimproved" land, it commissioned a number of geological, geographical, and topographical surveys that engaged a small army of geologists in assembling an exhaustive inventory of the region's rich natural bounty.[41] The United States also helped underwrite the construction of a transcontinental railroad, which put formerly remote parts of the Rocky Mountain West within reach of major population centers, transforming an arduous journey of several months by horse and buggy into one that could be undertaken in a matter of days.[42]

Along with the telegraph, the railroad linked the supply of natural history specimens to emerging centers of demand on the East Coast.[43] Two men in particular were just beginning to earn a nationwide reputation for their interest in vertebrate fossils during the late 1870s. The first was Cope, from the Philadelphia Academy of Natural Sciences, and the second was Marsh, from the Peabody Museum at Yale. Both were young, rich, and ambitious, and by the late 1860s they were embroiled in a bitter rivalry that would have a profound impact on the course of American paleontology.[44] In their zeal to outdo each other, each raced to publish more papers than the other. They also fought over who could succeed in naming the largest number of new species. To maximize their own productivity, Cope and Marsh often relied on local informants, cultivating relationships with anyone who could supply valuable fossils. Since both men were independently wealthy, they could spend a considerable portion of their inherited fortunes on the purchase of specimens. By 1888, Cope had spent about $100,000 amassing his collection, whereas Marsh was even more profligate, spending over $200,000 between 1868 and 1882 alone.[45] Roughly equivalent to over $2 million and $4 million today, respectively, these were considerable sums of money, certainly enough to capture the attention of a fossil hunter like Reed.

The commercial market in dinosaur bones was forged out of numerous factors working in tandem. An abundant supply of vertebrate fossils in the American West was linked to the demand for research material among wealthy paleontologists in the East by the construction of an expansive communication and transportation infrastructure. But what really brought collectors and researchers together was the culture of America's mineral industry. A fossil hunter like Reed did not understand himself to be making a contribution to science when he realized the commercial potential of Como Bluff. Rather, his behavior had much more in common with that of a mineral prospector who has just uncovered a promising vein of silver or a valuable seam of coal. Of course, Reed understood that there was a difference between fossils and other resources. But given the way that he chose to approach a prospective buyer like Marsh, and especially given the way the two subsequently dealt with each other, there is no mistaking the fact that Reed's behavior was informed by his experience in mining.

A principal difference between fossil collectors in the American West and their predecessors in England was an almost wanton disregard for traditional norms of respectable conduct. British collectors like Anning were certainly interested in making money, but they also coveted the more intangible forms of credit one could receive by having one's contribution to knowledge publicly recognized. This led them to adopt a modest approach when dealing with naturalists, helping them to establish identities as trustworthy partners in science. Similar social relationships predicated on a shared sense of mutual obligation and reciprocity obtained during the early years of American paleontology as well. Leidy, for example, attached the specific name *foulkii* to *Hadrosaurus* in honor of Foulke's magnanimous decision to donate his specimen to the Philadelphia Academy of Natural Sciences.[46] In much the same fashion, when J. C. Vorhees donated the bones of *Laelaps aquilunguis* to that institution several years later, Cope explicitly acknowledged this contribution in print, writing that if only "all persons engaged in digging marl were equally interested in the preservation of bones which come under their notice, we might have been far nearer an elucidation of this, one of the most extraordinary faunae which have been placed upon our planet."[47] In striking contrast, such acts of public munificence were practically nonexistent between fossil hunters from the American West and paleontologists from the East.

Because he was a transitional figure, Arthur Lakes serves as an especially revealing case in point. Of the three men who each independently discovered a dinosaur quarry in 1877, Lakes was by far the best educated. After studying theology and natural history at Queen's College, Oxford, he moved to the small mining town of Golden, Colorado, in hopes of improving his financial prospects. Before long, he became fascinated by the region's geology, especially once he came into contact with Ferdinand Hayden's United States Geological and Geographical Survey during the mid-1870s.[48] When Lakes found several vertebrate fossils weathering out of a hillside near his home, he wasted no time in contacting Marsh to announce the discovery of a "gigantic saurian."[49] Initially, Lakes wanted to "present" some of the fossils to Marsh in exchange for taxonomic information about them, but it was not long before these feelings of generosity gave way to more mundane concerns.[50] Complaining about his meager salary, he asked whether Marsh might be induced to pay for additional specimens. "I am very much interested in these discoveries," he wrote, but "whilst I am thoroughly imbued with enthusiasm about such pursuits . . . & should greatly like to continue them I have not the pecuniary means to do so." This meant that "some terms of remuneration must be agreed upon." Lakes went on to close this awkward missive with an expression of hope that his confession would not make his motives "appear unduly mercenary" or indicate he did not "truly partake of the spirit of scientific research." "I hope the time may come," he added, "when I can honestly afford to follow such pursuits for the pure love of them."[51]

The roundabout way in which Lakes asked for payment illustrates his reluctance to treat paleontological discoveries as objects of economic exchange. Given his educational background, he would have known that selling his quarry would diminish the credit an eminent naturalist such as Marsh could bestow, and he must have been especially demoralized when Marsh failed to reply. Rather than change his mind about selling the specimens, however, Lakes turned to Cope in search of a more receptive trading partner. After sending several small skulls and teeth for Cope to inspect, Lakes wrote to inform Marsh of what he had done. "I told [Cope] and I think I told you that my circumstances obliged me to sell the specimens," he explained, announcing that he "had no preference but they should go to the highest bidder."[52] This finally spurred Marsh into action, and he immediately shot off a telegram offering to pay one hundred dollars "to cover the expense, labor, [and] time incurred" to dig up any remaining fossils. But Lakes informed Marsh he would have to do better, since Cope had already offered "$150 to $175 per month for the purpose to provide myself with tents, wagon & outfit."[53] Determined that Lakes's bounty should not go to his competitor, Marsh quickly dispatched a trusted assistant to Colorado, which led to a deal being struck that saw Lakes work in Marsh's employ over the course of several field seasons, digging for dinosaurs until he eventually secured more gainful employment at the Colorado School of Mines.[54]

Lakes's transition from fossil hunting to mining was no coincidence, as the two occupations were intimately linked at the time. An article that appeared in *The Cosmopolitan* around the turn of the century offers a particularly good illustration, explaining that paleontological "collecting is similar" to "mining for precious metals" because in both cases "the earth seems to withhold her secrets jealously."[55] At the same time, a great deal of vocabulary from mining made its way into paleontology. Bone hunters routinely referred to the discovery of a particularly rich quarry as having hit "pay dirt" or even "treasure dirt."[56] Moreover, there was a high degree of back-and-forth movement of people engaged in both occupations. A miner named Fred Brown, for example, made the transition to vertebrate paleontology when he began working for Marsh in the early 1880s.[57] But the movement of people went in the other direction, too, since a solid background in geology was indispensable for those hoping to invest in mines. Anyone who was a proven expert in the field was thus highly sought after by those in the industry, and academic geologists were not immune to the promise of wealth.[58] But even the best education and training could not guarantee success.

Henry Augustus Ward, a well-known dealer in natural history specimens and professor at the University of Rochester, furnishes a revealing account of the challenges that anyone who sought to do business in the mineral industry would have faced.[59] Although Ward's early reports to his financial backers about the mineral prospect that he had acquired were full of promise, stating that "never before" had he been "met with gold lodes in any part of the world which gave even one half so large a yield," he soon realized these initial impressions were overly optimistic.[60] In retrospect, it became clear that Ward had mistaken surface indications as a representative sample of what lay underground, and he admitted that he could "know nothing about any of the lodes for more than one or two yards ahead of our digging," as each vein "widens and narrows suddenly and irregularly without any rule, logical or empirical, which one can take for a guide."[61] Within a few months, the company's prospects looked so bleak that Ward was forced to admit, "I have been entirely, totally and completely deceived." As a result, his investors in Rochester quickly lost faith in the project and Ward was sent orders to cease operations after selling the land and machinery he had acquired over the previous year.[62]

Brief though it turned out to be, Ward's foray into mining is instructive on at least two counts: beyond demonstrating that academic scientists were actively involved in the mineral industry, it also showcases the difficulty of turning a profit in mining.[63] The act of valuing a mineral claim was especially tricky because doing so required simultaneously solving for two unknowns. First, it was crucial to determine how far and how wide a lode extended back into a hillside. Second, it was equally important to know how much of the ore contained in a lode consisted of precious metals and what percentage was waste. Reliable numbers for either were not easy to come by. The quality of an ore had to be ascertained by chemical assay, which required a great deal of equipment in addition to specialized training and expertise. Worse still, the exposed ore did not always provide a representative sample, often failing to reflect the quality of the whole lode. For similar reasons, the full extent of a lode could not be determined with any degree of certainty from surface indications alone. Its thickness might increase as one followed it into the hillside, or it might disappear altogether.[64]

Due to these difficulties, the value of a mine could not be reliably ascertained before concluding a purchase.[65] It was only *after* expending considerable resources to open a mine that anyone really knew whether it would pay to work the locality. Deep shafts had to be sunk to access the ore, a dense latticework of wooden beams installed to prevent cave-ins, large pumps brought in to remove groundwater that seeped into the mine, and steam-powered stamp mills erected to crush quartz and extract precious metals. The scarcity of reliable information before making these investments created conditions that were perfect for unscrupulous hucksters to take advantage of naive investors. To make matters worse, the mineral industry was quickly becoming a notoriously speculative bubble. The promise of striking it rich meant the American West was crawling with prospectors, imparting to the region its famously heady and volatile character. By and large, these were people of small means who could not afford the investment of capital required to develop their claims. Having made a promising discovery, their only hope was to sell out at the highest possible rate. However, the sheer abundance of people looking to do so made it difficult for investors to know whom they could trust with their money. Richard Stretch, who authored one of the period's ubiquitous instructional guides, advised holding out for claims that promised "a large ore body" of "fair average quality" with "good working facilities and reasonable access."[66] But this was far easier said than done, as exuberant miners sank shafts and dug tunnels with such abandon that even the richest localities contained only a few efforts that eventually paid off. The hustle and bustle of boomtowns therefore brought with it not just competition but also confusion.

Because no one really knew how much a promising claim was going to yield, prospectors and investors had to negotiate under conditions of relative ignorance. To make the most of their claim, prospectors tended to paint their discoveries in the best possible light. For that reason, all mining manuals stressed the importance of prudence, admonishing investors to rely on informed, expert judgment when evaluating a new claim. Failure to do so could have disastrous consequences, especially if one considered the frequency with which unscrupulous miners dug holes in the ground "on the merest pretext of indications to catch the ignorant, adventurous tenderfoot capitalist purchasers, or 'suckers,'" as Lakes warned in another instructional guide. Worse still, investors had to contend with the possibility that a mine had been actively tampered with. "Take care you ain't salted," Lakes advised, for "so clever are the miners that cases are on record where a most experienced expert has been taken in, and comparatively or wholly valueless properties sold for large sums, the purchase followed later by woeful dismay and surprise when dividends were called for and did not appear."[67] For that reason, negotiations between potential buyers and sellers tended to be extremely long and drawn out, not to mention contentious, and they were marked by strong feelings of mutual distrust and suspicion.

Precisely the same market dynamics occurred in vertebrate paleontology too. Just like an investor looking to purchase a mine, a paleontologist could not assess the value of a new dinosaur quarry with certainty until its contents had been dug out of the ground. For that reason, not only were business transactions highly competitive, but success also hinged on the ability to manage the flow of scarce and unevenly distributed information to one's advantage. That said, there were also significant differences between a dinosaur quarry and a mineral prospect. The most important was the relative size of each market. Having discovered a promising deposit of gold, silver, or coal, a mineral prospector could expect to find a number of potential investors. The same was not true for fossil hunters, at least during the 1870s and 1880s. In the early years, Marsh and Cope were nearly the only game in town (with the possible exception of Louis Agassiz at Harvard). This meant the supply of bones far outstripped the available demand, providing paleontologists with a considerable advantage.

Carlin and Reed provide a particularly illuminating example of the way business was practiced in late nineteenth-century American paleontology. After the initial exchange with Marsh, negotiations over the specimens they had found at Como Bluff would go on to last over a year, during which both parties worked hard to extract the best possible deal. For all those involved, ignorance was by far the most salient obstacle to be overcome. Marsh, Carlin, and Reed were all in the dark about what the Como discovery might be worth, but for different reasons. Marsh, who was located in New Haven—nearly two thousand miles away from the dig site—had to contend with a problem of access, both to the fossils as material objects and to reliable information about their completeness, state of preservation, and abundance. This made it difficult to evaluate the quality of Reed's discovery. The two collectors out in Wyoming faced a different problem. Although they had direct access to plenty of bones, neither Carlin nor Reed possessed the expertise to determine their taxonomic position, age, or rarity. But this was precisely what was required to know how much someone like Marsh would be willing to pay for these objects, and so the two men frequently (if somewhat naively) asked Marsh for help in identifying the fossils in their quarry.

Having been stationed at Wyoming's largest coal-mining operation before Como, Reed was well acquainted with the mineral industry. Once he realized that his discovery had monetary value, Reed was inclined to treat dinosaur bones like any other scarce resource one could dig out of the ground. The manual Lakes published after transitioning out of paleontology was geared to investments in mining, but he could just as well have had fossil hunting in mind when he cautioned investors that typical reports of a new discovery "give the most favorable view" and thus "must be taken '*cum grano salis.*'" In their initial letter to Marsh, Carlin and Reed made sure to include a range of information that reflected well on their find. First, they guaranteed sole proprietorship of the quarry to Marsh, writing, "We have said nothing to anyone else as yet." Next, they talked up the specimen's remarkable size, declaring that its shoulder blade measured "four feet eight inches (4 ft. 8 in.) in length." Finally, they advertised their own credibility—something they knew was in short supply—by offering to ship a few samples to New Haven as "proof of our sincerity and truth." Clearly, all of these details were carefully chosen to entice Marsh into making a purchase.[68]

As newcomers to the fossil trade, Carlin and Reed also committed several tactical errors. Most damaging to their interests was the admission they had stumbled on something whose value they were not fully competent to assess, referring only to "a large number of fossils, supposed to be those of the Megatherium, although there is no one here sufficient of a Geologist to state for a certainty."[69] It is telling that Marsh did not help Carlin and Reed to identify the bones in their quarry. At first, he promised to send on a taxonomic key. But for mysterious reasons, it never reached Wyoming. Carlin and Reed repeatedly pushed Marsh to send them another copy, but he never complied. At one point they reported waiting "anxiously . . . for the pamphlets," admitting that Marsh's reluctance to send information was irksome because it deprived them of the means to ascertain "whether the bones we have found are rare or not."[70] But Marsh was a shrewd negotiator, and he never gave in to this demand.

For his part, Marsh was familiar with the mineral industry too, having held an appointment as a vertebrate paleontologist with the US Geological Survey. Thus, one of the first things he did upon hearing from Reed was to try to get a better sense of what Reed had found. His initial letters to Wyoming were primarily designed to elicit sample specimens, even providing a detailed set of instructions on how fossils should be packaged for travel.[71] But since he had just purchased the dinosaur quarry discovered by Lakes, Marsh was in a position to bide his time. Rather than rush into what must have appeared an uncertain and expensive proposition, he maintained a blasé attitude, ignoring Carlin and Reed's request that he pay for the sample specimens up front. As a result, it was three months before the first shipment of fossils finally arrived in New Haven. Upon receipt of the bones, however, Marsh suddenly showed a clear interest in acquiring ownership of the quarry. On the very same day, he rushed to send off an expensive telegram with instructions to "send rest with all small pieces."[72] He also posted a letter to acquire additional information. Samples now in hand, Marsh knew enough to determine that the bones were of high quality, but several questions remained. For example, he did not know whether he had been sent scattered fragments or pieces of an articulated skeleton. "Were the bones all found close together?" he asked. Moreover, even if whole skeletons could be exhumed, he had no idea what kind of animal they might represent. "Did the vertebrae go with the largest or next to the largest thigh bone?" he wanted to know. And with "what bones were the teeth found?"[73] All of these factors would have had a bearing on the discovery's value.

Besides the high quality of Reed's specimens, Marsh was spurred into action by the threat of competition. In a letter they wrote shortly after sending the samples, Carlin and Reed warned Marsh, "[Although] we are keeping our shipments of fossils to you as secret as possible, . . . there are plenty of men looking for such things and if they could trace us they could find discoveries which we have already made."[74] His interest now having been piqued, Marsh was not about to relinquish what might well turn out to be an unusually significant find. He immediately telegrammed his assistant, Samuel Wendell Williston, to go to Wyoming, where he was instructed to "collect and learn all possible."[75] When he arrived at Como Station, Williston was to superintend the new quarry, protecting his employer's interests by serving as Marsh's hands, eyes, and ears on the ground. Perhaps most important was to influence events so that his boss would prevail in the ongoing negotiations. In his first letter back to New Haven, Williston advised Marsh to hurry and get as many bones shipped as quickly as possible. The best strategy, he felt, was to stall making a final settlement and avoid paying outright for the discovery: "The point will be to get into & ship as soon as possible, for the men have got pretty exaggerated notions of what [the fossils] are worth." By purchasing specimens on a crate-by-crate basis, Marsh could hedge his bets, save time, and gather more information to help him determine the quarry's exact value. "I have engaged Reed to go to work," Williston explained, "guaranteeing an increase of wages on what he is now making [at the railroad] ($60) but not saying how much."[76] Caution, he stressed, must rule the day: "Should you publish any description of this I would suggest that you do not send C + R this publication," Williston advised. Otherwise they "will think [the locality] more valuable." This had to be avoided at all cost, seeing as how "their only object is money."[77]

Why would Carlin and Reed have agreed to work for piecemeal wages rather than insist that Marsh purchase their discovery outright? They too appeared to have stalled in the hopes of strengthening their bargaining position. Before Williston arrived in Como, Carlin had raised the possibility of negotiating a final contract in person rather than through the mail. He claimed to have already made plans to travel east sometime in November or December, providing an opportunity to stop over in New Haven.[78] But then, just before leaving Como several weeks later, Carlin informed Marsh that he was entertaining counteroffers from an unnamed third party.[79] Later, Carlin also sent Marsh a local newspaper article that announced the discovery of the Como dinosaurs, valuing them around $2,000.[80] Carlin and Reed must have been hoping to drum up additional interest in their dinosaur quarry, given that auctioning off their discovery would have effectively forced Marsh to match what a knowledgeable competitor was willing to pay. In the end, though, the tactic did not meet with success and the two prospectors failed to secure another potential investor.

Once Carlin reached New Haven and was presented with a concrete offer by Marsh, he was dismayed to find that it was much less than he expected. It did not take long for negotiations to break down completely, and Carlin ordered Reed to cease any further shipments until a final contract was signed.[81] Angry and frustrated, Reed accused Marsh of having taken advantage of his good will. "Your proposal . . . was not what I had expected nor what you led us to believe you would do," he complained, adding, "We have put all confidence in you and now you have the advantage and you do not seem slow to take it." Apparently, when push came to shove, Marsh had refused to offer the two men more than fifty dollars for the discovery, plus a monthly salary slightly above what they earned at the railroad.[82]

It turned out that Carlin and Reed made a fatal mistake by agreeing to begin shipping specimens to New Haven before signing a contract. Their strategy failed to take into account a crucial difference between mineral prospecting and fossil hunting: so long as it continued to yield coal, the marginal value of a mine did not greatly diminish with time, but the same could not be said of a dinosaur quarry. Once a locality's type specimens had been secured, the remainder was duplicate material and the quarry's value plummeted precipitously. Reed realized that he had come to understand this distinction too late, and accused Marsh of having deceived him by promising to "pay us well if we dealt only with you and in that we have complied to the letter. You now have so many bones," Reed continued, "that to another man [the quarry] is of very little value [and that] is a very poor way of doing business and I do not approve of it."[83] Still, having already shipped several crates of specimens to New Haven, Carlin and Reed were effectively locked into the transaction, which left them with few options besides acceding to Marsh's miserly terms. And so it came to pass that, in January 1878, Reed signed a contract and entered Marsh's employ at a rate of ninety dollars per month (thirty dollars more than he made at the Union Pacific). "The discovery and work done before any arrangements were made I had thought were of more value than you put upon them," he complained, "but as it is settled now I will work for your interest to the best of my ability."[84] Carlin eventually signed a contract as well, and both men got back to work excavating the bones Reed had found.

Conclusion

Dinosaur fossils in the American West were brought to the attention of learned naturalists such as Marsh because entrepreneurial frontiersmen like Reed came to regard them in much the same way as any other scarce resource one could dig out of the ground. It was the promise of pecuniary gain that motivated fossil collectors to make contact with paleontologists on the East Coast, circulating their discoveries within the scientific community.[85] For their part, Marsh and Cope had no qualms about negotiating with commercial collectors. Railroad ties and telegraph lines linked their two worlds together, allowing both parties to exchange information and material goods with comparative ease and speed. But while the new transportation and communication technologies made long-distance transactions possible, it was money that accomplished the real act of translation. Marsh and Reed were able to strike a deal because both recognized dinosaur bones as objects of economic exchange. The main point of contention was not whether fossils ought to be bought and sold but exactly how much they were worth.

The negotiation between Reed and Marsh provides a revealing look at the way scientific exchanges were conducted on North America's late nineteenth-century mining frontier. Here, as elsewhere, value was primarily measured in dollars and cents. Nonetheless, the price of a dinosaur remained hard to establish. Because the extent and quality of a new discovery was extremely hard to determine, especially from afar, the threat of deception and fraud loomed in the background of many transactions. Hence, rather than trust prices to converge on an equilibrium point that effectively mediated between the desires of both parties, fossil hunters and paleontologists adopted business practices that were commonly found in the mineral industry. In turn, this gave rise to negotiations that were long, drawn out, and intensely competitive bordering on acrimonious. More importantly, success in these contentious negotiations largely turned on the ability to manage the flow of scarce information to one's own advantage, profiting from the difficulty of acquiring reliable knowledge about the quality of a new prospect.

The conditions that structured scientific exchanges on the mining frontier proved to be short lived. At the beginning of the twentieth century, a wave of mergers and acquisitions swept across the American business landscape, resulting in the consolidation of many small, individually owned and operated firms into heavily capitalized and professionally managed corporations. Before long, the tycoons who oversaw the corporate reconstruction of the US economy realized they could temper the risks of doing business under conditions of asymmetrically distributed information by instituting a process of vertical integration, internalizing the market for raw materials and taking control of their own distribution networks. In the decades that followed the 1877 discoveries, natural history underwent a similar transformation, with the creation of several large, philanthropically funded museums in cities such as New York, Chicago, and Pittsburgh. As these museums began to acquire their specimens through in-house expeditions, fossil hunting came to be treated as a profession, and independent collectors like Reed would be largely, if not entirely, pushed out of the field.[86]

Taking place, as it did, in the narrow window of time after the transcontinental railroad was built but before the vertically integrated corporation had proliferated across the industrial economy, the negotiation between Reed and Marsh provides us a glimpse of a short-lived world. The exchanges that took place between them relied neither on personal relationships that could guarantee goodwill from the bottom up nor on powerful institutions capable of enforcing compliance from the top down. It is to the rise of precisely these institutions in natural history that we now turn.

2

Tea with *Brontosaurus*

On February 16, 1905, some five hundred New York notables gathered beneath the towering skeleton of *Brontosaurus* in the American Museum of Natural History. As the headline of one newspaper described it, this "mammoth" creature measured some sixty-seven feet in length, and it had a "stomach cavity as big as the kitchen of a Harlem flat" (Figure 2.1).[1] Everyone who was anyone seems to have attended the specimen's unveiling, including George B. McClellan, the city's mayor, and Morris K. Jesup, the museum's president. The financier J. P. Morgan was in attendance as well, joined by representatives of the city's chamber of commerce. New York's wealthy and powerful mingled with the museum's scientific and curatorial staff, feting the opening of its new dinosaur hall while the mayor's wife and Corinne Roosevelt Robinson (President Theodore Roosevelt's sister) served them all tea.[2]

We are used to seeing the rich and powerful attend an exclusive soirée to commemorate a new wing of an art museum or the premiere of a symphony. But a similar gathering held under a dinosaur in a natural history museum? Far from being an incongruous one-off event, this party represents a triumphal moment of metropolitan self-fashioning several decades in the making. Its exclusive guest list also reveals the culmination of a strategic alliance between a new generation of wealthy capitalists and America's learned naturalists, especially those who specialized in the excavation, study, and exhibition of vertebrate fossils. The story of exactly how and why that alliance was forged speaks to the precise ways in which the culture of capitalism and the history of paleontology continued to inform one another during the decades that

LOS ANGELES HERALD SUNDAY SUPPLEMENT.

By Walter L. Beasley

Figure 2.1. "The Giant Brontosaurus in Central Park," from the *Los Angeles Herald*, 1905.

followed Reed's 1877 discovery, and it forms the topic of this chapter, bringing us from the nineteenth century into the early twentieth century.

But it was not only wealthy elites who came into contact with *Brontosaurus*. Throngs of visitors numbering in the hundreds of thousands streamed through the museum's newly opened dinosaur hall each year after it opened in 1905. While America's largest and most spectacular dinosaurs hailed from remote parts of the country, it was in densely populated urban centers that vertebrate fossils were introduced to a more popular audience. For that reason, whereas the previous chapter examined the way that vertebrate paleontologists first learned about dinosaurs from the American West, this chapter asks how a much larger and less circumscribed group of people came into contact with these creatures. This too required putting American dinosaurs into circulation, albeit in a different way. Rather than moving specimens through physical space, it involved making them travel across social space. This boundary-crossing primarily took place at the museum, where dinosaur fossils were mounted into spectacular exhibitions capable of attracting a large and socially diverse group of visitors.[3]

More so than anywhere else, it was at the museum that dinosaurs from the Rocky Mountain West were brought to the attention of a popular audience. But not all museums engaged in this process equally. In the nineteenth-century United States, a complex array of institutions exhibited natural history specimens. By far the most numerous were commercial "dime museums," which ranged from small storefront operations to ambitious and well-publicized entertainment venues like the one run by P. T. Barnum in New York City. On the other end of the spectrum were research museums catering to the community of learned naturalists, which were often associated with universities or scientific societies. But as a rule, neither was in a position to mount a spectacular, free-standing dinosaur fossil. Whereas dime museums lacked access to specimens, most research museums were loath to pander to popular tastes by assembling the bones of creatures whose anatomy remained insufficiently well understood to mount a credible and authoritative exhibit. It was not until a new kind of museum emerged during the last third of the nineteenth century—one that combined a willingness to engage in popular spectacle with a large staff of trained scientists and an extensive collection of specimens—that exhibits such as the *Brontosaurus* display in New York began to proliferate.

Dinosaurs from the American West primarily entered more widespread circulation in the context of a new institutional framework: the philanthropically funded museum of natural history. One of the first and arguably the most influential of these was the American Museum of Natural History in New York. But there were others as well, most notably the Carnegie Museum in Pittsburgh and the Field Museum in Chicago. Initially conceived as elaborate municipal gifts, they were much larger and more lavishly funded than research museums. They also differed from older institutions in that they were managed and run by a group of wealthy trustees instead of the curators themselves. Finally, and perhaps most importantly, because philanthropically funded museums were created to show off the civic liberality of their benefactors, it was crucial for them to attract a large audience. This led them to embrace a new conception of the museum's institutional mission with particular gusto. Initially articulated in nineteenth-century England and often described as the New Museum Idea, it held that public museums ought to combine original scientific research with popular spectacle and education. The New Museum Idea therefore resembled dinosaurs themselves, in that it originated in England but took off in the United States, where it offered a model for how naturalists and philanthropists could forge a mutually beneficial relationship.

Going back to the Renaissance Cabinet of Curiosities, collections of the rare and the wonderful have long served as a means to demonstrate one's elite social status, functioning as material evidence of wealth, power, and mastery.[4] However, the long nineteenth century saw museums became increasingly open, accessible, and responsible to an emerging conception of the democratic citizen.[5] This gave rise to a conflict between two very different conceptions of what museums ought to accomplish. Should they serve a forum in which a culture's core values are debated, contested, and at times even overturned, or a temple for the veneration of sacrosanct objects, ideas, and persons?[6] Because it articulated a hybrid mission that mediated between these competing visions, the New Museum Idea helped to quell this debate. It also refigured the museum into educational institution whose goal was to edify and uplift the public. This was especially so in the United States, where philanthropic museums proliferated during the Long Gilded Age. Perhaps even more so than in Europe, philanthropic museums in the United States sought to expose working people to the awesome achievements of industrial capitalism, placing extraordinary faith in the power of material objects to turn unruly audiences into responsible citizens.[7]

During the last third of the nineteenth century, American capitalists began to show off their civic munificence and republican virtues by making large collections of spectacular objects available for public consumption. To borrow vocabulary from the contemporary social critic Thorstein Veblen, such highly visible displays of lavish magnanimity can be described as acts of conspicuous generosity, for they were clearly designed to reflect the high-minded liberality of the urban elite. And what could possibly be more conspicuous than a massive new *Brontosaurus* display? But wealthy capitalists were interested in more than just mass public appeal. To impart these exhibits with an air of legitimacy, they enlisted the participation of respected naturalists. The trustees who ran philanthropic museums began hiring curators of vertebrate paleontology who could fill their public galleries with authoritative renderings of these towering creatures.[8] As this happened, dinosaurs came to be valued for new reasons by a new group of people. During the 1870s and 1880s, fossil hunters like William Harlow Reed treated specimens as objects of economic exchange, whereas paleontologists such as Marsh saw them as material traces of the history of life on Earth. Now, at the turn of a new century, dinosaurs acquired a third meaning, becoming a preferred means for wealthy philanthropists to display their elite social standing while simultaneously demonstrating their civic liberality.

Albert Bickmore Encounters the "Paleozoic Museum"

The extractive economy whose development we encountered in the previous chapter was of far more than just local consequence. In addition to completely transforming the social and economic geography of the American West, it helped fuel the expansion of an industrial economy across the United States. To take just a single example, between 1869 and 1880, American steel production jumped from 35,000 short tons to nearly 1.5 million, only to exceed 15 million tons during the first decade of the twentieth century.[9] This caused industrial output to soar, and real gross domestic product grew at nearly an exponential rate, more than doubling at least once every twenty-five years between 1860 and 1930.[10] Such a rapid pace of industrialization had far-reaching social and cultural implications, including a sharp rise in inequality. Whereas the wealthiest 1 percent of US households claimed 15 percent of the country's income at the time of its founding, that number exceeded 25 percent by 1890 and approached 50 percent before dipping back down again with the stock market crash of 1929.[11] As the wealthy grew richer, the poor saw their wages stagnate and at times even fall, with predictably dire—at times violent—consequences for labor relations. But not everyone who reaped the fruits of industrialization already belonged to the social elite. The last third of the nineteenth century witnessed a new group of financial and industrial capitalists join older and more established merchant families at the top of the economic hierarchy. And before long, this powerful new social class rose to dominate America's cultural institutions as well as its business life.[12]

A comparative anatomist named Albert Bickmore understood these developments with particular clarity. After completing his training with Louis Agassiz at Harvard, he hatched a plan to found a new museum during the early 1860s. With a shrewd eye for economic opportunity, Bickmore settled on New York City as an auspicious site for the venture. Given that "science does not appear to create wealth directly," he reasoned, it "must depend upon the interest which rich and generous men may take in it." And since "New York is our city of the greatest wealth," it seemed obvious to him that it was "the best location for the future museum of natural history for our whole land."[13] Unfortunately, while he encountered no shortage of goodwill among the city's wealthy elite, he was dismayed to learn that all the available funds were tied up in support of the Union army. Disappointed by this initial lack of success, Bickmore embarked on an expedition to bide his time and collect specimens for the imagined museum.

When Bickmore returned to New York in 1867, the Civil War had been won and the nation's economy was booming. "Millions of dollars came to be regarded in the same manner that hundreds of thousands had been before," Bickmore recalled.[14] His goal finally seemed within reach. The ambitious young naturalist contacted William E. Dodge Jr., whose family had made a fortune in mining, to inform him that the wife of a recently deceased European taxidermist sought to sell off her late husband's extensive collection of stuffed mammals and birds. Estimating that the lot of valuable specimens could be had for about $25,000, Bickmore appealed to Dodge's vanity, stating that if he could "furnish the needed sum," he "would always be regarded as the founder of the Institution, even if other gentlemen should give $50,000 or $100,000 afterwards."[15] Although Dodge demurred, he did provide Bickmore with an introduction to Theodore Roosevelt, an importer of plate glass whose son would eventually go on to become president of the United States. Based on Dodge's recommendation, Roosevelt agreed to support Bickmore's proposal and even sent him to see two other members of New York's emerging philanthropic community: the corporate lawyer Joseph Choate and J. P. Morgan. It was not long before another sixteen men had signed their names to the undertaking, including the department store magnate Alexander T. Stewart, a banker named Morris K. Jesup, the corporate lawyer William T. Blodgett, and Dodge's brother, Anson Green Phelps Dodge. These men held several private meetings to finalize their plans and, on December 30, wrote to the Board of Commissioners of New York's Central Park to inquire about the feasibility of placing a natural history museum on public grounds.[16]

The ease with which Bickmore was able to secure funds for his museum venture demonstrates how important it was for New York's wealthy elite to be seen as culturally sophisticated. Although industrial capitalists such as Dodge quickly rose to great power and influence, their integration in high society was not guaranteed. The city's traditional merchant families in particular looked down on their upwardly mobile neighbors, especially industrialists with relatively humble, artisanal backgrounds, who were often derided as rude and uncultivated. Without a shared social background to bind them together, wealthy capitalists engaged in ritualized performances of social distinction to find common ground. These ranged from a visible appreciation of fine art and music to the display of erudite modes of speech and refined styles of dress.[17]

American capitalists could be surprisingly unimaginative in how they signaled their elite status to themselves and each other. Often, they simply adopted the trappings of an older, European aristocracy to perform their bourgeois identity. To that end, they frequently embarked on grand tours to purchase extensive collections of beautiful paintings and sculptures. As one dinner guest at Leland Stanford's opulent mansion in San Francisco remarked, the railroad magnate's house "looked as if the old palaces of Europe had been ransacked of their art and other treasures."[18] The mania for collecting everything from Renaissance paintings and classical statuary to rare books and valuable furniture reflected a widespread enthusiasm for material culture that pervaded nineteenth-century society. But just as important was a much older tradition of surrounding oneself with opulent goods as evidence of one's ability to make fine-grained distinctions in matters of quality, a skill that continued to matter for success in a mercantile economy.[19]

Lavish collections could indicate business acumen and high social standing, but they were increasingly used to show off one's liberality and public munificence too. Whereas the display of a judiciously chosen set of paintings had long served to distance oneself from the more vulgar elements in society, the period after the Civil War saw this practice become infused with a republican spirit as well. Eventually, this gave rise to a boom in the construction of public museums explicitly modeled on older institutions such as the Louvre in Paris or the British Museum in London. But whereas European museums tended to be state-run institutions, America's largest and most impressive examples were usually created through a philanthropic bequest.[20] Hence, by the time the social observer Thorstein Veblen poked fun at what he called the "conspicuous consumption" of the "leisure class," members of high society were already engaging in acts of conspicuous generosity, offering their most prized possessions for public exhibition in an effort to legitimize their over-the-top spending habits at a time of increasing social and economic unrest.[21] Choate put it especially well. "Think of it, ye millionaires of many markets," he bellowed during a speech to celebrate the opening of a new building for the Metropolitan Museum of Art, "what glory may yet be yours if you only listen to our advice, to convert pork into porcelain, grain and produce into priceless pottery, the rude ores of commerce into sculptured marble." In this way, Choate argued, the "rage of Wall Street . . . to convert all baser things into gold" would be counteracted by "the higher ambition to convert your useless gold into things of living beauty that shall be a joy to a whole people for thousands of years."[22]

Among their many philanthropic activities, American capitalists were particularly keen supporters of popular science. This was partially due to the decisive role science played in driving the period's economic expansion, but it was also because science was seen as an especially reliable means to cultivate personal attributes, such as objectivity and disinterestedness, that meshed well with traditional bourgeois values. The epistemic virtues of the modern sciences were tailor-made for a class of people who valorized personal discipline and moral restraint in addition to learning, and wealthy philanthropists were keen to instill these ideals among working people as well.[23] Of all the modern sciences, natural history was especially well regarded. Because it consisted of drawing conclusions based on the careful inspection of material objects, it was believed to cultivate the faculty of attention, which emerged as a core pedagogical preoccupation at the time.[24] Natural history also taught object lessons in Christian theology, illustrating the benevolence and omnipotence of a supernatural creator whose presence was immanent in the natural world.[25] The second half of the nineteenth century therefore saw natural history become a favorite target of philanthropic largesse among American capitalists eager to establish their cultural bona fides. Not only did the study of nature elevate the acquisition of knowledge that underpinned America's extractive economy to a nobler calling, it was also seen to promote desirable habits of mind, domesticating an otherwise unruly working class by filling their newly acquired "leisure time" with wholesome, pious, and uplifting pursuits.[26]

The widespread enthusiasm for natural history among wealthy philanthropists helps to explain why Bickmore found such a receptive audience among elite New Yorkers. But it also accounts for the warm response that his museum venture elicited among the commissioners of Central Park, which was itself a creation of the city's economic elite. As the park's comptroller, Andrew Haswell Green, put it, the commissioners "entirely concur in the desirability of the establishment of a Museum in the Park" and would "very gladly receive the collection to which you allude."[27] Indeed, although Green did not mention it at the time, there had been talk of establishing a museum on the park's premises for some time. But whereas Bickmore envisioned something resembling the Museum of Comparative Zoology at Harvard, Green and the park commissioners planned to build an extravagant "Paleozoic Museum" that would showcase sculptural reconstructions of the extinct beasts that once roamed North America's prehistoric landscape.

Central Park was designed to establish New York as a world-class metropolitan center by supplying it with a grand public space that could rival the Bois de Boulogne in Paris and Hyde Park in London. In addition to helping

Figure 2.2. A lithograph of Hawkins's proposed design for the "Palaeozoic Museum" in Central Park, 1870.

New York overcome its perceived lack of refinement, the park was also supposed to provide a place for the city's elite to take their Sunday promenade while the working classes enjoyed the salubrious effects of fresh air, escaping the crowded conditions and industrial pollution of the immigrant neighborhoods farther downtown.[28] But it was soon decided that green space alone did not suffice. As early as 1863, the board of commissioners lamented the absence of an establishment "for popular amusement and instruction" in the natural sciences on the park grounds.[29] Thus, when the celebrated British artist Benjamin Waterhouse Hawkins crossed the Atlantic to deliver a series of lectures on the relationship between art, science, and religion, Green jumped at the chance to inquire whether he might be for hire. Hawkins had gained considerable notoriety for the three-dimensional models of antediluvian monsters that he made for the Crystal Palace exhibition in London during the 1850s, and Green was convinced that a series of similar sculptures would make a fitting addition to Central Park.[30]

Elaborate reconstructions of prehistoric creatures were well suited for an institution designed to help New York overcome its deficiencies compared to Europe. Anxious about their lack of refinement in the pursuit of high culture, intellectuals from the United States routinely turned to the natural world as a source of national pride. Of course, the Old World had its share of forests, rivers, and mountains as well, so American patriots tended to emphasize the wild and untamed nature of their country's interior as an indication of its superior status. By transporting a bit of the wilderness into the city, the Paleozoic Museum served an important ideological function. While the park had its brambles, rolling hillocks, and peacefully grazing flock of white sheep, the sketches that Hawkins prepared for his proposed museum clearly emphasized America's wild natural heritage instead (Figure 2.2). In so doing, Hawkins echoed an old tradition of mining the deep past for especially potent symbols of American exceptionalism. The mastodon, for example, had long served as material proof the young nation's exceptional power and vitality, and when the French naturalist Georges Buffon described North America's climate as enfeebling and its fauna as degenerate, Thomas Jefferson held it up as material proof of the New World's superiority over the Old, arguing that it "should have sufficed to have rescued the earth it inhabited, and the atmosphere it breathed, from the imputation of impotence."[31] But of all the ferocious and fearsome creatures that once roamed across the American interior, none could rival the recently discovered dinosaurs. In addition to being much larger than prehistoric creatures from Europe, American dinosaurs were often depicted as especially violent creatures. Thus, instead of sculpting replicas of the bulky, passive, and slow-moving creatures he had previously designed for the Crystal Palace in England (see Figure 1.2), Hawkins's designs for the Paleozoic Museum foregrounded two hadrosaurs rearing up on their hind legs while a pair of *Laelaps* brandished their imposing teeth not far behind.[32]

Besides boosting New York's metropolitan status and playing up well-worn themes of American exceptionalism, dinosaurs also promised to attract a large popular audience. The commissioners were especially eager to discharge their "duty of education." In the first instance, this entailed training young minds in painstaking observation and judicious comparison by bringing them "into direct relation with things in themselves." Second, the commissioners also sought to induce viewers to use their imaginative and analytical faculties simultaneously. Finally, they realized that, as a rule, people were more receptive to lessons that mixed entertainment with edification. Dinosaurs excelled on all three counts because nothing could be expected to exercise audiences' minds quite like the realization that "generations of the most gigantic and extraordinary creatures lived through long geological periods, and were succeeded by other kinds of creatures equally colossal and equally strange." The new field of vertebrate paleontology was therefore seen to supply some of the best material for popular education.[33]

By the end of the 1860s, the completion of the Paleozoic Museum seemed close at hand, but municipal politics got in the way. Green had already spent some $30,000 in public funds to lay the museum's foundation when the elections of 1870 swept a controversial political machine that was associated with the Democratic Party and known as Tammany Hall that especially catered to immigrant workers into office. Since the park's board of commissioners had been appointed by the city's Republican Party, the new administration quickly replaced them with a Department of Public Parks staffed by its own political appointees. In its first annual report, the new department estimated that at least $300,000 would be needed to complete the Paleozoic Museum, which was judged far too expensive for a science that, "however interesting," remained in their eyes "yet so imperfect as not to justify such a great public expense."[34] Tammany Hall Democrats therefore ordered an immediate halt to the project. Hawkins was incensed by the decision, not least because it meant forfeiting the balance of his salary. In addressing the New York Lyceum, he even insinuated that corruption and greed motivated the controversial decision to scrap his museum, to which Tammany Hall responded by raiding his studio and smashing the half-finished sculptures, whose remains are presumably still buried somewhere in the park.[35]

These municipal controversies made Bickmore and his financial backers anxious about the fate of their institutional project. They were particularly afraid that Tammany Hall Democrats would try to seize control of their museum venture and redirect it to meet their own ends. Bickmore and his colleagues therefore decided to seek the protection of corporate personhood. It is no small irony that the museum's founders called on none other than Tammany Hall's charismatic leader, William "Boss" Tweed, to help them acquire an official state charter. Although there was not much love for Tammany Hall among the city's wealthy elite—William E. Dodge Jr., for example, described them as "greedy, corrupt, and treacherous"—the two groups did forge an uneasy alliance at times.[36] So long as Tammany Hall helped to maintain control over the flood of poor immigrants entering the city and pushed for expansionist municipal projects that benefited wealthy landowners financially, the latter begrudgingly agreed to support its patronage-based politics.[37] Thus, when Bickmore went to see Tweed in Albany, he was gratified to hear Tweed promise that he would "see [the] bill safely through," and when it came up for discussion in committee the following day, "no hint of any question was made," clearing the path for easy passage before the state legislature.[38] In this way, America's first philanthropically funded civic museum of natural history was officially incorporated during the spring of 1869, forged out of a complex social network that tied an ambitious young naturalist to a group of wealthy capitalists and the Tammany Hall Democrat William "Boss" Tweed.

From a Respectable Museum to a "College of Discovery"

From the moment that the American Museum of Natural History was founded, its leaders faced a tension between their desire to see the museum become a respectable institution of science and the need to attract a broad audience. Initially, the museum's founders primarily sought to acquire rare, expensive, and spectacular objects of natural origin while hosting a series of exclusive parties. As the *New York Times* described it, the first of these gatherings featured "excellent music" to accompany the "*élite* of the City" as they "promenaded up and down inspecting the numerous cases, and filling their minds with science, while their ears were filled with the soft strains of LANNER and STRAUSS."[39] At the same time, however, the museum's founders wanted to draw in a more socially diverse group of visitors, hoping that an exposure to nature would edify and uplift the city's working poor, whose daily routine was otherwise confined to an exclusively urban environment: the factory floor, cramped apartments in tenement houses, and the dingy streets of the Lower East Side. Unfortunately, it soon became clear that the museum was failing to draw large numbers of working people into its exhibition halls. Not only did the city's system of elevated railways fail to extend far enough uptown at the time, but the museum had not yet cornered the market on its particular brand of cultural goods.[40]

Part of the problem stemmed from stiff competition. Despite a constant refrain among the city's elite that New York desperately needed a new museum, there were a whole range of institutions that already exhibited natural history specimens. The largest and most successful of these belonged to P. T. Barnum, who was arguably the most famous American showman of the nineteenth century. By coincidence, Barnum's museum was also named the American Museum, so it was Bickmore's good fortune to see it burn (for a second time) in a catastrophic fire only a few years before his own institution finally opened its doors. Barnum decided to move on and try his luck running a traveling circus, but his museum continued to cast a long shadow over the city's amusement offerings, and it was survived by a large number of similar institutions. Although smaller and less well known, places such as the Eden Musée, Wood's Museum, and many small storefront operations of a similar vein continued to attract large audiences all over the city.[41]

These dime museums (so called because of the customary admission fee) were in stark contrast to the kind of institution that Bickmore and his financial backers envisioned. Whereas Bickmore's museum was a philanthropic endeavor, dime museums were commercial operations meant to enrich their proprietors by attracting a large number of paying customers. Again, it was Barnum who primarily set the tone for the genre, describing his own museum as an explicitly capitalist undertaking. In his best-selling autobiography, he openly confessed, "I liked the Museum mainly for the opportunities it afforded for rapidly making money."[42] But just because dime museums could be lucrative undertakings does not mean they were necessarily seen as lowbrow. On the contrary, they catered to the widespread enthusiasm for "rational recreation" among the middle and working classes alike, a characteristically nineteenth-century form of entertainment that was designed to educate as well as delight. To that end, Barnum continually updated his exhibits by purchasing "genuine" specimens of live and stuffed animals from all over the world—"regardless of cost," he boasted—in addition to other "curiosities" that included mechanical contrivances such as the magic lantern. Like many other institutions of its genre, Barnum's museum also featured a lecture hall that could seat over a thousand people, in which he showcased "philosophical demonstrations" alongside a constantly changing suite of "transient attractions," including "educated dogs, industrious fleas, automatons," and "ventriloquists," as well as "Albinos, fat boys, giants, dwarves," and "rope-dancers," not to mention an "English Punch and Judy" show.[43]

Barnum's real stroke of brilliance was to lure paying customers out of their homes and into the city by emphasizing the moral and educational benefits of his museum. To that end, he took care to advertise that he had abolished "all vulgarity and profanity from the stage," insisting that all "the attractions of the Museum are intended for the moral and the intelligent, in contradistinction to those who seek unwholesome excitement."[44] Similarly, a visitors' guide from the mid-1860s described all the different animals on display in extensive detail, informing readers that there was "no study that is more important to the youth of a rising generation, or to adult age, than that of Natural History," for it "teaches man his superiority over brute creation, and creates in his bosom a knowledge of the wisdom and goodness and omnipresence of a supreme and all-wise Creator."[45] Barnum clearly sought to project an image of his museum as a respectable institution, one that catered to a popular but morally chaste and socially upright audience. He even boasted of hatching an ambitious plan during the mid-1860s that would have seen his museum transformed into "a great free institution, which would be similar to and in some respects superior to the British Museum in London."[46] Needless to say, that plan never came to fruition, and given how forthright he was about wanting to maximize profits, there is good reason to doubt its sincerity. Still, Barnum routinely called on the authority of well-known naturalists such as Louis Agassiz, who publicly attested that Barnum could be trusted to exhibit "animals as nearly as possible in their state of nature."[47] Toward the end of his life, Barnum even endowed a philanthropic museum of his own, at Tufts University in Massachusetts.[48]

Barnum's claims that his museum primarily served an educational mission notwithstanding, the moral and epistemic value of dime museums was often called into question. Not only did many commercial entertainment venues feature lurid exhibits of so-called monstrous births and other aberrations of both the social and the natural orders—one newspaper account derided their tendency to exhibit everything from "the shadowy form of the 'Living Skeleton'" to "the hirsute wealth of the bearded lady," as well as "the India rubber man, the two-headed boy, and other odds and ends of nature's handiwork"—but the city's numerous medical museums even invited curiosity seekers to ogle wax models of human genitalia ravaged by venereal diseases such as syphilis.[49] To make matters worse, commercial museums were often rumored to engage in the clandestine sale of alcohol. On one occasion, the city brought suit against a purveyor of popular spectacle called "Sandy" Spencer for permitting "liquors to be sold in the auditorium while a strange performance was in progress."[50] Perhaps worst of all was that dime museums were widely accused of sheltering pickpockets and other petty criminals, some of whom were even said to be prone to violence. A tourist guide from the mid-1860s explicitly advised visitors to beware the "bodily danger" one could incur in such "places of amusement," and an 1885 article in the *New York Times* reported that two men had been assaulted by the cashier of a Bowery museum when they refused to purchase a "bottle of blood purifier."[51]

Commercial museums also resorted to playful deceptions to lure visitors in off the streets. Most famous among these was Barnum's "Feejee Mermaid," a curiosity that combined the anterior half of a monkey and the posterior end of a fish. The authenticity of this notorious exhibit was actively called into question by Barnum himself, who gleefully urged audiences to see if they could discover any stitching between the mermaid's two halves.[52] While antics like these got him branded a cheat and a charlatan, Barnum relished the controversy, claiming it only "helped to advertise me, and I was willing to bear the reputation."[53] But not everyone was in on the joke. An 1866 article published in the *New York Herald,* for example, denounced the city's "humbug show concerns" as "disgusting cheats," and a few months later, the same newspaper elaborated that "to be of sterling value," a museum "must be a public institution." Lambasting Barnum as the "charlatan general of showmen whose sole reputation is based upon the shameful exposure of frauds of which he himself is the chronicler," the *Herald* judged dime museums to be "of no possible value." Similarly, an editorial in the *New York Times* opined that what the city really needed was "a well gotten-up museum" with a truly educational and scientific collection, one whose exhibits "might be made as attractive even to country folks and youngsters as a paltry collection of preposterous things which Mr. Barnum so justly styles humbugs."[54] In addition to being seen as morally lax and permissive, then, dime museums were widely considered intellectually suspect as well.

Deception and fraud were a part of business as usual in the nineteenth-century United States, and the problem of authenticity was on everyone's mind. This was true not only in the American West but in large urban centers also. As a result, the confidence man became a favorite of nineteenth-century literature, offering readers a chance to fantasize about being fleeced in a low-stakes and fictional world.[55] Something similar could be said of the dime museum, whose exhibits openly challenged visitors to decide for themselves what was authentic and what was fake. Thus, the experience of visiting a commercial museum can be likened to making a purchase in a major metropolis. The "complex system of urban market exchange . . . looked to many contemporary observers like a kind of humbug too," the historian James Cook explains, and museum visitors appraised a dubious specimen such as the Feejee Mermaid in the same spirit as they did "a heavily promoted patent medicine, or cut-rate goods at a Chatham Street auction."[56] In other words, Barnum's exhibits were enjoyable because they elicited the active participation of viewers, inviting audiences to hone the very same faculties that were required to navigate a dynamic and notoriously unstable marketplace populated by what often seemed to be untrustworthy strangers.

For Bickmore's museum to succeed at its core goal of reflecting well on the city's wealthy elite, it had to distinguish itself from the competition. A failure to do so would have invited comparison with for-profit amusement venues that exhibited questionable morals and even more questionable knowledge, which is precisely what Bickmore's financial benefactors sought to avoid. This was not just because the open and democratic nature of New York's marketplace had come to signify all that was mean and vulgar about the city. It was also because their visions for what a museum ought to accomplish differed radically from those of someone like Barnum. In the eyes of its founders, Bickmore's museum was not going to challenge visitors to determine the meaning of an exhibit for themselves. Quite the contrary, it would offer them an authoritative appraisal of the latest and most trustworthy science. To that end, its founders quickly moved beyond simply mounting lavish exhibits and holding exclusive soirées and began to hire their own team of naturalists who could add to the stores of reliable knowledge.

The stark difference between Barnum's and Bickmore's museums was immediately apparent from their sharply contrasting architectural styles. Located in a commercial neighborhood on the busy downtown intersection of Broadway and Anne Streets, Barnum's museum was designed to entice visitors inside with a colorful and inviting exterior. Besides a variety of large flags and banners, the facade was adorned with bold illustrations of the animals one could expect to encounter within. In addition, Barnum often hired a band to play loud music on the front balconies in the hopes of capturing the attention of passersby. In contrast, the American Museum of Natural History was deliberately located in a more exclusive and residential part of the city. During the 1870s, it moved from its original, temporary location in the Arsenal Building in the southeastern portion of Central Park to its current home on the Upper West Side. The museum's location not only reflected its origins as a philanthropic project of the city's wealthy elite, who had been moving farther uptown for decades; it also ensured the museum would be far removed from the city's commercial thoroughfares. The new building was designed in a Victorian Gothic style, and it was intended to form part of a much larger architectural schema that would feature a central, octagonal crossing covered by a massive, five-story dome. By the 1890s, the museum's wealthy trustees commissioned Josiah Cleveland Cady to design an even more ambitious plan, which led to the construction of a new southern wing that enveloped the first building completely. Whereas Barnum's museum had been busy and colorful, this stately and majestic structure was imposing and fortresslike, adorned with a series of Richardsonian turrets and towers that made this institution appear more intent on protecting its precious contents than welcoming visitors in off the street.[57]

Despite the building's imposing architectural style, Bickmore and his wealthy benefactors were genuine in their desire to attract a broad audience. They recognized that to succeed in its mission of disciplining the working classes and educating the general public, their museum had to draw in large numbers of ordinary New Yorkers. Thus, while they were careful to avoid sacrificing their institution's seriousness of purpose, they were not above engaging in a bit of Barnumesque showmanship. As one trustee confided to another, "We must sprinkle our wholesome bread with a little sugar."[58] To do so, they acquired attention-grabbing displays such as an infamous piece of theatrical taxidermy by Jules Verraux that showed an Arab courier on a camel being attacked by a stuffed Barbary lion. At one point during the 1880s, the museum's president, a successful banker named Morris K. Jesup, even sent Barnum a flattering letter in which he admitted that, "If we can succeed in a measure in interesting the public to a small extent of what you have done, we shall consider our efforts a success."[59] That said, the museum's trustees also recognized they were walking a fine line, and unlike Barnum, who deliberately played with the difficulty of distinguishing fact from fiction, they instead sought to reinforce the boundary between science and popular knowledge.

In search of a strategy to balance the museum's elite status with the desire to attract a large audience, its founders carefully calibrated the exhibitions to communicate on a number of registers simultaneously. For that reason, its paleontology displays not only featured large and eye-catching show specimens, such as the mastodon, the Irish elk, and the giant moa, which were prominently placed in the exhibition hall's central isle. It also flanked these on either side by rows upon rows of glass cases containing more sober displays. The latter were arranged taxonomically, and each specimen was accompanied by a printed label that featured its binomial designation in Latin. To make the point even clearer, a guide to the museum's paleontology halls admonished visitors to pay careful attention to the minute details characterizing each specimen, as well as the relationships between them. Visitors were directed to inspect each printed label, as well as the objects themselves, in hopes that they might get a sense of the complex classification scheme that was used to organize everything in accordance with the most up-to-date scientific standards.[60]

In mounting exhibits that mixed sober pedagogical elements with more sensational showpieces, the museum's founders borrowed a page from Barnum's playbook, yet they took it several steps further. Whereas dime museums sometimes invited respected naturalists such as Agassiz to deliver lectures for paying customers, Bickmore's wealthy supporters gradually transformed their museum into a bona fide research institution. The decision to pair rational recreation with original scientific research was far from unique, inspired as it was by the New Museum Idea, which was gaining widespread traction at the time. As the British Museum's keeper of zoology, John Edward Gray, wrote in an influential polemic, it was vital for public museums to distinguish between two very different purposes: they were responsible for "the diffusion of instruction and rational amusement among the mass of people," but they also had "to afford the scientific student every possible means of examining and studying the specimens of which the museum consists." For a museum to reach its full potential, neither purpose could be neglected or folded into the other. Rather, each had to be recognized as a priority in its own right, so much so that Gray pushed for a sharp separation between a museum's display specimens and what came to be called its "study collection." Whereas the former should be arranged to capture the attention of ordinary visitors, the latter should strive for completeness and "admit of the most minute examination."[61] In this way, the public museum could balance the needs of its several constituencies, mounting spectacular exhibitions without sacrificing its scholarly mission.[62]

Bickmore's philanthropic partners embraced the New Museum Idea wholeheartedly. Although their institution began as a public exhibition space, it soon acquired a research mission, especially after moving out of the Arsenal Building and into a purpose-built structure in Manhattan Square during the 1870s. Whereas the original site only had room for exhibits, the new structure was designed to combine "large palatial saloons for the public" with "spacious and well-lighted rooms" in which naturalists could "pursue their favorite studies."[63] Because the museum did not yet employ its own scientific staff, these workrooms were primarily intended for use by "distinguished scientific men from abroad," although anyone capable of "adding to the existing stock of knowledge" would be admitted, with the telling caveat that all "charlatans and pretenders will be excluded."[64] This was welcome news to members of the scientific community, many of whom argued that even more should be done in support of their work. In 1874 the secretary of the Smithsonian Institution enjoined wealthy New Yorkers to "endow this Park Museum with a college of discovery" that was capable of "interrogating nature and discovering new facts," as well as "expounding established and known truths." Only then, he concluded, could it truly hope to rival the great European centers of learning in Paris, London, and Berlin.[65] Similarly, at a public ceremony to celebrate the museum's new building in 1877, the paleontologist Othniel Charles Marsh used his speech to challenge "the old idea of a museum" as a mere "showroom," arguing that its trustees ought to embrace the "modern idea" that "makes it a workshop as well." Playing on the vanity and the insecurities of his audience, Marsh even predicted that their project would fail "to achieve more than local influence unless the work-rooms above [the exhibits] are made the most important feature of the whole."[66] The scientific community was determined to convince the museum's wealthy trustees to expand on their philanthropic ambitions and turn their museum into a genuine research institution.

This not-so-subtle advice did not fall on deaf ears among a group of wealthy philanthropists who coveted the legitimacy that naturalists such as Marsh could bestow on their efforts. As a result, they agreed to transform the museum into an internationally renowned center of scientific research. In 1881 they began publishing a regular periodical, the *Bulletin of the American Museum of Natural History*. Around the same time, the museum also began to amass a study collection of specimens not meant for public display. Perhaps most important was the decision to place several respected naturalists on the museum's payroll. In 1885, for example, the founders lured a promising young ornithologist from Harvard, Joel A. Allen, to New York with the title of professor and a salary of $3,000 per year.[67] These changes represented a crucial development for the museum, allowing it to begin forging reciprocal relationships with peer institutions all over the world. By engaging in the nonmonetary exchange of knowledge, specimens, and publications, the museum slowly succeeded in joining the international scientific community. As the American geologist John S. Newberry wrote to Jesup, the decision to invest in a research program would "give fame and influence to the Museum," clearly signaling that its trustees would "not permit it to be diverted from its original purpose, and become a mere show-room of natural curiosities." To that end, Newberry especially praised Jesup's willingness to invest "large sums for the purchase of material that does not appeal to the vulgar or even an uneducated interest."[68]

Although the museum's first well-known research scholar was an ornithologist, it soon became clear that the new science of vertebrate paleontology was particularly well suited to furthering its complex institutional goals. As the creators of the ill-fated Paleozoic Museum had recognized several decades before, the huge and often strange-looking creatures discovered by paleontologists were especially popular with the public. Just as important, however, was the fact that that geology and paleontology increasingly ranked among the United States' most prestigious branches of science. This was not only due to their enormous importance for the mineral industry, which generated large amounts of wealth during the period, but also because paleontologists supplied some of the most convincing evidence for the controversial hypothesis that all species of biological organisms had evolved from a common ancestor. As Marsh pointed out in his speech to celebrate the completion of the museum's new building in 1877, the fossil discoveries coming out of the American West spoke directly to the "great problems" of the age, helping to elucidate "the origin of life itself."[69] Marsh was not alone in this view. The British evolutionist Thomas Henry Huxley made sure to visit the Yale Peabody Museum in New Haven in person, stating, "There is nothing in any way comparable . . . for their scientific importance, to the series of fossils which Professor Marsh has brought together."[70] Perhaps even more impressive was that Charles Darwin praised Marsh for uncovering some of "the best support to the theory of evolution, which has appeared within the last 20 years."[71] For an American natural history museum keen to develop a research mission, it was therefore all but inevitable that vertebrate paleontology would take center stage.

During the late 1880s, the American Museum of Natural History began to develop an ambitious program in vertebrate paleontology. As one curator informed Jesup, "Your magnificent museum . . . will need some representation of the giant vertebrate fauna which Marsh and Cope and Leidy have made known to the world."[72] Jesup clearly agreed, and he hired a young paleontologist from Princeton University named Henry Fairfield Osborn. Osborn was an obvious choice. Not only was he an up-and-coming star in the field, but his uncle was none other than J. P. Morgan, one of the museum's founding trustees. Having been extensively courted by Jesup, Osborn decided to take up a teaching position at Columbia University in the spring of 1891, proposing to spend half of his time at the museum and promising to make New York into "a center for exhibition, publication and research" in paleontology, a branch of science "in which American leads the world." The son of a wealthy railroad magnate, Osborn asked no remuneration for his own "services in supervising the work," but he did request that the museum "appropriate the sum of $5,000" a year to hire assistants and cover operating expenses.[73] Within less than three months, the board of trustees voted to create a paleontology department under Osborn's direction to help further "the cause of science, education, and popular interest" at the museum.[74]

Only a few years after Osborn arrived in New York, he began shipping dinosaurs from the American West to the museum. Although he initially focused his efforts on mammalian fossils, Osborn was convinced to change tack in 1897 when a young collector named Barnum Brown discovered a rich trove of dinosaur bones while digging in one of Marsh's old quarries at Como Bluff. Brown's initial discovery turned out to be a magnificently preserved *Diplodocus,* a large plant-eating dinosaur with an exceedingly long neck and tail. But there were plenty of other specimens too, including the bones of the closely related *Brontosaurus.* Before long, it was clear that despite all the work Marsh's collectors had done there during the 1870s and 1880s, Como Bluff still contained plenty of dinosaur bones. Brown even went so far as to describe the new quarry as "a veritable gold mine," adding, "I have been in bones up to my eyes."[75] In response, Osborn directed a whole crew of field assistants to join Brown in Wyoming, where they would continue to excavate dinosaur bones over the next several years.

Almost immediately, Osborn was captivated by the idea of mounting one of the newly found dinosaurs as a freestanding display. Hopes ran high that Brown's first *Diplodocus* specimen might be sufficiently complete to furnish such an exhibit, but further digging into the hillside revealed that the forelimbs, neck, and skull were all missing. Luckily, the next several field seasons continued to bring spectacular new discoveries, including a rich site the museum's field crew named Bone Cabin Quarry because a local shepherd had used the region's superabundant fossils to build a small hut for himself nearby (Figure 2.3). Then, in the summer of 1899, a field assistant named Walter Granger found yet another quarry just a few miles from Como that contained a second *Brontosaurus* specimen. This gave Osborn and his assistants the idea of fabricating a so-called composite mount by combining elements from several individuals in a single display. "I am very much pleased with your discovery of [the] Brontosaur," Osborn wrote to Granger, saying that "if it is of the right size" to match up with the Como specimen, "it is a great hit, in fact, the very greatest you could have made."[76]

Despite his initial interest in mammalian fossils, Osborn had become completely captivated by dinosaurs. Citing the broad popular appeal they could be expected to command, he encouraged museum trustees to fund the construction of a new exhibition hall. Although they were "more difficult to find

Figure 2.3. Excavating a sauropod dinosaur at Bone Cabin Quarry, Wyoming, 1898.

and more expensive to collect" than mammals, dinosaurs promised to become the museum's biggest draw, "representing more ancient and less known types of life, more widely different from those of the present day, and in many respects far more extraordinary than animals shown in the Hall of Fossil Mammals."[77] By 1903, a rectangular space in the museum's East Wing had been reserved for the exhibition of dinosaur bones. At first, specimens were displayed individually on long tables and in glass cases for detailed inspection. But from the start, Osborn's plan was to create synthetic and lifelike displays that emphasized the dinosaurs' colossal dimensions by combining individual pieces into a larger assemblage. To that end, he had the vertebrae of a huge *Brontosaurus* "laid out on a series of tables in as nearly as possible their natural relations, giving a much more vivid idea than has heretofore been possible of the gigantic size of these animals."[78] Around the same time, he instructed technicians to mount the hind limbs of three different dinosaurs upright along the exhibition hall's wall. As Osborn explained in the museum's annual report, this arrangement was sure to make a powerful impression on visitors, providing them with a vivid sense of these creatures' functional anatomy.[79] While these early exhibits were remarkably successful, Osborn had something far more ambitious in mind, and he soon set to work assembling fragments of the museum's largest and most spectacular dinosaur, *Brontosaurus,* into a single, freestanding display.

Osborn had good reason to believe that a fully articulated dinosaur skeleton would capture the public's imagination. Similar exhibits had done so before. Most notable was a plaster cast of *Hadrosaurus* that Benjamin Waterhouse Hawkins assembled for the Philadelphia Academy of Natural Sciences during the late 1860s. After Hawkins had been commissioned to make a series of dinosaur sculptures for the ill-fated Paleozoic Museum, he decided to familiarize himself with the fossil remains of *Hadrosaurus* and *Laelaps* in Philadelphia. To thank Joseph Leidy for granting him access to the academy's collections, Hawkins assembled a cast of *Hadrosaurus* into a freestanding skeleton in the academy's exhibition hall. This was the first time that anyone had mounted a dinosaur in this way. When it was unveiled to the public, the new exhibit proved a spectacular success. Attendance at the academy's museum immediately shot up, nearly doubling in the space of a year. But not everyone was pleased with these developments. The academy's scientists in particular complained that the huge crowds only got in the way of "those who would really wish to examine the collections."[80] As an exhibition guide from 1876 made abundantly clear, the academy's members were anxious to maintain an institutional distance from popular showrooms "in which are exhibited chiefly animal monsters and effigies of strange things, . . . in a word, whatever a wondermonger can collect to allure the curious and idle many to amusement at small individual cost to them but lucrative to the showman."[81] As a result, the academy began charging an entrance fee of ten cents, which its members judged sufficiently dear to "moderate the crowds" without turning away students who had a truly abiding interest in natural history.[82]

Not only did the leaders of research museums worry about pandering too much to popular tastes, but curators also felt anxious that dinosaurs remained too little known to furnish material for an authoritative, scientific display. When

Marsh was asked about the feasibility of exhibiting a copy of *Hadrosaurus* at the 1876 Centennial Exposition in Philadelphia, for example, he objected in no uncertain terms, stating, "I do not believe it possible, at present, to make a restoration of any of the more important extinct animals of the country that would be of real value to science or to the public." Because it would "certainly end in serious mistakes," Marsh insisted that any attempt to assemble a dinosaur would do a signal disservice to paleontology, "as error in such a case is very difficult to eradicate from the public mind."[83] A couple of decades later, Marsh offered similar misgivings in print. Given the wildly erroneous ways in which European dinosaurs were often depicted, he joked that they had "suffered much from both their enemies and their friends." They "were destroyed and dismembered long ago by their natural enemies," and now "their friends have done them further injustice in putting together their scattered remains." He especially singled out Hawkins's sculptural reconstructions at the Crystal Palace for derision, arguing, "There is nothing like unto them in the heavens, or on the earth, or in the waters under the earth."[84] Thus, while dinosaurs had the power to capture the attention of large and diverse audiences, research museums generally preferred to exhibit them in a state that was accessible to direct observation—namely, as scattered and disarticulated fragments of fossilized bone.

In contrast, the New York museum was willing to use popular spectacle to attract visitors, boldly assembling the *Brontosaurus* fossils collected in different parts of the American West into a single "composite" skeleton. The result measured nearly seventy feet in length and over fifteen feet in height. Osborn described its completion as "the most noticeable event in the work of this department," boasting that it "attracted a great deal of attention in the press," which resulted in a 25 percent increase in attendance.[85] When the doors to the new dinosaur hall were thrown open in 1905, *Brontosaurus* was at the very center of the new exhibition hall. To help audiences make sense of this monster, the museum also exhibited a painting and scale model underneath its long neck depicting the way curators imagined the creature would have looked when alive (Figure 2.4). Over the next several decades, the museum continued to mount increasingly lifelike and extravagant dinosaur displays. These included a large and ferocious *Allosaurus* that was shown in the

Figure 2.4. The *Brontosaurus* display in the Hall of Fossil Reptiles, American Museum of Natural History.

act of feasting on a section of *Brontosaurus* tail, as well as a pair of duckbilled dinosaurs evading a predator near the shores of a lake. These met with an enthusiastic response, drawing throngs of people numbering in the hundreds of thousands into the museum each year to catch a glimpse of its latest dinosaur.[86]

Despite its undeniable success, not everyone agreed that the new *Brontosaurus* befitted a serious museum of natural history. The *New York Times* quoted a "professor with large glasses" who derided the decision to host a tea party in the new dinosaur hall as "an absurdity," adding that such publicity stunts were "in bad taste in a place devoted to science." However, the professor's companion apparently had a more sanguine view, saying, "I don't like [the spectacle], but I must excuse it" for having "drawn all these people here, being a splendid bit of advertising." Regardless, it was clear to the *Times* reporter that whatever all those people got out of their visit, authoritative knowledge about the history of life on Earth ranked near the bottom of the list. "The poor beast if alive would not have recognized his scientific name in the many variations it took," the newspaper reported. "Some wanted to see the 'dino,' others the 'diorso,' and among other designations were 'the octopus,' and 'His Nibs, Old Boney.'" In the end, everyone seemed to take away from the exhibit just what suited his or her fancy. A butcher, for example, speculated that such an animal would "burst the Meat Trust in a week."[87]

Museum curators knew they would court controversy by appealing to popular tastes, but they concluded that the risks were outweighed by the rewards. In a widely read article that appeared in *Science* magazine, the anthropologist Franz Boas argued that as a public institution devoted to popular education, the museum had to engage in more than just specialized scientific research. It was also responsible for creating "healthy and stimulating surroundings" where ordinary New Yorkers could "employ their leisure time," thereby serving to counteract "the influence of the saloon and of the race track." To draw in these people, the museum had no choice but to stage exhibitions that, "first of all," were "entertaining." Few among the city's idle masses would seek out the genuine moral and educational benefits that natural history had to offer, Boas suggested, had these not first been made widely palatable. "The people who seek rest and recreation resent an attempt at systematic instruction while they are looking for some emotional excitement," he reported, singling out dinosaurs as an especially valuable public relations asset. "When the installation of a new immense mounted skeleton of some extinct animal is announced," he wrote, "people will flock in crowds to the museum to see the specimen."[88] By all indications, Boas was right. In 1905, for example, the museum reported an "unusually large increase in the number of visitors," amounting to more than half a million in all. This "gratifying" development was chalked up to "the opening of several striking exhibits, particularly the huge *Brontosaurus*."[89] With the aid of a spectacular new dinosaur, the museum had successfully implemented the New Museum Idea, drawing large numbers of people into its exhibition hall, where they could be uplifted by exposure to the latest scientific research being conducted by curators in house.

Figure 2.5. Dinner inside the *Iguanodon* mold at the Crystal Palace in 1853, from an original watercolor by Benjamin Waterhouse Hawkins, 1872.

Conclusion

Thirty years after Andrew Haswell Green pushed for the construction of the Paleozoic Museum, New York finally had a complete dinosaur on display. In many ways, the *Brontosaurus* exhibit was modeled on its Victorian predecessors, and there were numerous parallels between it and the dinosaurs sculpted by Hawkins. Not least was Osborn's decision to host a tea party under the museum's newest star specimen, which called to mind a similar gathering held in December 1853 to celebrate the completion of Hawkins's Crystal Palace dinosaur sculptures. In concert with Richard Owen, Hawkins invited about twenty prominent men of science, art, and commerce to join him for a New Year's Eve dinner inside the hollow mold of the *Iguanodon*. Owen presided over the event from his place at the head of the table, and judging from a watercolor that Hawkins drew later, it must have been a tight fit (Figure 2.5). Despite these close quarters, the evening was a rousing success, featuring an elaborate menu of mock turtle soup, *turbot à l'hollandaise,* and *nougat à la Chantilly,* to name just a few of the delicacies on offer that night, accompanied by a selection of sherry, Madeira, port, and claret. So taken were people with the event that it received generous coverage in newspapers and magazines across England, including a satirical piece in *Punch* that congratulated Owen and Hawkins "on the era in which they live; for if it had been an earlier geological period, they might perhaps have occupied the Iguanodon's inside without having any dinner there."[90]

But the dinner hosted by Hawkins also differed from Osborn's tea party a half century later in many respects. The most striking were the two institutional settings. Although both gatherings took place in a space that was devoted to rational recreation, the principle of their organization could hardly have been more different. First, whereas the American Museum of Natural History was understood as a philanthropic endeavor, the Crystal Palace pleasure ground was a commercial venture that promised to pay financial dividends to its shareholders.[91] Second, the American Museum of Natural History also distinguished itself by combining the goals of rational recreation and the furtherance of original scientific research, whereas the Crystal Palace only sought to "diffuse" the latest geological knowledge among a broad audience. To be sure, the New York museum was hardly unique in this regard, and museums in Paris, Berlin, and London all developed similar hybrid missions during the last third of the nineteenth century. However, the conspicuous generosity of its wealthy benefactors ensured that the New York museum had access to far more lavish funds than did its European counterparts. This, in addition to its proximity to the American West, quickly allowed it to distinguish itself in the collection, study, and exhibition of dinosaurs.

During the decades that followed, a number of additional philanthropic museums sprang up in the United States, and they began to compete with each other for access to the largest dinosaurs. Arguably most important was a museum founded by the wealthy steel magnate Andrew Carnegie in Pittsburgh, which soon began manufacturing dinosaur displays on an industrial scale for export abroad.

3

Andrew Carnegie's *Diplodocus*

In May 1905, only a few months after the public unveiling of *Brontosaurus* in New York, an even more sumptuous ceremony took place under another long-necked plant-eating dinosaur. This time, however, the gathering was held at the British Museum of Natural History in London's South Kensington neighborhood. While the New York museum could boast of an impressive guest list of notable civic and business leaders, the British Museum assembled an even more august group of dignitaries, including members of the nobility and Parliament, several high-ranking military officials, and men of science with international standing. But that does not mean America's paleontological dominance was on the wane. Quite the contrary, for the dinosaur being unveiled that afternoon hailed from the American West, and the guest of honor was none other than America's richest and best-known industrialist, Andrew Carnegie. Carnegie had personally donated this specimen to the museum, and it was so closely tied to the famous steel magnate that it even bore his name, having officially been christened *Diplodocus carnegii* by vertebrate paleontologists.[1]

To many who were in attendance, Carnegie's dinosaur seemed a fitting symbol for the United States and its entrepreneurial zeal. As one of the evening's speakers noted, "It is appropriate that such a monster as this should have lived on a great continent like North America," blessed as that region was with such "wonderful resources."[2] The satirical magazine *Punch* put it even more bluntly, observing that "even in the earliest periods, our American cousins did things on a more colossal scale."[3] These sentiments were only

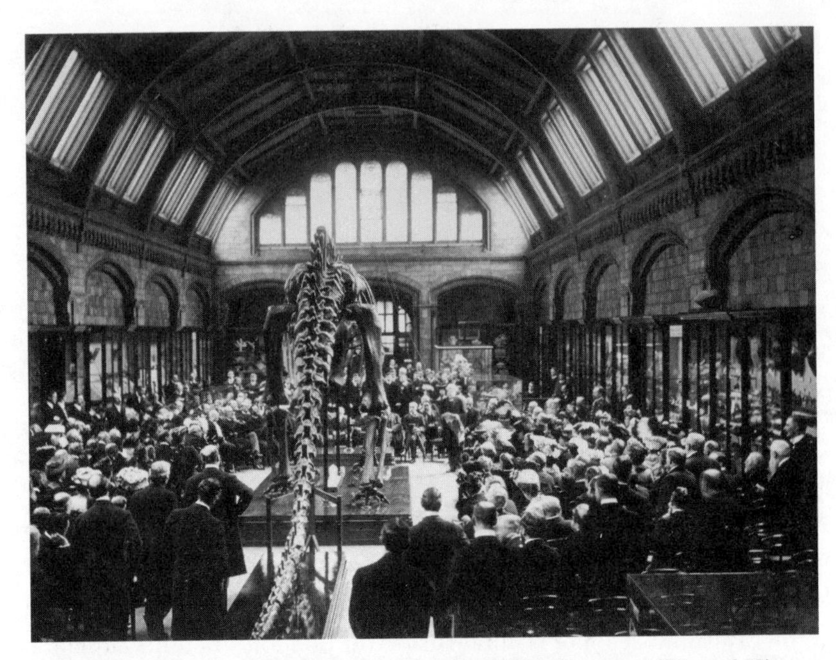

Figure 3.1. Unveiling *Diplodocus carnegii* at the British Museum of Natural History, May 12, 1905, with Lord Avebury speaking in the background.

amplified by the fact that Carnegie's dinosaur was not even a real fossil. It was only one of about a dozen identical copies the American steel magnate had fabricated in plaster. Whereas the original specimen was to go on display in Pittsburgh, the replicas were donated to some of the world's oldest and most respected natural history museums. Besides London, this included museums in Berlin, Paris, Vienna, Bologna, Saint Petersburg, Madrid, and Munich, as well as La Plata and Mexico City. Everywhere that it went, the fossil caused a sensation, and it remains one of the most recognizable dinosaurs to this day.[4]

The story of *Diplodocus carnegii* demonstrates that the widespread enthusiasm for popular science among wealthy capitalists was far more than a mere sideshow to their core business interests. In the case of someone like Carnegie, philanthropy was nothing less than a life's work, an all-embracing social philosophy, and the evolutionary apex of modern capitalism all rolled into one. Although he worked hard to give the impression that his life was divided into two distinct parts—whereas his youth was spend amassing a huge fortune, his old age was devoted to giving nearly all of that money away—Carnegie's grand philanthropic ambitions and considerable business acumen were perfectly of a piece with each other. This makes him a particularly clear illustration of the degree to which ostentatious displays of generosity were part of, rather than separate from, the cycle of accumulation that dominated American culture during the Long Gilded Age.[5]

For Carnegie, building a natural history museum and running the world's largest and most profitable steel-manufacturing company were not fundamentally distinct enterprises. Rather, they functioned as two sides of the same coin, complementing each other in perfectly dialectical fashion. Whereas Carnegie's business ventures supplied the funds for his philanthropic endeavors, his philanthropic projects helped to legitimize the unrivaled profits that he made in the steel industry. This led to a widespread suspicion that Carnegie's true motives were not nearly so selfless as he liked to profess, and contemporary critics like Harold Joseph Laski argued that philanthropy was just a veneer used to cover up unsavory business practices. Moreover, because it offered a means to circumvent the political process and shape civil society directly through private bequests, philanthropy was, and remains, fundamentally antidemocratic.[6] Critics like Laski were right, but this chapter will argue that Carnegie's desire to give the bulk of his fortune away did not have to be insincere or purely self-serving to further his personal interests. On the contrary, business tycoons such as Carnegie were no less keen to accrue a large stockpile of intangible or "symbolic" capital than they were to maximize profits, in some cases perhaps even more so. Hence, conspicuous acts of generosity did not have to yield financial dividends to be highly lucrative, even in the traditional sense of the word. Much more important was the extent to which they helped to enlarge one's social network, augment one's reputation, and leave a lasting legacy behind.[7]

Unlike fossil hunters on America's late nineteenth-century mining frontier, people like Carnegie depended on tight-knit social networks forged out of personal bonds of trust to make their ambitious projects succeed. This was especially true in the high-stakes world of international finance. Often, the most lucrative investments were also the most difficult to assess and police using formal procedures, legal obligations, and transparent accounting standards. Access to financial credit was therefore primarily mediated through personal credit, which rested on one's good name and reputation. As Carnegie himself recognized, "Nothing is more essential to young business men than their untarnished credit, credit begotten of confidence in their prudence, principles and stability of character."[8] Active participation in a shared culture of exclusivity was thus about much more than just keeping up with appearances. It also performed an important economic function, helping to broadcast moral qualities, such as honesty, respectability, and trustworthiness, that were essential to building one's reputation as a desirable partner in commercial affairs.[9]

The Most Colossal Animal on Earth

The story of Carnegie's dinosaur begins with a familiar character: William Harlow Reed, the fossil collector who first discovered the famous Como Bluff dinosaur quarry in the summer of 1877. Reed eventually sold that quarry to Othniel Charles Marsh for just fifty dollars, much less than he believed it to be worth. Over the next several years, Reed would remain in Marsh's employ, although the relationship between the two men continued to be marred by friction, in large part due to Marsh's habit of failing to remit wages when they were due. Partially for that reason, when word of the rich diggings at Como spread through the paleontological community, Reed's onetime partner, William E. Carlin, did not hesitate to defect and begin selling fossils to whomever was willing to pay, including Edward Drinker Cope in Philadelphia and Louis Agassiz at Harvard. In 1881 Reed accused Carlin of operating a "Bone Company" in Wyoming.[10] Before long, the competition among field crews working the area grew so intense that Reed even resorted to demolishing any bones that Marsh did not want just to keep them from falling into his rival's hands.[11]

Despite staying loyal to Marsh, Reed bristled at his employer's overbearing demeanor. By the early 1880s, he had decided to quit paleontology altogether and try his luck raising sheep. "I regret leaving the Bone business," he confided to Marsh, but he expected to do much better financially by changing course, insisting, "I think it is my duty to look at my own interests first."[12] Unfortunately, disaster struck down Reed's plans when an especially hard winter killed off most of his flock and left him financially destitute. With nowhere else to turn, he wrote to ask Marsh whether there might still be a job for him collecting dinosaur bones. "I have lost everything I had," he complained, "and of the poor I am the poorest."[13] Marsh agreed to lend Reed some money but refused to take him back as a salaried employee, saying that he would now only consider purchasing individual specimens on a crate-by-crate basis.[14] Knowing what a hard bargain Marsh was likely to drive for each fossil, Reed refused, preferring to spend the next decade or so doing odd jobs such as making hay and working for the railroad.[15]

Finally, in the summer of 1894, Reed's luck began to change. The reason for his newfound good fortune was a significant expansion in the market for American dinosaurs. During the 1870s and 1880s, only a handful of paleontologists had both the means and the desire to purchase a dinosaur quarry. But over the next several decades, a growing enthusiasm for vertebrate paleontology led to the creation of several new museums, which began to compete with one another for specimens. Before long, demand caught up with supply, and collectors like Reed could achieve much greater rewards for their work.

The first indication that things were starting to look up for the veteran fossil hunter was a decision by the University of Wyoming to establish a collection of vertebrate fossils. During the early 1890s, the university hired a young geologist named Wilbur Knight. Initially, Knight's duties were primarily aimed at spurring economic growth in the area by developing the state's mineral industry. In addition to teaching skills such as assaying that were required to work in the mines, Knight published numerous surveys of the region's geology. He also wrote about half a dozen letters a day, mostly responding to requests for information from people looking to evaluate a promising claim.[16] To make the most of a cash-strapped institution, however, the university's president—Albinus Johnson—decided to expand Knight's responsibilities to include dinosaur paleontology. Noting that more and more "prominent Scientific men" were asking for samples of the state's "fossils and minerals," he ordered Knight to begin collecting the material for a museum. Johnson estimated that by taking full advantage of its location, the frontier university could build up a collection "worth $5,000" within a single field season.[17] To help him find the most valuable specimens, Knight hired Reed as a field assistant. At first, Reed worked as a simple day laborer, but Knight feared that he might be tempted to "take the best things he knows of and ship them to eastern colleges."[18] He therefore convinced Johnson to hire Reed on a salaried basis,

Figure 3.2. The "bone room" at the University of Wyoming in Laramie, 1897, with William Harlow Reed in the background.

paying him $1,000 per year in exchange for a promise that all his discoveries would go to the university. By all appearances, the plan was successful, and in 1897 the university boasted that its "bone room" was so large that it rivaled Marsh's extensive collection at Yale (Figure 3.2).[19]

Reed was a shrewd businessman, and Knight's fears that he might defect at any moment were well founded. When he discovered a new marine reptile before being placed on a regular salary, Reed wrote to tell Marsh that "the bones are for sale and will go to the highest bidder."[20] In his response, Marsh was characteristically hard-nosed, stating that while he "should like to have this specimen," its "value, as you well know, depends on how perfect [it] is, and whether it is new or not," and he could not tell this "without seeing the specimen itself."[21] Given the increasing demand for Wyoming fossils, Reed was not inclined to send Marsh the fossils on spec, saying, "I do not think you value them as high as other parties are willing to pay." An experienced negotiator, Reed also cautioned Marsh that a bidding war was in the offing by strategically disclosing that "I am corresponding with several museums and colleges," including Marsh's onetime assistant Samuel Wendell Williston at the University of Kansas.[22] Although Marsh did not take the bait, Williston had offered to pay $250 for Reed's discovery, which Knight eventually matched to acquire it for the University of Wyoming.[23]

Even after the university hired him on a salaried basis, Reed was determined to grab the emerging bull market by its horns. Within a year of signing an exclusive contract, he was already thinking of quitting his job. In the fall of 1898, he confided to Marsh, "Our university is so poor that I am thinking of leaving it and selling my fossils in Europe or to some other American Museum."[24] His confidence must have been buoyed when, acting on the advice of Knight, Henry Fairfield Osborn ordered Barnum Brown to begin collecting dinosaur bones in Marsh's old quarries at Como Bluff.[25] Reed even accompanied Brown on some of his prospecting trips, showing him a quarry "where the bones were literally packed one on top of another."[26] However, rather than continue to share hard-won intelligence about the location of valuable claims freely, Reed judged that he could do better on his own. To increase his chances of success, he sought out a public endorsement from Marsh, asking him to issue an open letter guaranteeing that "the bones here are no. 1 in quality and some of them are monsters."[27] By November of that year, the idea had captivated him completely. "I have found 36 quarries within one mile of my camp," he wrote to Marsh, and "am writing to several large museums . . . in hopes to work up a market for this material before spring."[28] Clearly, Reed set his ambitions much higher than working as an assistant for Knight at the University of Wyoming.

Always the prospector at heart, Reed dreamed of striking it rich by selling an especially grandiose dinosaur. To create a buzz around his discoveries, he lured a reporter from Denver to publicize his success in the field.[29] The plan worked, and news of his exploits soon reached newspapers as far away as New Zealand.[30] Later that winter, a particularly gripping account appeared in William Randolph Hearst's *New York Journal* (Figure 3.4). Under a banner headline that read, "Most Colossal Animal Ever on Earth Just Found Out West," the *New York Journal* article's subheads claimed, "When It Walked the Earth Trembled under Its Weight," "When It Ate It Filled a Stomach Large Enough to Hold Three Elephants," "When It Was Angry Its Terrible Roar Could Be Heard for Ten Miles," and

Figure 3.3. William Harlow Reed posing next to a dinosaur femur, 1898.

"When It Stood Up Its Height Was Equal to Eleven Stories of a Sky Scraper."[31] This sensational account was illustrated with a number of drawings and photographs, including a portrait of Reed posing next to the gigantic femur of a large, plant-eating sauropod dinosaur, as well as an imaginative rendering of the huge beast as it was thought to have appeared in life, rearing up on its hind legs to peer through a window atop the recently completed New York Life Building in Manhattan.

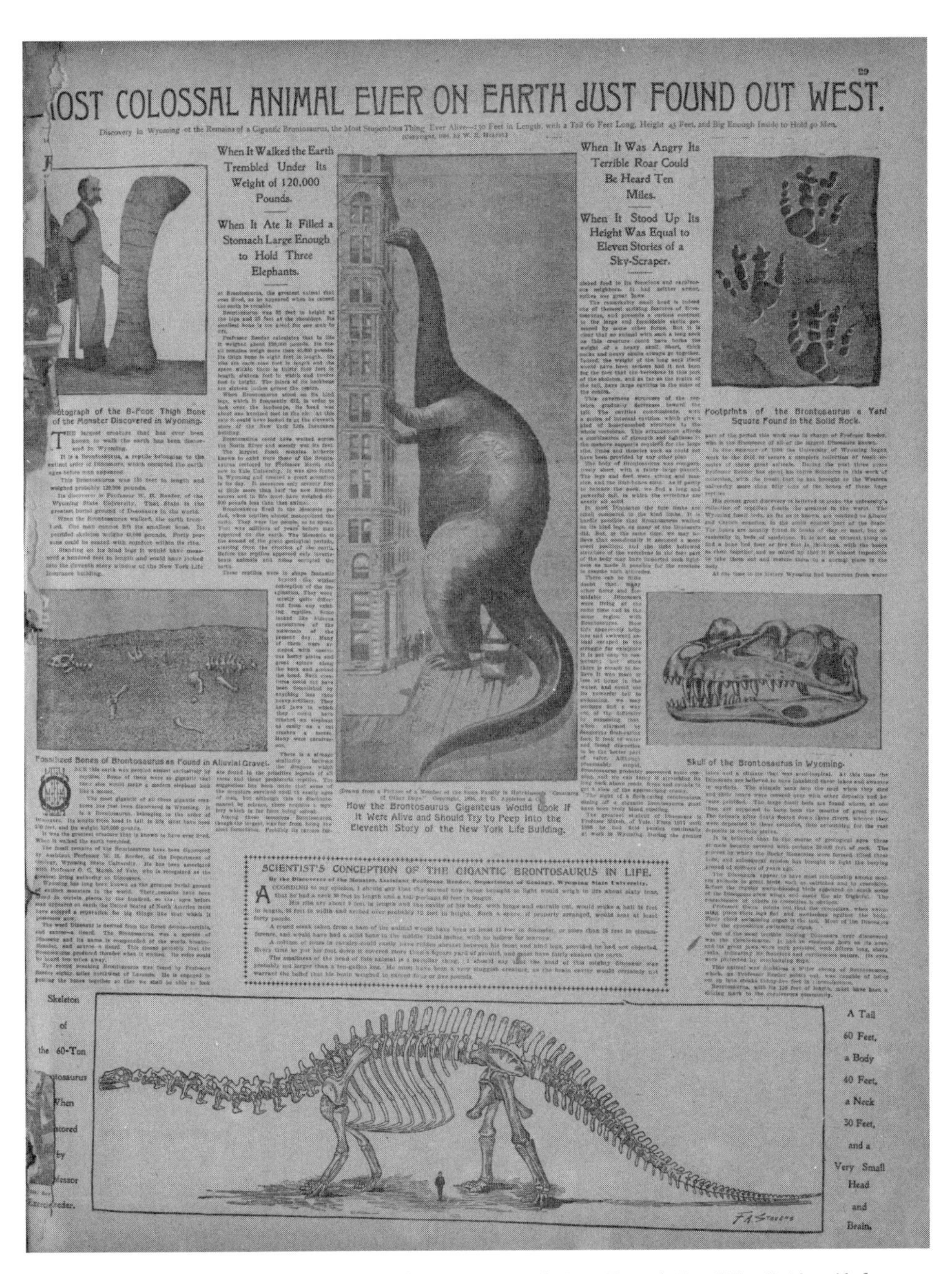

MOST COLOSSAL ANIMAL EVER ON EARTH JUST FOUND OUT WEST.

Discovery in Wyoming of the Remains of a Gigantic Brontosaurus, the Most Stupendous Thing Ever Alive—130 Feet in Length, with a Tail 60 Feet Long, Height 45 Feet, and Big Enough Inside to Hold 40 Men.

When It Walked the Earth Trembled Under Its Weight of 120,000 Pounds.

When It Ate It Filled a Stomach Large Enough to Hold Three Elephants.

When It Was Angry Its Terrible Roar Could Be Heard Ten Miles.

When It Stood Up Its Height Was Equal to Eleven Stories of a Sky-Scraper.

Photograph of the 8-Foot Thigh Bone of the Monster Discovered in Wyoming.

Footprints of the Brontosaurus a Yard Square Found in the Solid Rock.

How the Brontosaurus Giganteus Would Look If It Were Alive and Should Try to Peep Into the Eleventh Story of the New York Life Building.

Skull of the Brontosaurus in Wyoming.

Fossilized Bones of Brontosaurus as Found in Alluvial Gravel.

SCIENTIST'S CONCEPTION OF THE GIGANTIC BRONTOSAURUS IN LIFE.

Skeleton of the 60-Ton Brontosaurus When Restored by Professor Reeder.

A Tail 60 Feet, a Body 40 Feet, a Neck 30 Feet, and a Very Small Head and Brain.

Figure 3.4. "Most Colossal Animal Ever on Earth Just Found Out West," *New York Journal*, 1898.

The comparison between the United States' overgrown dinosaurs and its massive skyscrapers was becoming a conventional trope at the end of the nineteenth century, and it was often invoked to emphasize the colossal size of these creatures. Bigness was clearly their most remarkable feature, and the *New York Journal* article repeated Reed's claim that his find measured "130 feet in length and probably weighed 120,000 pounds" in the flesh. It was so large, readers were told, that "one man cannot lift its smallest bone," and the skeleton as a whole was estimated to weigh a total of 40,000 pounds. Readers were further informed that these creatures were "so gigantic" that they "would make a modern elephant look like a mouse," ensuring that even "before man appeared on the earth the United States of North America must have enjoyed a reputation for big things like it possesses now."[32] In other words, just as the United States came to see itself as surpassing Europe in the growth of its cities and the productive capacities of its factories, its dinosaurs too were being described in equally superlative terms.

All this publicity had the desired effect. News of Reed's discovery spread quickly, even capturing the attention of Carnegie, who had recently endowed a new natural history museum in Pittsburgh, the city in which he had earned most of his fortune. Used to doing things on a grand scale, Carnegie was determined to make his museum the most talked about institution of its kind. The humongous remains of a long-extinct reptile whose image was splashed across the yellow pages of New York's newspapers fit Carnegie's vision perfectly. When he saw another article announcing that Reed had discovered the "biggest thing on earth," he hastily scrawled a message to his new museum's director across the margin: "Can't you *buy* this for Pittsburgh," he snapped, adding, "Wyoming State University isn't rich—*get an offer*."[33] Reed's dream of hitting pay dirt finally seemed within reach.

Preaching the Gospel of Wealth

While the frontier collector valued his fossil in economic terms, Carnegie was far more interested in the symbolic value of dinosaurs, coveting Reed's specimen for the social prestige that it could confer. To understand Carnegie's obsession with the accumulation of dinosaurs therefore requires taking a step back to examine how he developed into one of the era's most famous philanthropists. The story of Carnegie's rise as an industrial capitalist reveals how much of his business success derived from the social relationships that he cultivated. These friendships, in turn, were strengthened by Carnegie's reputation as an honored, respected, and valued member of the community, a reputation he assiduously tended by engaging in acts of conspicuous generosity.

One of the late nineteenth century's most adept social climbers, Carnegie was born into an impoverished family of handloom weavers from Dunfermline, Scotland. Despite these humble beginnings, Carnegie learned how to cultivate his personal reputation among a growing circle of friends from an early age. When his parents moved to Pittsburgh in search of a brighter future, the young Carnegie almost immediately capitalized on the family's extended social network, finding work in the city's large Scottish community. At first, he kept accounts for a bobbin manufacturer and dreamed of becoming a bookkeeper, but after his uncle recommended him to a friend at the O'Rielly Telegraph Company, Carnegie was hired to run messages across town. Before long, he was promoted to a position with greater responsibility and more pay, where he earned the admiration and trust of a fellow Scotsman named Thomas A. Scott. When Scott was promoted to superintend the Pennsylvania Railroad's western division, he took Carnegie on as his personal secretary and chief assistant, even loaning him money to begin purchasing privately traded stock. By the time Carnegie was twenty-four, he counted the railroad's president, J. Edgar Thompson, as a personal friend, which helped him earn a promotion to run the company's entire Pittsburgh division. Soon thereafter, Carnegie branched out into his own business ventures, first in oil but then also bridge building, high finance, and, eventually, iron and steel manufacturing.[34]

None of these early triumphs would have been possible had Carnegie relied on his energy, his intelligence, or his capacity for hard work alone (although he certainly possessed all of those). Rather, in every case, his success rested in one way or another on the help of his friends. This is especially clear in his decision to move from iron manufacturing into steel, where he would earn most of his great fortune. Having seen the revolutionary new Bessemer process in action while traveling through the United Kingdom, Carnegie decided to make the capital-intensive improvements in his own plants. But doing so required a massive investment of cash. "Carnegie knew nothing of steel," an economic historian explains, but "he did know how to bring together men with finance."[35] At the time, formal capital markets could not supply sufficiently large blocks of finance to make the costly conversion, which helps to explain why the transition to Bessemer steel had eluded previous iron manufacturers in the United States. Carnegie therefore turned to his social network for financial support, raising about $700,000 by entering a partnership with a friend from the oil business and other close personal acquaintances, including Carnegie's brother Tom. The group of investors even named their new operation the Edgar Thompson Steel Works after Carnegie's old boss, correctly predicting that such flattery would result in lucrative contracts from the railroad.[36]

The adroit way that Carnegie leveraged personal relationships for financial gain illustrates how the friendships we form provide opportunities for further advancement in much the same way that money can be made to beget more money. As Carnegie understood, access to the right social network could be just as important to commercial success as a large stockpile of financial assets, in some ways perhaps even more so. Social and cultural resources thus constitute a valuable form of what sociologists often call "symbolic capital."[37] Carnegie himself recognized the power of this analogy when he urged workers in one of his steel mills to avail themselves of the free libraries he built to "lay up intellectual capital that cannot be impaired or depreciated."[38] In most cases, however, the analogy between different manifestations of capital was far less transparent. Because symbolic capital derives much of its value from the fiction that it operates at a remove from the marketplace, it was crucial to draw a clear demarcation between the pure and disinterested realms of art, science, and culture, on the one hand, and the self-interested, acquisitive, and transactional world of commerce, on the other.[39]

Carnegie learned that social capital could be turned into economic capital, and vice versa, but the conversion could rarely, if ever, be executed in a straightforward way. Thus, although Carnegie knew how to leverage his friendships for personal gain, he also knew they could not be acquired for that reason directly. Instead, Carnegie spent much of his life working to earn the esteem of his colleagues by building a reputation as someone whose aspirations extended far beyond commerce. To that end, he frequently visited Europe, attending exclusive soirées, going to museums, and generally taking in all that he could. He also styled himself as a man of letters, authoring numerous books and articles that showed off his wit and considerable erudition.[40] But perhaps most important of all was Carnegie's turn to philanthropy. As early as 1868, he had resolved to begin cultivating "an idol" beyond "the amassing of wealth," which he considered "one of the worst" and "most debasing" "species of idolatry." Beyond the $400,000 he had saved up thus far in his life, Carnegie vowed he would "make no effort to increase my fortune, but spend the surplus each year for benevolent purposes."[41] However, despite (or rather, because of) these lofty ambitions, Carnegie's fortune continued to grow, so much so that by the time he retired from the steel business in 1901, he was far and away the wealthiest person alive. In practice, then, the distinction between the philanthropist's generosity and the robber baron's rapaciousness was far from clear-cut. On the contrary, both worked in tandem to enlarge Carnegie's reputation among an ever-increasing network of friends.

Carnegie's turn to philanthropy not only paid dividends by increasing his social and cultural capital, it also helped him to reconcile his business success with his working-class roots. As a boy in Scotland, Carnegie had been steeped in a radical brand of classical liberalism that sought to overthrow Britain's aristocratic class system and undo its strict social hierarchies. This led him to embrace the egalitarian ideals of his adopted homeland with fervor, and he grew fond of lecturing anyone who would listen that America's "government of the people" surely provided the most secure "foundation of individual growth and of national greatness."[42] But Carnegie also had to contend with the fact that, in reality, working people saw their social and economic mobility undergo significant erosion during the late nineteenth century. Indeed, the period's increasing inequality was largely a result of precisely the kind of industrial capitalism of which Carnegie himself was a principal architect. Despite his outspoken claim to be a friend of the working man, then, Carnegie soon found himself at odds with the burgeoning labor movement. This dissonance became especially hard to ignore in the wake of the infamous Homestead Strike of 1892. When negotiations during a bitter contract dispute at one of Carnegie's steel mills came to a standstill, his deputy, Henry Clay Frick, not only imposed a lockout but also hired a private security force of some three hundred Pinkerton "detectives" to intervene. During the violent confrontation that followed, several workers were shot and many more Pinkertons taken hostage.[43] Rather than rejoice when the strike was finally broken and the union expelled from Homestead, however, Carnegie remained haunted by these events for the rest of his life. "Nothing," he recalled in his autobiography, "wounded me so deeply" as the violence at Homestead.[44] Although he steadfastly refused to give up his political ideals, Carnegie had to acknowledge that a century of industrial growth had yet to result in widespread prosperity.

Carnegie found a compelling way to address the problem of inequality in the writings of Herbert Spencer. One of the most influential philosophers of his day, Spencer developed an all-encompassing theory of universal progress rooted in the period's most cutting-edge science, biology. In an early essay first published in 1852, Spencer postulated a single "law of progress" that directed the development of everything under the sun, including human society.[45] This law was based on the observations of embryologists such as Karl Ernst von Baer, who had found that a simple, fertilized egg advanced "from homogeneity of structure to heterogeneity of structure" as it developed into a complex, adult organism. Drawing a parallel between the individual and the society, Spencer projected the basic trajectory from simplicity to complexity, from chaos to organization, onto all aspects of life, including the evolution of language, culture, commerce, and art. "From the earliest traceable cosmical changes down to the latest results of civilization," Spencer argued, "progress essentially consists" of a "transformation of the homogeneous into the heterogeneous."[46] Why should the same Law of Progress not also hold true for the evolution of industrial capitalism?

Spencer's progressivist theory of universal development is often dismissed as crude social Darwinism.[47] But Spencer's development hypothesis actually predated the publication of Charles Darwin's *On the Origin of Species* by nearly a decade. Moreover, whereas Darwin and Alfred Russel Wallace both identified natural selection as the driving force of evolutionary change, Spencer held that it constituted just one among a whole range of responsible mechanisms, including the Lamarckian idea that useful adaptations acquired during an organism's lifetime could be passed down to its offspring. Hence, although he did not discount the benefits of competition—indeed, it was Spencer who coined the phrase "survival of the fittest"—neither did he emphasize it nearly as much as is often supposed.[48] That said, Spencer identified many parallels between the social and the natural worlds, arguing that both were characterized by the same basic processes. Besides postulating that human history mirrored the development of an individual organism, he also drew on the ideas of Adam Smith to suggest that a physiological "division of labor" could explain the emergence of heterogeneous complexity out of homogeneous simplicity, arguing that evolution would favor individuals who specialized in a particular task because they could discharge their function more efficiently.[49] By "changes as insensible as those through which a seed passes into a tree," Spencer concluded, "society has become the complex body of mutually-dependent workers which we now see."[50] Thus, while Spencer did not discount the role of competition entirely, he primarily saw evolution as an organizing force that worked to produce complex assemblages whose constituent elements cooperated to form an integrated and harmonious whole.[51]

Spencer was far from the only late nineteenth-century thinker to extend the evolutionary worldview to human society, and faith in the notion of universal progress grew so widespread that his ideas became a kind of cliché during the period.[52] But it was his work in particular that had a strong impact on Carnegie. When the steel magnate learned that his intellectual hero planned to visit the United States in 1882, he secured a letter of introduction from a mutual friend and booked passage on the same ocean liner from England. He was so eager to meet Spencer in person that he even arranged to be seated at the philosopher's table.[53] The ploy proved successful, and the two struck up a friendship that Carnegie treasured for the rest of his life, showering Spencer with lavish praise, expensive gifts, and, eventually, an anonymously endowed pension. But the relationship was hardly one-sided. As Carnegie recalled in his autobiography, Spencer repaid the wealthy industrialist many times over by supplying a productive new philosophical outlook. "'All is well since all grows better' became my motto, my true source of comfort," Carnegie recalled toward the end of his life, having grown certain there was no "conceivable end" to humanity's "march to perfection."[54]

Armed with Spencer's evolutionary optimism, Carnegie turned the critique of industrial capitalism as an engine of inequality on its head. Because the division of labor would invariably give rise to a distinction between producers and consumers, rulers and citizens, labor and capital, the period's fast-growing inequality was actually a sign of progressive development. As Carnegie put it in his most famous essay, "The Gospel of Wealth," the stark "contrast between the palace of the millionaire and the cottage of the laborer" should inspire pride rather than envy, for it "measures the change which has come with civilization." Thus, while Carnegie recognized that industrial capitalism could be "hard for the individual," he insisted that everyone "must accept and welcome" the "great inequality of environment" and "the concentration of business" that it engendered as "essential to the future progress of the race." Yet he was not blind to the fact that Spencer's "law of progress" was supposed to yield a smoothly functioning, cooperative society, not one that was racked by fractious and at times violent labor disputes. Hence, something more was required to fulfill the evolutionary potential of modern capitalism. For Carnegie, that was philanthropy.[55]

Whereas the first stages of industrial capitalism had concentrated society's wealth in the hands of the few, Carnegie argued that its future development required the wealthy to make civilization's greatest achievements accessible to all parts of the social organism. In this way, philanthropy could help engineer a capitalist society in which "the ties of brotherhood may still bind the rich and poor in harmonious relationship." For Carnegie, philanthropy was an evolutionary engine of progress, supplying the motive power that would propel the United States into a new stage of capitalist development. Indeed, he was so confident that philanthropy was "the true antidote for the temporary unequal distribution of wealth, the reconciliation of the rich and the poor," that he even began to describe his worldview as "evolutionary socialism," which he contrasted with the "revolutionary socialism" of more radical thinkers like Mikhail Bakunin and Karl Marx.[56]

Eager to put theory into practice, Carnegie supported all manner of philanthropic causes during his lifetime, ranging from the donation of church organs to the creation of a "Hero Fund" for war veterans. But given his fascination with Spencer's evolutionary philosophy, he was particularly keen to promote science. In addition to the huge sums he donated for the construction of libraries and a technical university aimed at the diffusion of useful knowledge, he also endowed the Carnegie Institution for Science to fund cutting-edge research. But perhaps most impressive was his vision for a large and expansive museum complex in Pittsburgh. Initially conceived as a lending library, the plan for this ambitious project was allowed to lie fallow for over a decade due to the city's unwillingness to appropriate enough funds to cover maintenance costs. By the time a deal was finally hammered out,

Carnegie had enlarged the size of his gift and considerably expanded the scope of his vision. It now encompassed an ambitious cultural multiplex that combined a circular music hall seating two thousand listeners, a richly stocked reference library, an art gallery, and a natural history museum all under one roof.

Visually dominated by two large towers, Carnegie's museum projected a seriousness of purpose befitting an august temple of social and cultural progress.[57] Before long, natural history had established itself as the institution's centerpiece. In no small part, this was due to Carnegie's growing enthusiasm for dinosaur paleontology. From the very beginning, the museum embraced the New Museum Idea that we encountered in the previous chapter. Like its counterpart in New York, the Carnegie Museum was to be a "laboratory of research" that simultaneously served as "an inspiration to the student."[58] As Carnegie himself put it, if his museum did not "attract the manual toilers, and benefit them, it will have failed in its mission."[59] Because of their popularity among working people, Carnegie made sure to assemble an extensive collection of large and spectacular fossils in his museum, including a mastodon, a *Megatherium,* and an Irish elk.[60] But as the museum's director, William J. Holland, argued that it was still lacking a grand dinosaur hall, a place "in which we will put the big things," including Reed's "biggest thing on earth"[61] From almost the moment that it was founded, Carnegie's museum was therefore especially eager to mount a large and impressive dinosaur for display.

Wrangling a Dinosaur

Carnegie was keen to acquire Reed's dinosaur for his museum, but doing so proved to be easier said than done. The steel magnate's haughty and overbearing demeanor posed an especially significant obstacle when dealing with frontier settlers who resented the power of moneyed elites in the East to shape the region's political economy from afar. The aggressive way that his deputies dealt with local authorities did not help matters much, and Holland's excessive self-confidence must have come off as especially arrogant. From the vantage point of Wyoming, Carnegie's philanthropic pretensions were seen as pompous and self-aggrandizing, and the frontier institution was hardly content to relinquish Reed's dinosaur just because Carnegie wanted a centerpiece for his museum in Pittsburgh.[62]

When Holland initially contacted Reed in December 1898, he wanted to make sure the specimen was as impressive as advertised. To that end, he informed Reed that his boss "very much" wanted "one of these huge saurians," "preferably the biggest specimen that has ever been discovered." But Holland also worried that Reed could not deliver as promised, reminding him that he was "in the position of a miner who has located a 'pay-streak,' but how wide and how long it may be is impossible to tell until you have completed the work of excavation."[63] In response, Reed assured him that while he did not "pretend to see further into the ground than other men," he could say unequivocally "that this is a good prospect, the best I have seen for many years." But trouble was already looming on the horizon. Reed's superiors at the university informed him that he was "under obligations to stay with this institution."[64] Apparently, the university had leveraged Reed's discovery to convince the state legislature to appropriate funds for a new building in which to display its growing fossil collection. This prompted Holland to tell Carnegie that he would have to act fast or risk losing their "chance to get this monster."[65]

To get a jump on the matter, Holland rushed out to Wyoming and met Reed in person, offering to pay him an annual salary of $1,800 for a minimum of three years if he joined the Carnegie Museum staff in Pittsburgh.[66] This was a considerable sum at the time, about $200 a year more than Reed's boss, Samuel H. Knight, earned at the university.[67] But almost immediately after Reed signed a contract, Holland was disheartened to learn that he could not guarantee ownership of the fossil. Reed had previously filed a so-called grubstake claim to protect the specimen from falling into the hands of competitors. This was a mineral claim in which two or more parties agreed to share the labor, expense, and effort of locating and developing a new prospect.[68] It conferred an official ownership stake in the dinosaur quarry on the University of Wyoming. Infuriated, Holland offered the university's board of regents $2,000 for Reed's specimen, but they flatly refused. "It is the biggest thing on earth," one of the regents reportedly told him, "and we think that it is worth a hundred thousand dollars."[69] In response, Holland retained a lawyer and purchased the land on which Reed's discovery was located in fee simple, hoping this would divest any mineral claim the university may have filed.

Although Carnegie continued to yearn for Reed's dinosaur, such controversies were not to his liking. Worried that his name might be tarnished if word of Holland's aggressive tactics leaked to the press, Carnegie urged Holland to adopt the "method of concession and kindness." This, in turn, prompted Holland to make the slightly absurd gesture of donating Reed's specimen back to the university "in the interests of science."[70] Holland also sent a copy of the letter informing the university of Carnegie's magnanimous decision to the local newspaper.[71] Behind closed doors, however, Holland continued to scheme for ways to acquire Reed's dinosaur. He even began meeting with the state's governor and other local politicians in hopes of enlisting them as allies in the protracted dispute, assuring Carnegie, "We shall ultimately get possession of our coveted monster."[72] For his part, Carnegie mostly stayed out of the fray, preferring to let Holland get his hands dirty, much as he had left the job of handling the bloody strike at Homestead to his deputy several years earlier.

The controversy over Reed's dinosaur called into question Reed's honesty and reliability as a partner in science. Although Holland had once said there was no "better man for this work," he now felt it was necessary to "get the services of some thoroughly expert paleontologist to supervise the scientific portions of the enterprise."[73] To that end, he hired Marsh's onetime assistant, Jacob Wortman, as well as one of Henry Fairfield Osborn's most skilled fossil preparators, Arthur S. Coggeshall.[74] Writing to instruct Reed that he would no longer have free rein, Holland tried to put the best spin on the situation, telling him to cooperate with Wortman as Holland would expect Reed to cooperate with him "were I myself personally present."[75] But once the weather warmed up enough to send a field crew out to Wyoming, Wortman was horrified to discover that Reed's fossil had been all but destroyed. Wilbur Knight from the university had torn down Holland's ownership claim and opened a large cut right down the quarry's center, taking out most of the fossils and destroying the rest.[76] Back in Pittsburgh, Holland flew into a rage, accusing Knight and the university of incompetence while he fumed that "they apparently do not know how to meet manly men in a manly way, but are as full of little narrow, petty jealousies as an egg is of meat."[77] Holland's plans having been totally dashed, there was now a real danger that Carnegie's dream of acquiring a spectacular dinosaur might never materialize.

To make matters worse, the competition for American dinosaurs escalated considerably that summer. In addition to the crews that Holland and Osborn sent into the field, scores of other geologists descended on the area to prospect for vertebrate fossils. Their interest was piqued by a publicity stunt from the Union Pacific Railroad designed to turn southern Wyoming into a tourist destination. In collaboration with Knight, Edward Lomax from the railroad's Passenger Department sent invitations to over two hundred scientists from across the United States, offering free passage to anyone who wanted to see the region's abundant "reptilian fossils of enormous size" with their own eyes.[78] The goal was to acquire material for an attractive pamphlet featuring numerous photographs and articles written by participating geologists, which was subsequently distributed at no cost to passengers making the long journey on the overland route.[79] The call having been issued, dozens of scientists accepted, including a team led by Elmer Riggs from the newly established Field Columbian Museum in Chicago (later renamed the Field Museum of Natural History). The Union Pacific's Fossil Fields Expedition was a success, resulting in the discovery of many new specimens and attracting the attention of newspapers from all across the country.[80]

In spite of the intense competition, Holland remained undeterred. "It is . . . of the utmost importance that our Museum should succeed in obtaining a fine display of showy things," Holland lectured his field crew, adding that "Mr. Carnegie has his heart set on Dinosaurs—'big things'—as he puts it."[81] Luckily, the museum's perseverance paid off. Reed managed to locate another promising claim before the summer was out. It was near Sheep Creek, a small tributary to the Little Medicine Bow River northwest of Laramie, and it contained two separate skeletons, one of which was a large and remarkably complete *Diplodocus*. After the fossil had been shipped back to Pittsburgh and prepared for more careful inspection, Holland rushed to tell Carnegie the good news. Describing it as "the most perfect and the biggest skeleton of its kind ever found," he was exceedingly pleased that all talk of "the biggest show on earth" could finally be put to rest.[82] However, just as Holland was celebrating the new discovery, Wortman decided to quit his position at the museum. Still certain that Reed could not be trusted to work on his own, Holland immediately embarked on a trip to Princeton University, where he hired another young geologist—John Bell Hatcher—to take Wortman's place.

Figure 3.5. The Carnegie Museum field crew (with Holland in the front left) near Sheep Creek in southern Wyoming, 1899.

Reed may have discovered a promising new dinosaur quarry, but his standing at the museum continued to deteriorate. Only a few weeks into the 1900 field season, Hatcher and Reed had an intense falling-out. Realizing that his reputation had suffered a considerable blow after the previous summer's controversies, Reed offered to give the museum a valuable series of mammal skulls he had located several years earlier. But after two weeks of intensive searching, the fossils remained nowhere to be found. Recalling Reed's failure to deliver the dinosaur that initially captured Carnegie's interest during the previous summer, Hatcher began to question whether he could be trusted at all. "I fear the entire story is a fiction," he wrote to Holland, adding, "I am afraid Mr. Reed's veracity is not to be depended upon & for my part I do not like to be needlessly fooled or tricked by one of our own men."[83] He even began to cast aspersions on Reed's skill in the field, saying, "Frankly I do not want such work done for us, for in many instances it simply means the destruction, rather than the preservation of rare & important material."[84] So when Reed had the temerity to disagree with Holland on a technical matter, Holland decided that enough was enough. As he wrote to Hatcher, "It seems to be of the highest importance that someone who has better scientific training and who can adhere a little more rigidly to the demands of strict truthfulness should be on the ground."[85] Reed's reputation was in tatters, and Hatcher ordered another assistant to begin overseeing his work in the field. But when the veteran fossil hunter refused to submit to the authority of a much younger and less experienced man, he was summarily dismissed.

The termination of Reed spoke volumes about the Carnegie Museum's institutional culture. Especially telling was the extent to which criticisms of Reed centered on a perceived deficiency in his character. It was impossible to deny that Reed knew the fossil fields of Wyoming better than anyone. Having discovered some of the region's most productive dinosaur quarries, he could hardly be faulted for failing to supply high-quality specimens. But Reed did not meet expectations of how a member of the museum's research staff ought to comport himself. According to Holland, Reed's frontier background was to be blamed for his inadequacy as a dependable employee, and Holland stated that a dignified institution of learning could not afford to associate itself with such "partially educated and so-called 'self-made' men."[86] For his part, Reed also felt ill at ease in his new institutional home. Just weeks after leaving Wyoming, he began to complain that not only was there "too much smoke and damp weather to suit" him in Pittsburgh, but, more to the point, he also felt awkward playing the educated scientist. Asked to "deliver a lecture on geology to a naturalist club," he sheepishly demurred, saying, "It does not take me long to tell all I know and I might run out of material."[87] The humble and unassuming demeanor that served Reed well in Wyoming was now seen as a personal failing. Holland therefore instructed Hatcher to find a replacement "who has had the benefit of a liberal course of education," adding, "We wish no more bumptious, verdant youths in the list of our employees."[88]

In the institutional context of philanthropically funded civic museums, the conventional division of labor between collectors and naturalists began to break down. Just like its cousin in New York, the Carnegie Museum was designed to stand as "a monument to the broad philanthropy and sagacity" of its "noble founder," as an editorial in the *Annals of the Carnegie Museum* explained.[89] Thus, whereas the first generation of American paleontologists did not expect fossil hunters to supply them with anything but the raw material needed to conduct their research, philanthropically funded civic museums wanted more from their employees. In addition to delivering dinosaurs, Reed found himself called on to help augment the reputation of the museum's generous founder. Unfortunately, Reed's acute business sense made it difficult for him to succeed in that role. Resulting in an acrimonious contest over who would eventually secure ownership of his "coveted monster," the transactional relationship that Reed learned to cultivate in the 1870s and 1880s now made him a suspect partner in science just a few decades later. This was not only because his behavior was informed by the commercial ambitions that also characterized dime museum impresarios such as P. T. Barnum. It was also because Reed's desire to make the most of his discoveries undermined the trust that could be placed in him to be scrupulously honest in all of his dealings. Now that he was on the museum's payroll, these fears not only threatened the institution's financial solvency. They also cast doubt on its status as a serious research institution. Beyond causing embarrassment, then, Reed's identity as a commercial collector undermined the museum's credibility as a source of reliable knowledge.[90]

Dinosaur Diplomacy

Whereas Reed found it hard to fit into his new institutional home, Hatcher thrived there, especially after he discovered a second *Diplodocus* specimen near Sheep Creek in April 1900. Although somewhat smaller than the first, it contained many of the portions missing from Reed's specimen. This suggested the possibility of combining the two into a composite mount, similar to the one being prepared in New York. To that end, the new specimen was immediately collected and shipped via the Union Pacific to the museum, where its bones were cleaned up by removing what remained of the surrounding rock matrix (Figure 3.6). All of this happened so quickly that, by 1901, Hatcher was able to publish a detailed description of *Diplodocus,* which later appeared in the very first installment of the *Memoirs of the Carnegie Museum of Natural History.* Based on a few morphological differences between the Sheep Creek specimens and earlier ones collected by Osborn and Marsh,

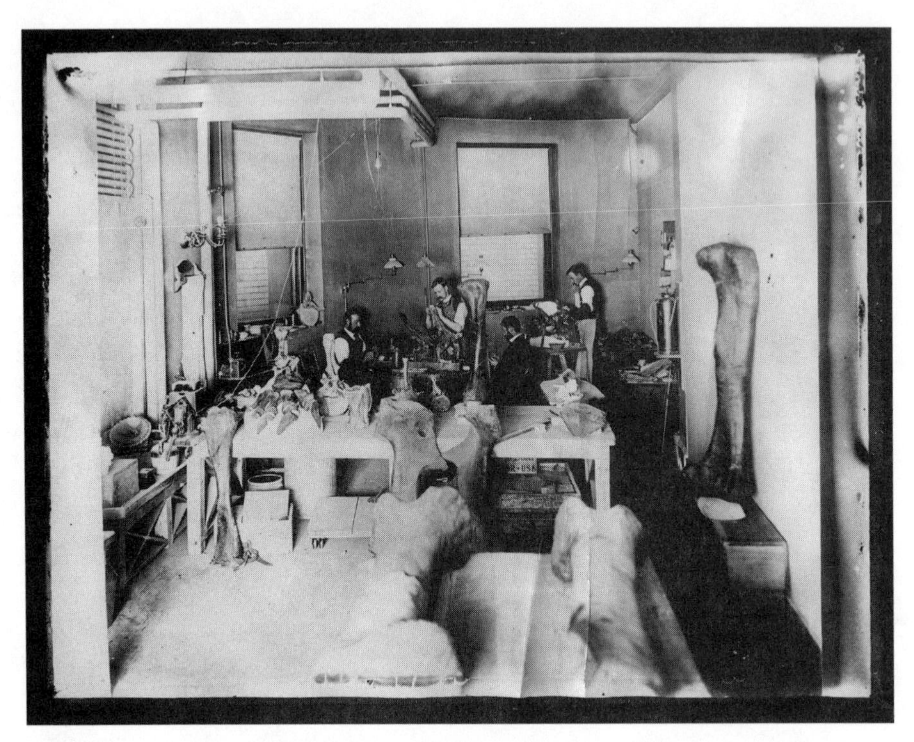

Figure 3.6. Fossil preparation laboratory at the Carnegie Museum of Natural History, ca. 1900.

Hatcher concluded that the Carnegie Museum had discovered a new species of dinosaur. In a shrewd move, he named it *Diplodocus carnegii* to commemorate the generosity of his museum's benefactor.[91]

Holland immediately sent a framed lithograph of the skeleton (Figure 3.7) to Carnegie's summer castle in Scotland, explaining that the new species of dinosaur had been named in Carnegie's honor and that it "is a bigger beast" than the first *Diplodocus* described by Marsh several decades earlier. A couple of weeks later, he wrote again to ensure that Carnegie understood the gravity of what had transpired: "Now the biggest thing on earth of its kind bears your name; so you are sure of immortality in the annals of science."[92] This was precisely what Carnegie wanted to hear: whereas a lifetime's worth of accumulated financial capital was fleeting, the honor and esteem earned from Carnegie's philanthropic investments would last forever. He was beyond thrilled, and he immediately donated the money required to construct a new building

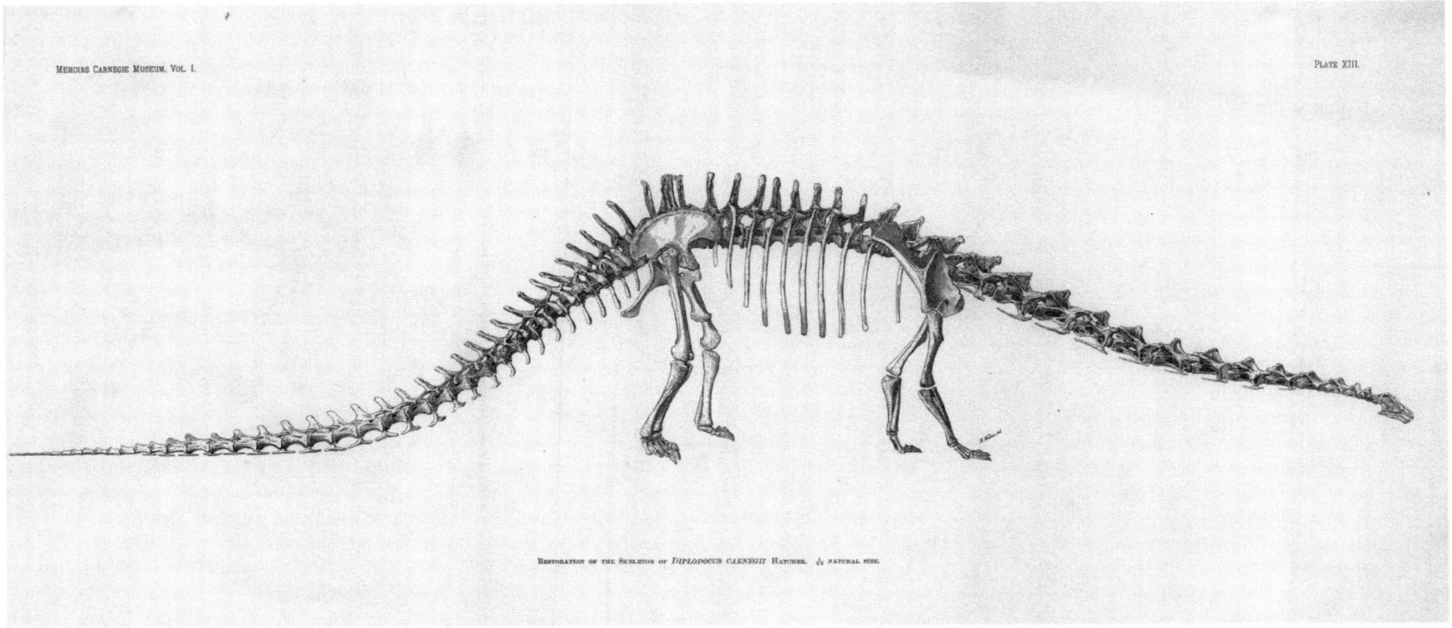

Memoirs Carnegie Museum, Vol. I.

Plate XIII.

Restoration of the Skeleton of *Diplodocus carnegii* Hatcher. natural size.

Figure 3.7. Skeletal restoration of *Diplodocus carnegii*, by John Bell Hatcher.

for Reed's dinosaur, whose remains were too big to fit anywhere in the original museum complex.[93]

Although its first home was in Pittsburgh, Carnegie's dinosaur was destined to travel beyond the United States. When England's King Edward VII visited Carnegie during the fall of 1902 and took note of Hatcher's framed lithograph on the wall, he adjusted his glasses and exclaimed, "I say Carnegie, what in the world is this?" Bursting with pride, Carnegie explained that this "namesake of mine" just so happened to be the "hugest quadruped that ever walked the earth."[94] King Edward was so impressed that he asked whether a similar "monster" could be acquired for the British Museum.[95]

Carnegie loved the idea of donating an American dinosaur to England, and he wrote to ask Holland whether a second specimen could be found. Knowing how difficult it would be to procure a skeleton of comparable size and completeness, Holland proposed making a plaster cast replica of the original. Carnegie agreed, especially after Holland explained that an "indefinite number" of identical casts could be fashioned from the same set of molds, which would allow him to send copies all over the world, earning "a royal return" for the wealthy philanthropist while guaranteeing a "fine harvest of paleontological material" in the form of exchange specimens from other museums as well.[96] Much like Carnegie's expanding network of friends and his growing reputation for cultural connoisseurship, his museum's rich storehouse of valuable specimens thus functioned as yet another form of capital, the accumulation of which could be leveraged for intellectual as well as reputational gains.

While Carnegie's *Diplodocus* ranks as the most widely distributed plaster cast of a dinosaur, the medium had long been in use to make valuable objects more accessible to the widest possible audience. During the nineteenth century, the practice had become so widespread that most large museums routinely circulated casts of their most important and sought-after specimens. Plaster casts played an especially important role in American art museums, which initially preferred to exhibit faithful replicas of well-known European masterpieces rather than locally created originals of far less renown. However, once wealthy philanthropists began to donate original artworks in large numbers, a new conception of authenticity began to emerge. Ironically, whereas the status of plaster casts as purely mechanical reproductions had once lent them the authority to stand in for the original, that very same feature now marked them out as aesthetically dubious. As a result, the first third of the twentieth century saw most American art museums relegate their plaster casts to the basement, place them in storage, or worse, destroy them outright.[97]

In light of the medium's contested museological status, Carnegie's decision to donate a plaster cast of his dinosaur to the British Museum served as a demonstration that America's cultural achievements could rival its economic success, to the point of overshadowing those of Europe. In this respect, natural history still differed considerably from the fine arts. When Carnegie's museum complex first opened in 1895, its sculpture galleries were filled with casts taken from Europe, including a copy of the famous Parthenon frieze and two of Luca Della Robbia's renowned mural marbles.[98] Similarly, the shelves of his libraries were overwhelmingly stacked with literature from abroad, and the most acclaimed music performed in his concert halls had been written by French, German, and Italian composers. Only the natural history museum offered a way to reverse the flow of cultural goods, providing a rare opportunity to export quintessentially and recognizably American objects to Europe. It must therefore have been a real coup that even such a renowned institution as the British Museum would agree to accept a mechanical reproduction whose original stood in Pittsburgh (Figure 3.8).

Because Carnegie's gift came with considerable strings attached, it did not meet with unvarnished gratitude. Not only did Carnegie's ostentatious displays of generosity demand recompense in the form of a public acknowledgment, but the fact that his gifts could not be repaid in kind only further highlighted the considerable asymmetries that distinguished him from the beneficiaries of his largesse. The symbolism of a Scottish-born factory owner presenting a stupendously large dinosaur to King Edward VII would have been clear to everyone who attended its unveiling at the British Museum in March 1905. For that reason, the reception of Carnegie's dinosaur offered an opportunity for people to voice their misgivings about America's industrial juggernaut quite generally. British newspapers loved to point out that although it was immensely large and fearfully powerful, *Diplodocus* was also a slow-moving and unintelligent reptile. It was a "great digesting machine" whose "brain cavity was no bigger than a walnut," one journalist joked.[99] Another delighted in describing it as "an aimless, stupid sort of reptile" with "an enormous mouth that, even in the bone, is suggestive of an expansive and inane

Figure 3.8. The fossil remains of *Diplodocus carnegii* mounted for public display in the Dinosaur Hall of the Carnegie Museum of Natural History, ca. 1910.

smile."[100] And a third journalist stated that although he was "sure Mr. Carnegie means well" and his "intentions are undoubtedly benevolent," he would nonetheless have preferred the *Diplodocus* be returned to its original owner, as "America is the proper sphere for such unreasonably developed monsters."[101]

Others insisted on pointing out that dinosaurs originally hailed from England, where they were first discovered, not the United States, which remained a relative newcomer to the science of paleontology. The *Westminster Gazette* even borrowed the language of political economy to argue its case, writing, "The great American Continent has by no means the monopoly of things colossal."[102] Tellingly, such sentiments were hardly confined to the popular press. In a speech that he gave at the unveiling ceremony, the British

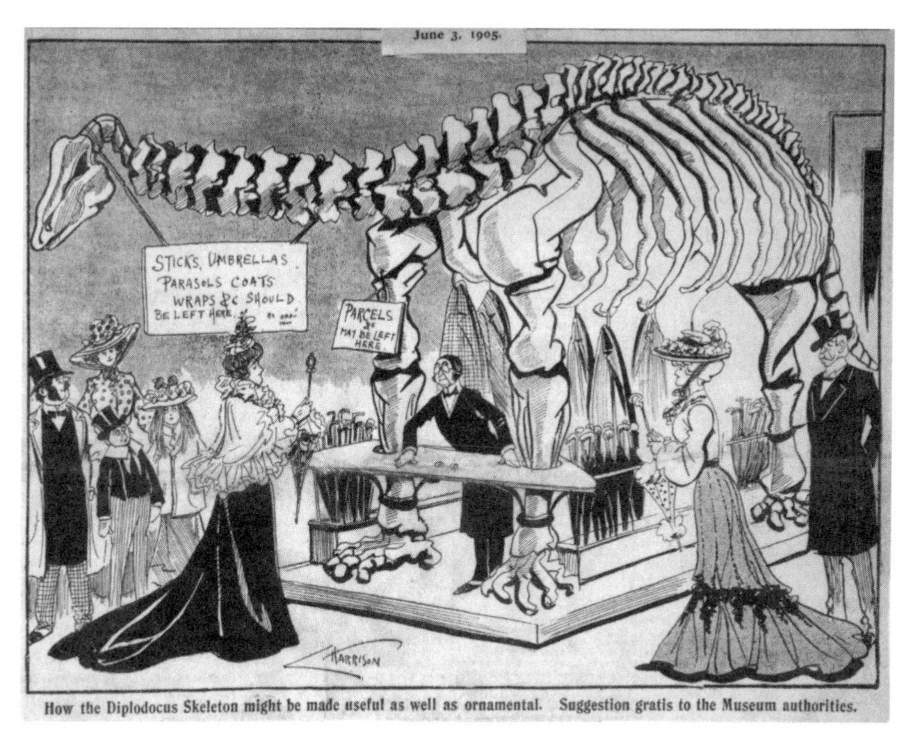

Figure 3.9. "How the Diplodocus Skeleton might be made useful as well as ornamental." *Morning Leader*, 1905.

Museum's director made sure to point out "that all the great progress which has been made in the American Republic has been founded upon ideas, which have germinated, and inventions, which have been really conceived, in England." Dinosaurs were no different, he went on to say, rendering *Diplodocus* no more than "an improved and enlarged American form of an English creature."[103] The implication of these comparisons would have been hard to mistake, especially given Carnegie's reputation for ruthlessness in the face of labor unrest: the United States may have overtaken Great Britain as the world's most productive economy, but only because industrial capitalism had been allowed to evolve unchecked there, developing into a monstrous behemoth that was too large and too powerful for its own good.

Confident in the knowledge that Great Britain remained the world's most formidable empire, English newspapers could accept Carnegie's generosity with a wink and a smile, easily poking fun at the pretensions that inhered in his gift. But things were different in Germany, for which the United States represented a more menacing rival. Although it too was a relatively young nation, having only been formed into a politically unified state during the 1870s, Germany was already becoming a formidable economic power. By the eve of the First World War, it produced about a quarter of the world's coal and much of its steel, and firms such as Bayer, BASF, Siemens, Bosch, Zeiss, and Krupp translated the latest scientific research into precision-engineered manufactured goods. Taken together, these developments caused the young nation's gross domestic product to skyrocket, doubling about once every fifteen years between 1880 and 1914.[104]

Alongside this economic growth came a desire for increased military and political power. In this respect, Germany's ambitions may even have exceeded those of the United States, and it fervently participated in the scramble for Africa by laying claim to large portions of present-day Cameroon, Nigeria, Tanzania, Rwanda, Namibia, Togo, and Ghana. To amass and protect its overseas colonies, Germany also invested heavily in the creation of a formidable navy, which grew sufficiently powerful to threaten Britain's dominance of the seas. Needless to say, all of this saber rattling contributed to the sense of crisis that became more acute by the day, culminating in a premonition that it was only a matter of time before Europe descended into war.[105]

As an outspoken proponent of world peace, Carnegie worried that Germany's military buildup would destabilize the fragile balance of power that kept the global economy humming. In hopes of averting war, he supported the creation of the League of Nations and even attended an annual sailing regatta in Kiel, where he tried to convince Kaiser Wilhelm II to embrace a scheme of compulsory arbitration enforced by an international police force in order to avoid open hostilities.[106] But none of his efforts met with success. He was therefore delighted when the kaiser sent a delegation to witness the opening of his museum's new wing—featuring the original *Diplodocus* fossil—in Pittsburgh during the spring of 1907. As Holland would later describe the encounter, they came bearing "a right royal gift, consisting of books, engravings, and photographs, illustrating the arts and material progress of the Empire."[107] Carnegie took this to be an auspicious development, because the scientific community's cosmopolitanism was widely regarded as one of the most sustained and successful examples of international cooperation. For Carnegie, the promise of peace and the practice of science went hand in hand, and he wasted no time to seize the opportunity, instructing Holland to reciprocate by offering a cast of the famed American dinosaur to the kaiser's delegation.

Carnegie hoped that his dinosaur would help foster a closer relationship between Germany and the United States. But when Holland and his team arrived at Berlin's *Museum für Naturkunde* in the spring of 1908, they met with an even more ambivalent response than in London. Curators at the Berlin museum were less than thrilled to devote a large portion of their main gallery to the cast of an American fossil, especially since German scientists doubted the competence of their overseas colleagues, questioning whether *Diplodocus* had been assembled correctly. The controversy was especially bitter because many of the specimen's bones were bereft of clearly defined articulating surfaces that showed how they fit together. Moreover, the cartilage that formed an important part of its joints was not preserved in the fossil record. These issues, plus the fact that *Diplodocus* lacked a close living relative to serve as a model, meant the exhibited dinosaur embodied a great deal of conjecture, which ensured that doubts about the display would eventually surface. It also meant they would be difficult to dismiss.[108]

The main point of contention was Holland's decision to mount Carnegie's *Diplodocus* on stiff, columnar legs, which made it resemble an oversize elephant with an extraordinarily long neck and tail. Holland reasoned that such an immense creature must have stood fairly erect or its legs would have buckled under the animal's weight. For similar reasons, most paleontologists suspected that *Diplodocus* spent most of its life partially submerged in the water to lessen the load on its feet. But not everyone agreed. Just as Holland was preparing to set up a cast of *Diplodocus* in Germany, an American paleontologist named Oliver Perry Hay issued a critique of the idea that such a massive animal could have lifted itself up off the ground, primarily citing the enormous weight of these creatures as "a strong argument against their having had a mammal-like carriage."[109] Hay envisioned an animal that, rather than extending its legs straight under its belly, laboriously dragged itself along on the ground with its legs splayed out to the side. A similar view was adopted by German paleontologists, who argued that curators from Carnegie's museum had committed a grave error in mounting his dinosaur as they did.

Almost immediately after Holland had finished erecting Carnegie's dinosaur, a well-known herpetologist at the Berlin museum named Gustav Tornier subjected its assembly to withering criticism. Point by point, he refuted the anatomical conjectures of Holland's exhibit in a remarkably pugnacious and condescending tone. First and foremost, he objected to Holland's choice of animal models. Given its evolutionary relationship to primitive reptiles such as the crocodile, Tornier reasoned that *Diplodocus* must have resembled the chameleon or monitor lizard more than elephants. Second, he took issue with the way Holland positioned the dinosaur's outer extremities. Tornier argued that its tail had been mounted at an incorrect angle, so that it curved up from the ground far too aggressively, thinking it far more likely that it extended back in a relatively straight line, as it does in modern lizards. Conversely, he felt that the shape of the fossil's cervical vertebrae indicated that the neck probably took the form of a sharp S curve, meaning the dinosaur held its head high in the air. The fact that Holland had earlier endorsed a similar view led Tornier to accuse him of knowingly fabricating an erroneous mount so the fossil's skull would be closer to visitors on the ground. But Tornier reserved his most vehement criticism of Carnegie's *Diplodocus* for its mammalian posture. Like Hay, he too insisted it must have crept on the ground in a lizard-like crawl. To claim anything else, he believed, was to dismiss all that was known about the functional anatomy of reptiles.[110]

Not everyone concurred with Tornier. A paleontologist in Frankfurt could not imagine how his American colleagues, "who mount dinosaurs by the dozen," could have made so many mistakes.[111] The well-known Austrian naturalist Othenio Abel also challenged Tornier's conclusions. In a lengthy and detailed analysis, he argued that *Diplodocus* appeared to have walked on its toes (as elephants do) rather than the balls of its feet (as lizards and crocodiles usually do). He also rejected Tornier's phylogenetic claims by pointing out that dinosaurs were anatomically related not just to lizards but in some ways even more so to birds, which stand upright on rectilinear legs.[112]

Not surprisingly, Tornier's arguments gained even less credence in the United States. Holland in particular was incensed at the suggestion that he had made an anatomical blunder, and he issued an incendiary riposte in which he cast doubt on Tornier's capacity to weigh in on matters of paleontological controversy. Dismissing him as a herpetologist who lacked personal experience

Figure 3.10. Heinrich Harder's Germanic interpretation of *Diplodocus,* in Wilhelm Bölsche's *Tiere der Urwelt,* 1910.

digging for dinosaur bones, Holland accused Tornier of being a mere "closet-naturalist" who had given free rein to his "brilliantly illuminated imagination." And when Carnegie donated a copy of his dinosaur to the Saint Petersburg museum, Holland even experimented with assembling it in the way Tornier imagined, just to show how unlikely the end result really was.[113]

Despite the heated debate they elicited, Tornier's unorthodox views became widespread and even gave rise to a recognizably "Germanic" tradition in the visualization of dinosaurs. In large part, this was a result of Tornier's tireless efforts to promote his ideas in popular science venues.[114] Especially visible was the work of Heinrich Harder, who arguably did more than anyone to disseminate Tornier's ideas among a broad audience in visual form. Harder collaborated with the well-known science writer Wilhelm Bölsche to produce a collection of several dozen full-color illustrations featuring prehistoric animals for the German chocolate manufacturer Kakao Compagnie Reichardt (Figure 3.10). Issued for promotional purposes, this series brought large

Figure 3.11. *Brachiosaurus brancai* towering over *Diplodocus carnegii* in the Berlin Natural History Museum, 1938.

numbers of people into contact with a vision of *Diplodocus* as it "waddled" about on the ground.[115] But Tornier's influence made itself felt in three-dimensional media also. Just a few years after the Reichardt chocolate company issued its collecting cards, the Berlin Zoological Garden commissioned Harder to adorn the facade of a new aquarium building with several relief sculptures. Harder chose the occasion to reconstruct a number of prehistoric creatures, such as *Triceratops,* again giving them a distinctly reptilian pose. For a time there was even talk of remounting Carnegie's dinosaur in the Germanic style, which received Kaiser Wilhelm's official blessing.[116] Although a lack of manpower and funds prevented that plan from coming to fruition, when the Senckenberg Museum in Frankfurt remounted its own *Diplodocus* skeleton (which had been donated by the American Museum of Natural History in New York) during the mid-1930s, curators made an exhibit the deliberately dodged the question by giving the specimen a pose it was believed to assume when submerged underwater. However, a scale model was also fabricated to illustrate how the creature crawled on its belly as it ventured ashore.[117]

Carnegie's *Diplodocus* did not only invite controversy for the way it was posed. The dinosaur's arrival in Berlin also threatened to eclipse a tremendous new fossil that German paleontologists had recently discovered in Africa. About two years before Carnegie's *Diplodocus* arrived in Berlin, a mining engineer, acting on the advice of native informants, unearthed a quarry of Mesozoic fossils about three hundred miles south of Dar es Salaam, in one of Germany's colonial possessions. At precisely the time Carnegie's *Diplodocus* was being assembled, German paleontologists were scrambling to raise enough funds for a costly expedition to what is now Tanzania, and to ship large numbers of specimens back to Berlin. Over the course of four years, they succeeded in extracting nearly four hundred thousand pounds of fossil material at a cost that exceeded 180,000 German marks. In the process, a dozen species of dinosaurs were unearthed, including the immense bones of yet another long-necked sauropod dinosaur (Figure 3.11). Much as Hatcher had done, German paleontologists declared it to be a new species, which they named *Brachiosaurus brancai* in honor of the head paleontologist at the Berlin museum, Wilhelm von Branca.[118]

Although it was a close relative of the American *Diplodocus, Brachiosaurus* was even larger. Upon the fossil's arrival at the museum, German paleontologists immediately prepared its femur—more than twice the size of the corresponding bone in *Diplodocus*—for exhibition. An iconic specimen, this was the same type of bone that Reed had been photographed next to over a decade before (Figure 3.3), only much larger. And just in case the fossil's colossal size did not speak for itself, a museum guide also informed visitors that Germany's *Brachiosaurus* had "even more gigantic dimensions" than American dinosaurs, inviting visitors to make a direct comparison between the two creatures' limb bones so as to "better demonstrate this tremendous difference in size."[119]

In *Brachiosaurus,* Germany had found a powerful answer to Carnegie and the United States. As the paleontologist Charles Schuchert begrudgingly admitted, although his countrymen had long been "proud in the belief" that "America had reared the largest of all animals," that "honor may now go to Germany."[120] But plans to mount the spectacular specimen for exhibition were cut short by the outbreak of war, and they did not resume in earnest until years after Germany's defeat. By the time that *Brachiosaurus* was finally mounted for display nearly two decades later, a new regime had come into power, one that aimed to realize the imperial ambitions that had animated Germany's aggressive expansionism during the time of the *Kaiserreich.* This made the sight of a dinosaur that hailed from Germany's erstwhile colonial possession in East Africa (which had been lost during the First World War) towering over Carnegie's American *Diplodocus* highly symbolic. Not only did its immense size help to announce Germany's renewed national confidence, but the creature's sweeping pose could not have escaped notice (see Figure 3.11). As if to emphasize the majesty of the Third Reich, the body of *Brachiosaurus* was elevated high off the ground instead of being mounted in the somewhat ridiculous waddling style advocated by Tornier. Additionally, its neck was curved into an elegant S shape, which further emphasized the creature's immensity. Visually dominating the museum's spacious and sun-drenched atrium, Berlin's *Brachiosaurus* could therefore look down on Carnegie's much smaller *Diplodocus* a dozen meters below.[121]

Conclusion

The fierce debate that *Diplodocus carnegii* touched off in Germany clearly involved far more than just arcane questions about comparative anatomy and dinosaur osteology. It was also, and perhaps even more so, a fight over who had the authority to make credible claims about the deep past. Proud of their country's reputation as the land of thinkers and poets, German intellectuals had long derived great satisfaction from the renown of its national culture. After all, it was in Germany that modern research universities had been born, and a majority of prominent nineteenth-century American scientists pursued their postgraduate studies there, including both Cope and Marsh. This helps to explain why German naturalists such as Tornier so resented the notion that an American paleontologist could teach them their craft. It was one thing to import a material specimen from the New World, but quite another to accept the knowledge and interpretations that came with it.

As its name so plainly suggested, *Diplodocus carnegii* was far more than just an object of science. It also stood for its country of origin, its famous patron, and even the rapid expansion of American capitalism (with all that entailed). As Holland bragged in a letter from 1912, Carnegie's grand philanthropic vision had yielded a "*Diplodocus factory*" that was churning out dinosaurs "in a wholesale way."[122] For an ambitious young nation like Germany to be on the receiving end of Carnegie's high-throughput philanthropic endeavors must have been especially galling, particularly given the obvious symbolism that attended his dinosaur's prodigious dimensions. *Diplodocus carnegii* was the very picture of American ostentation and vulgarity. It also represented the young nation's increasing power, politically, economically, and, before long, militarily.

Besides placing the history of American capitalism within a global context, the story of *Diplodocus carnegii* also helps demonstrate the important role that philanthropy played in the broader cycle of accumulation that dominated America's Long Gilded Age. Visible displays of conspicuous generosity served as a powerful signal of one's commitment to the norms of an elite social class. Securing a legitimate place in high society was of material consequence because it provided access to the finances required to launch an ambitious industrial project off of the ground. In the bourgeois culture that Andrew Carnegie inhabited, intimate friendships with other members of the wealthy elite were indispensable to one's continued business success. But inside of his companies, things could hardly have been any more different. Rather than getting to know employees personally, Carnegie implemented formal tools of managerial oversight to keep track of expenditures and maximize profits. Thus, whereas deeply personal relationships based upon mutual bonds of trust ruled Carnegie's social world, formal regimes governed by paperwork held sway in the institutions he ran. This would turn out to be equally true in the practice of natural history as in the manufacture of steel.

4

Accounting for Dinosaurs

On May 1, 1893, President Grover Cleveland pressed down on a gilded telegraph key, inaugurating the World's Columbian Exposition by closing an electrical circuit that gave life to an astonishing array of engines, fountains, and lights. Organized by Chicago's business elite, the world's fair was ostensibly held to commemorate Christopher Columbus's voyages of discovery some four hundred years earlier. But it was also a spectacle of industrial progress and modernity. The *Official Guide* declared that fairgoers were living through "the golden age of American enterprise, American industry and American development." Time and again, they were reminded of the commercial potential of modern science, especially through demonstrations of electricity's awesome power. Over eight thousand arc lamps and 130,000 incandescent lights illuminated the fairgrounds at night, consuming more power than the rest of Chicago combined. The fair also showcased a wide range of newfangled contrivances that organizers described as "the most novel and brilliant exhibits of the Exposition." These included a giant Ferris wheel, a moving sidewalk, and an elevated railway line. In large part due to such stupendous exhibits, the fair was a huge success, drawing more than twenty million people (equivalent to almost a third of the United States' estimated population) to Chicago's South Side.[1]

The fair's organizers were not content with a temporary exhibit, however, no matter how grand. Instead, they heeded the advice of scientists such as Frederick W. Putnam, a Harvard professor of anthropology and the cousin of Othniel Charles Marsh, who counseled them to establish a permanent museum once the fair ran its course. Prominent members of the city's Commercial Club such as Edward E. Ayer immediately began to solicit funds for the endeavor, eventually securing $1 million from the department store magnate Marshall Field. Others soon followed suit—George L. Pullman and Harlow N. Higinbotham each donated $100,000, for example—and on June 2, 1894, the Field Columbian Museum was formally dedicated as a permanent memorial to the world's fair, as a demonstration of Chicago's metropolitan ambitions, and as a testament to the conspicuous generosity of its wealthy elite.

Although it was initially housed in the fairground's only permanent building, the Field Columbian Museum soon began distancing itself from the legacy of the Exposition. In 1905 it changed its name to the Field Museum of Natural History. A decade later, construction began on a new, purpose-built structure located closer to the city's downtown. These changes reflected a significant institutional transformation. Whereas the 1893 fair primarily served as a celebration of industrial progress, the museum quickly abandoned its focus on applied science and removed nearly all commercial exhibits dedicated to American industry from public display. It also launched an ambitious program of original scientific research, acquiring a research collection of study specimens and hiring several notable scientists to join the permanent staff. Before long, it stood alongside of the Carnegie Museum in Pittsburgh and the American Museum of Natural History in New York as one of North America's most notable and well-funded centers for the study of dinosaur paleontology.[2]

Just like its two main competitors, the Field Museum was overseen by a board of wealthy trustees. But the task of managing its day-to-day operations fell to the museum's director, Frederick J. V. Skiff, who came to the attention of Chicago's philanthropic community because he successfully oversaw a large and extensive exhibition on Colorado's mineral industry for the Columbian Exposition. Since the new museum's first order of business was to accession a huge number of objects from the fair to its permanent collection, the organizational skills Skiff demonstrated made him a natural choice to lead the endeavor.[3] However, while Skiff showed a great deal of administrative promise, he almost immediately clashed with the new museum's research staff.[4] According to curators, much of the trouble stemmed from Skiff's officious and domineering management style. Whereas the museum's trustees regarded him as a loyal and capable deputy, curators felt that he lacked the requisite background to manage an institution dedicated to the production and dissemination of knowledge. Skiff hailed from the world of commerce, and he could not understand why members of the scientific community guarded their institutional autonomy so jealously, nor why they resented the measures he implemented to manage their work from above. In January 1897, curators went into open revolt. They complained bitterly about the notion "that only a business man, and a business man only, can conduct the business of an institution—museum or otherwise," which they derisively called "The Chicago Idea."[5]

It is tempting to interpret the curators' revolt as a protest against the commercialization of science. But that would be a mistake. While curators objected to Skiff's businesslike attitude, they could not fault him for failing to promote the ideals of what was often called "pure science." Indeed, Skiff emerged as an especially outspoken proponent of shifting the museum's institutional focus by replacing exhibits that celebrated American commerce and industry with more sober, scholarly displays. Only one year after it threw open its doors to the public, for example, he triumphantly proclaimed that visitors who came to the museum "under the impression that it was a miniature World's Fair, have discovered their error," for, as he later explained to the institution's wealthy trustees, "Inappropriate and undesirable material is constantly disappearing to be supplanted by that which is nearer the [strictest scientific] standard."[6] Nor did curators resent the museum's identity as a philanthropic project of Chicago's business elite. If anything, they enthusiastically courted the financial largesse of wealthy capitalists with an interest in natural history, just as their peers in New York and Pittsburgh had done. Instead, curators objected to something far more specific than Skiff's cozy relationship with the American bourgeoisie, namely, his embrace of an administrative philosophy that stressed the importance of bureaucratic accountability in managing a large and complex organization. While scientists were glad to accept the financial backing of tycoons such as Marshall Field, they did not welcome the reams upon reams of paperwork that accompanied their philanthropic bequests. Thus, when museum curators complained about "The Chicago Idea," they were less worried about the commercialization of science than the notion that a museum could to be managed and run like a business.

During the decades around 1900, the political economy of American capitalism underwent a period of rapid consolidation, which resulted in many small, family-owned businesses being absorbed into much larger and corporately organized firms.[7] Historians often describe this period as the "Great Merger Movement" in American business, but the corporate reconstruction of the US economy had a profound impact on American culture and society too.[8] As industrial firms grew larger in size and more expansive in scope, people's relationships with their employers, their jobs, and one another were dramatically altered. Corporately organized firms grew so large and complex that a sole proprietor could not supervise their operation alone. For that reason, industrial behemoths were usually carved up into a dense latticework of quasi-autonomous units overseen by a cadre of paid administrators. This meant that ownership was increasingly divorced from management, and the latter became a recognizable (not to mention profitable) profession in its own right. In addition, because the capital requirements of large corporate firms could be astronomically high, ownership was often spread across dozens of shareholders, if not hundreds or more. Before long, even ordinary Americans came to regard themselves as investors with a stake in their nation's economy, an amorphous entity whose size and well-being was soon being measured using new statistical indicators like GDP. Taken together, all of these changes created a fervent desire for accountability, not only among corporate managers who answered to shareholders but also a growing sector of the broader public who were increasingly tracking the stock market's performance using convenient new bellwethers such as the Dow Jones Industrial Average.[9]

The corporate reconstruction of the US economy had an impact on museums like the one in Chicago as well. Philanthropists such as J. P. Morgan, Andrew Carnegie, and Marshall Field invested in dinosaurs to flaunt their cultural exclusivity while simultaneously proclaiming a genuine devotion to public munificence and civic responsibility. But they did not cease to be astute businessmen in the process, and they always remained anxious to get the most out of their investments. As Albert Bickmore wrote in a letter in 1873, his museum's "trustees like to be sure they are getting value for their money."[10] So when trustees hired salaried managers such as Skiff to take on the day-to-day operation of their museums, they expected them to ensure that their philanthropic bequests were put to the best possible use. Managers such as Skiff complied with these wishes by making themselves and the institutions they ran accountable through a suite of bureaucratic practices drawn from an industrial context. This led to the implementation of a hierarchical, rule-governed, and inflexible administration that placed a great deal of emphasis on the production, dissemination, and storage of paperwork.[11]

At philanthropically funded civic museums, dinosaur paleontology was bureaucratized. In exchange for a generous infusion of cash, the work of vertebrate paleontologists came to be ruled from above, with strict orders coming down the chain of command and information flowing back up. In this way, trustees could be satisfied that their money had been wisely invested without having to spend time and effort running these institutions themselves. Skiff's controversial leadership style thus did not represent a hostile take-over by business interests so much as a distinctly bureaucratic vision of the way that a large, capital-intensive, and organizationally complex institution ought to be managed. Because that vision involved forcing independently minded curators to submit to the authority of an officious administration, it is easy to understand why they rebelled.

Management at a Distance

Philanthropic museums may have been run like businesses, but that does not mean they engaged in commercial activities. Philanthropists coveted dinosaurs because they were large, impressive, and distinctly American. But those features alone did not yet suffice to serve as a mark of cultural exclusivity. For dinosaur fossils to aid in the performance of social distinction, they first had to be removed from the commercial specimen trade, ensuring that they would no longer be treated like other scarce natural resources one could dig out of the ground. Failure to do so would have invited comparison between patronizing a natural history museum and purchasing a coal mine, which is precisely what philanthropists sought to avoid. This provided another reason for philanthropically funded museums to turn to the power of bureaucratic authority. The complex administrative structures museums put in place at the beginning of the twentieth century not only helped wealthy trustees maintain oversight of their institutions. Bureaucracy also offered a means to control the movement of dinosaur bones. By restricting the contexts in which they circulated, dinosaurs could be removed from the commercial market, thereby conferring an added layer of exclusivity to the ownership of these spectacular creatures.

At the same time, members of the scientific community were anxious to insulate themselves from the marketplace by assiduously disentangling their work from commercial affairs. The demand that science be regarded as a noble calling—pursued for its own sake and not for material gain—was as old as the practice of science itself. But it became particularly vociferous toward the end of the nineteenth century. It was at this time that scientists most visibly celebrated (without always strictly adhering to) their value neutrality. And it was also at this time that they most clearly articulated a conception of objectivity as the denial of one's individual motivations, selfish desires, and personal biases. Paradoxically, however, that did not prevent them from joining forces with philanthropists who were eager to invest in the creation of cultural capital. In fact, exactly the opposite happened. Because the conversion of economic wealth into cultural capital was necessarily circuitous, if not in fact hidden, the campaign to "purify" science by protecting it from the world of commerce only made it an ever more lucrative site for philanthropic investment.[12]

The call to insulate science from the demands of the marketplace reached a crescendo during America's Long Gilded Age, but money's corrosive effects on the production of reliable knowledge had been a source of anxiety for some time. In 1838, the eminent physicist Joseph Henry confided in his colleague Alexander Dallas Bache that, absent a concerted effort to "raise our scientific character," the United States was in danger of succumbing to "charlatanism" and "quackery."[13] Initially, Henry and his ilk primarily sought to bolster the credibility of American science by policing the composition of their community, promoting what they called "abstract" knowledge produced by "theoretical" researchers over the work of "practical men" such as miners, mechanics, and inventors.[14] By the last third of the century, however, the personal motives of individual scientists also became a locus of intense scrutiny and concern. This gave rise to a clear terminological shift, and Gilded Age scientists began to adopt the language of purity for their purposes. Undoubtedly the most well-known and oft-cited example is the Johns Hopkins University physicist Henry Augustus Rowland, who delivered a famous "Plea for Pure Science" before the American Association for the Advancement of

Science in 1883, exhorting its members to create "something higher and nobler in this country of mediocrity" by cultivating a personal desire to "study nature from pure love."[15] Rowland encouraged fellow scientists to be steadfast and assiduous in avoiding all financial entanglements and developing disinterested habits of mind, for only thus could they shore up the foundations of truth. In addition to policing the boundaries of their community, then, Gilded Age scientists such as Rowland sought to control the private desires and motivations of individual researchers as well.[16]

The ethos of "pure science" developed from long-standing fears that commerce undermined the production of reliable knowledge. As the historian Paul Lucier puts it, "An appeal to 'pure science' bespoke a pessimism about the corrupting influence of money and materialism."[17] But that did not mean Gilded Age scientists shunned the rich and the powerful. In fact, they actively solicited the support of the business community. Rowland was remarkably candid in this regard, arguing that although scientists were engaged in "something more honorable than the accumulation of wealth," they nonetheless required "instruments and a library" to get on with their work, not to mention "a suitable and respectable salary to live upon." To that end, he called on America's financial elite to underwrite the production of knowledge, judging that only they had both the means and the "moral qualities" to carry out such an ambitious agenda. Rowland also instructed fellow scientists to fashion themselves into worthy recipients of conspicuous generosity, counseling them, "We must live such lives of pure devotion to our science that all shall see that we ask for money, not that we may live in indolent ease at the expense of charity, but that we may work for that which has advanced and will advance the world more than any other subject, both intellectually and physically."[18] This was a message tailor-made for its age, conforming precisely to what the likes of Morgan, Field, and Carnegie wanted to hear.

In a striking convergence, the same period also witnessed the emergence of a formal distinction between for-profit and nonprofit corporations. Earlier in the nineteenth century, corporate personhood had been understood as a privilege the state could bestow on behalf of its citizens, and anyone who wished to apply for a corporate charter had to demonstrate how their proposed business venture benefited the commonwealth. Early corporations were private enterprises understood to perform a public service that, for whatever reason, the state did not wish to offer directly. Often, these included infrastructure projects such as turnpikes, canals, and railroads. They also included charitable organizations, educational institutions, and similar efforts to promote the common good. As a result, the scope of an early to mid-nineteenth-century corporation was restricted to the specific purpose explicitly declared in its charter. Starting with Pennsylvania in 1874, however, one state after another began to enact a new set of general incorporation laws, which removed these restrictions in favor of a sweeping distinction between for-profit and non-profit enterprises. As a result, business corporations no longer had a specific and circumscribed purpose. Rather, they could engage in any commercial activity expected to yield revenue. In contrast, nonprofit corporations were exempted from having to pay taxes because they did not have a commercial intent. The general requirement that all corporate bodies must serve the common good was therefore replaced by a scrutiny of the private intentions that motivated their public behavior.[19]

Precisely the same period that saw an ethos of purity used to police the internal motivations of scientists also witnessed the creation of a new legal entity—the nonprofit corporation—that was designed to promote the philanthropic ambitions of capitalists who desired to perform good works. Indeed, the asceticism of pure science fit hand in glove with the noble high-mindedness that nonprofit corporations sought to project. This promoted a symbiotic, mutually beneficial relationship between Gilded Age naturalists and their benefactors. Precisely because the applicability of pure science was so hard to assess, its financial support allowed wealthy philanthropists to bolster the claim that they were motivated by genuine altruism. In turn, altruism was exactly the motivation that researchers like Rowland most valued in their patrons, because it helped to ensure they would not have to justify the value of their activities on narrowly utilitarian grounds. In other words, by insisting on an almost monastic devotion to truth, and by jealously guarding their institutional autonomy, advocates of pure science fashioned themselves into attractive targets for philanthropic largesse. At the same time, their visible devotion to pure knowledge allowed wealthy elites such as Field to argue that modern capitalism produced genuine public goods in addition to profits.

These developments help to explain why the division of labor in vertebrate paleontology changed so dramatically in the context of philanthropic museums at the beginning of the twentieth century. As we saw in the previous chapter, William Harlow Reed was dismissed from the Carnegie Museum within just a few years of discovering its most celebrated dinosaur because he could not fully shed his identity as a commercial fossil hunter. In a telling contrast, another freelance collector named Charles Hezelius Sternberg managed to make the transition that eluded Reed. In no small part, Sternberg succeeded because he was willing to play the part of a self-effacing servant to truth. Reflecting back on a long, arduous, and penurious life in his autobiography, Sternberg insisted that he was gratified to have contributed his "humble part toward building up the great science of paleontology." Noting that he could have done better financially by selling fossils to "showmen" or dealers, Sternberg stressed that it was the esteem of his peers that he especially coveted, arguing that credit from the scientific community constituted the highest reward for someone who "values his labor as something that cannot be measured by money."[20] Ironically, then, it was precisely the lack of a businesslike attitude that made it possible for Sternberg to do business with philanthropic museums for many years after Reed had been fired from Pittsburgh.[21]

The pure-science ethos was attractive to vertebrate paleontologists and their benefactors for another, more practical reason as well. Because of the vast expanse that separated urban museums such as the one in Chicago from the fossil fields of the American West, vertebrate paleontologists could not be confident in the quality of a new dig site without visiting it in person. This made it difficult to know whether freelance collectors such as Sternberg and Reed were honest about their latest discovery, and naturalists worried endlessly they might resort to outright fraud in the hopes of inflating its value. Since the acquisition and display of adulterated or otherwise fraudulent specimens could do significant damage to a museum's reputation and credibility, the ethos of pure science offered a valuable means for these institutions to protect themselves from the risk of being defrauded that ran so high on the mining frontier. However, the power to instill desirable moral qualities could not be taken for granted and always remained a matter of trust. In practice, there was no way of knowing with certainty that someone like Sternberg was sincere in his claim to be after more than just money, or whether he merely paid lip service to the ethos of pure science. Hence, paleontologists developed a number of other techniques to exercise a measure of control over events in the field. Often, these centered on the production and circulation of paperwork designed to allow naturalists in the museum to act at a distance and discipline the behavior of their collectors out in the field.[22]

One way that US paleontologists controlled events in the field was by issuing printed instructions. An early example was drawn up by Othniel Charles Marsh during the 1870s and came to be known as "Marsh's Fifteen Commandments." With it, Marsh sought to regulate all the minutiae of his collectors' day-to-day work. For example, he told them to bring "sacks, paper, cotton, and twine" when going out for the day, and to pack any specimens they might find in burlap "with a label inside, and a tag outside, giving locality, formation, collector, and date." These were then to be shipped to the museum in "boxes of moderate size" made of "one inch boards" and surrounded by enough "hay or straw" to ensure that their contents "cannot move when the box is turned over."[23] Continuing in like fashion for two pages, Marsh clearly did not trust even veteran fossil hunters with immense field experience to follow their instincts.[24]

Marsh's obsessive attention to detail was in part a response to the very real difficulty of reconstructing a lost world from its fragmentary remains. Since Georges Cuvier, paleontologists had understood themselves to practice an "archival" science whose "documents" had been broken into countless pieces and scattered across the globe. To puzzle out the detailed story of life's history, paleontologists sought to reconstitute the earth's archive within the more manageable confines of a museum, where its contents could be more readily assessed, compared, and interpreted. For that strategy to succeed, however, it did not suffice to ensure that specimens were transported to the museum unharmed. Just as important, but far more difficult, was to protect the *relationship* between them as well. This made it essential to preserve information about the precise location, depth, and geological horizon in which fossils were found, as well as their position and orientation relative to each other. Otherwise, there would have been no way to sort out the contents of a particular quarry, let alone reconstruct extinct and strange-looking creature from innumerable fossil remains that had been dug up in different locations, sometimes separated by thousands of miles.[25]

To reconstitute the fossil record from a huge mass of fragmentary remains inside the museum, paleontologists developed a suite of sophisticated information technologies. In Marsh's case, perhaps most important among these were printed labels that collectors filled out and affixed to the inside of each shipping crate. These labels gave Marsh a reliable record of exactly when and where each discovery had been made, as well as the rock formation in which it was found. Still, the task of keeping track of the specimens coming into the museum could be overwhelmingly difficult. As one of his laboratory assistants recalled, collectors during the 1870s and 1880s usually took specimens up in a "potato fashion," meaning that individual fossils were "picked up in pieces" much as "potatoes are dug in the field." Next, they were simply shipped off in whatever order they came out of the ground, leaving the task of reassembling a complete specimen entirely to the museum, which often proved to be "an utter impossibility, especially if the pieces were quite small and the fractures not characteristic enough to determine their position."[26] Hence, Marsh began to instruct his collectors to draw maps of each quarry, showing the exact location of bones that lay in close proximity to one another and indicating which ones were grouped into an individual shipment. Such practices were crucial to determining whether two pieces belonged to the same animal, and if so, what their anatomical relationship might have been. In some cases, a similar procedure was even used to "map" a single bone that was broken into innumerable pieces. When dealing with an especially fragile discovery, collectors were told to take a large piece of paper, lay it flat on top of the fossil, and trace out its overall shape, indicating where any major cracks, fissures, or other points of weakness were located. By assigning a number to each segment of bone and marking it on the drawing as well as the fossil itself, they made it easier for the specimen to be reassembled back in the museum.[27]

Faced with an avalanche of material coming in from the field, Marsh also devised an elaborate system of record keeping to be used by his clerical staff. By the mid-1880s, he had hired some three dozen assistants to make the Peabody Museum a sophisticated information-processing center. The goal was to create a physical database that could render the fossil record legible to the human eye by putting detailed, accurate, and up-to-date information about every specimen at Marsh's fingertips. Whenever a new shipment arrived on the railroad, it was immediately assigned an accession number that was duly recorded in one of a series of pocket-size receipt books. As the shipment was unpacked, each crate was assigned a unique box number that was recorded in a large-format ledger, along with relevant information from printed field labels and correspondence files. Together, accession and box numbers formed a hierarchical finding aid, making it possible for museum staff to keep track of each specimen, as well as the various maps, labels, and letters that its collection had generated, all of which were kept on file also. Further robustness was built into the system via redundancy by directly marking each fossil and the corresponding paperwork with the appropriate box and accession numbers. Cross-referencing everything to provide multiple points of entry into the system allowed Marsh and his staff to determine the provenance of a specimen even in the event that a ledger or receipt book was lost or destroyed.[28]

This elaborate system of paperwork did far more than just allow Marsh to keep track of everything that was coming into the museum. It also allowed him to manage events in the field without ever having to leave New Haven. Rather than trust collectors to make an informed judgment about the most valuable information to record, Marsh issued standardized printed forms that prompted them to report precisely what he most desired to know. Similarly, when he hired a promising student named John Bell Hatcher to assist Sternberg in Kansas during the mid-1880s, he made the young man keep a diary of the work that he did every day, which was condensed and sent to the museum on a regular basis, usually once a week.[29] Not long after leaving New Haven, Hatcher also devised an improved method to manage the excavation of fossils. Rather than merely making crude drawings to indicate where each bone was buried, Hatcher systematically divided his quarry into square sections of uniform size to act a guide when drawing a more comprehensive map. As he boasted to Marsh, "I work each section separately and make a large diagram of it and whenever I take up a bone I number it," being careful to "put the same number on the diagram that I put on the bone."[30] Within less than a year, other collectors working for Marsh had been instructed to use the same method as well. For example, the fossil collector Fred Brown kept track of the precise location in which specimens were uncovered when he took over the work of digging at Como from Reed during the 1880s (Figure 4.1). Notably, Brown's maps would later prove to be crucial in working out the complex osteology of *Stegosaurus*, whose functional anatomy posed a particularly difficult puzzle for Marsh.

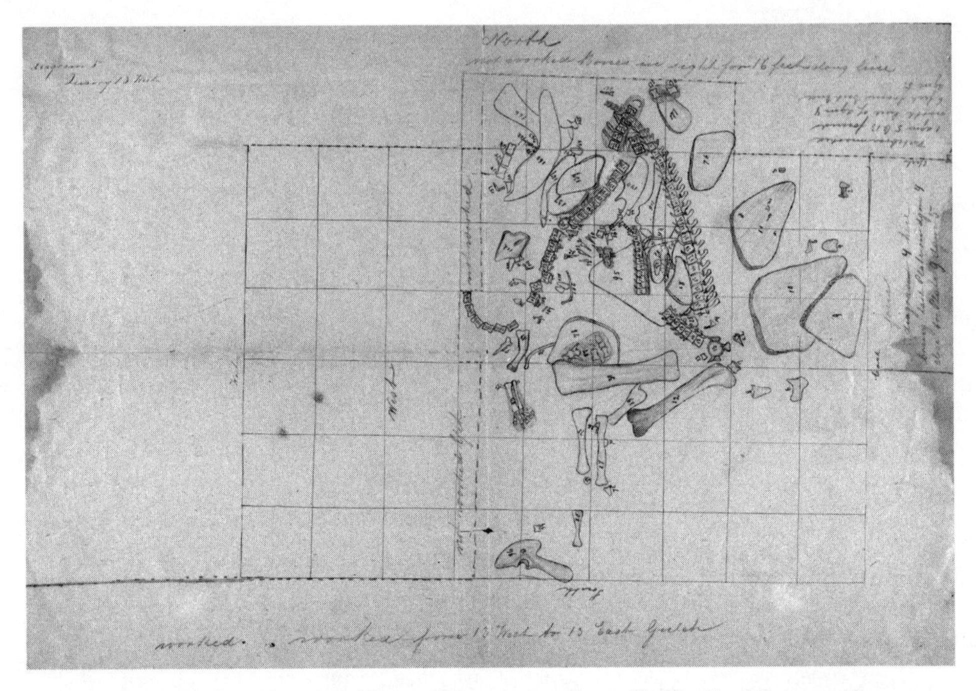

Figure 4.1. A diagrammatic grid map of Quarry 13 at Como Bluff by Fred Brown, 1885.

Not long after Hatcher began using a grid system to locate specimens on a more detailed map, he also developed a new way of taking specimens out of the ground. Whereas collectors had previously dropped individual bones into a burlap sack and then packaged these in a wooden box with plenty of straw, Hatcher began "bandaging" specimens with strips of cloth that had been soaked in a mixture of flour and water. In this way, large pieces of dinosaur bone could be protected en route to the museum. The new method spread quickly, and before long Marsh's assistants began to use plaster of Paris rather than flour. By the late 1890s, field crews working for Osborn at the New York museum were bandaging entire sections of rock without ever handling the fossils still buried inside (Figure 4.2). By digging down into the matrix alongside a specimen, they could tunnel beneath it and apply strips of cloth covered in plaster while the excavation was still taking place.[31]

These disparate practices and information technologies all evinced a pervasive desire among museum curators to control the process of specimen acquisition from afar. The rigid rectangular grid lines that Hatcher began using

Figure 4.2. Barnum Brown pouring plaster on a section of *Diplodocus* tail vertebrae still encased in the surrounding rock matrix at Como Bluff, 1897.

when drawing up maps of his quarry not only ensured that more detailed and accurate information about every specimen would be preserved. Insofar as they satisfied a more stringent requirement of realism, they also constrained his ability to exercise personal judgment about how to represent the location and relationship of each specimen in situ. Maps with grid lines therefore resembled a standardized printed form in that they transferred a measure of control over the collection process to the museum, virtually transporting Marsh from his office in New Haven to a dinosaur quarry several hundred miles away. The practice of bandaging up entire sections of rock matrix was an even more telling case in point. Instead of allowing collectors to decide the best way to excavate fossils, the process effectively left entire sections of a quarry intact, so that large portions of rock matrix could be transported to the museum untouched, with the bones still buried inside.[32]

Although it was certainly baroque and sophisticated, the elaborate bureaucracy Marsh devised at the Peabody Museum did not offer a total solution to the problem of trust. To the extent that it allowed Marsh to project his watchful eye into the field, it lessened the information asymmetries that plagued American paleontology during the late nineteenth century. But no matter how complex or intrusive, it could never eliminate them entirely. Insofar as he still had to trust his collectors to work conscientiously and with painstaking attention to detail, Marsh could not afford to place all his faith in the power of bureaucratic authority. For that reason, he also surrounded himself with a cadre of field and administrative assistants who were tasked with ensuring that excavations were carried out in accordance with his exacting standards. Unlike freelance collectors such as Reed, these assistants had a background in academic paleontology. Extensively socialized in the culture of pure science, they shared many of Marsh's expectations about the self-restraint and personal discipline that were required to produce reliable knowledge.[33] Even so, Marsh never fully trusted his assistants to work on their own, and most left the Peabody Museum having become disgruntled with their domineering employer.[34]

As a result of Marsh's obsessive and megalomaniacal exercise of control, he quickly alienated those underneath him and considerably hampered the creation of paleontological knowledge at his museum. Starting in the late 1870s, his ambition had been to set up a streamlined production process resembling a paleontological factory that was supposed to commence with the collection of huge numbers of fossils, proceed with a steady stream of preliminary reports drawn up as individual specimens were removed from the rock matrix, and culminate in the publication of no fewer than nine detailed and lavishly illustrated monographs. Issued in a large folio format, each was to be the last word on a particular group of extinct animals from North America. But before long, Marsh's plans faltered, largely due to his autocratic and domineering management style. As two early biographers described it, his usual practice was to assign a "series of fossil bones" to an assistant, along with "precise instructions as to what he wanted done." Then, as museum employees busied themselves with a series of specimens, he would make the rounds to check up on their work. More often than not, he only got in the way, actively interfering with the work of his preparation staff by insisting they never make any decisions without asking him first. As one might expect, this method did not deliver results very efficiently. According to his biographers, Marsh soon became "overwhelmed and confused by the very mass of his fossil riches, and by the effort required to direct his superabundant staff." Because he was so adamant about micromanaging everyone's work, his assistants frequently found themselves idling away several days with nothing to do, especially if their boss happened to be traveling away from the museum, which was often the case. As a result, of the nine monograph publications he had promised the United States Geological Survey, Marsh managed to complete only three.[35]

The Vertical Integration of Dinosaur Paleontology

The museological practices Marsh implemented at Yale were elaborate and intrusive, but they were hardly sui generis. His obsession with exercising managerial oversight from afar reflected a much broader trend in late nineteenth-century America that could also be seen in the process of vertical integration among industrial firms. By internalizing the market for raw materials and taking control of their own distribution networks, firms such as Carnegie Steel and Singer Sewing Machine Company found they could reduce the cost of doing business over long distances. Rather than continue to deal with independent and untrustworthy third parties, vertically integrated firms enforced compliance within a company's ranks through coercive measures and incentive regimes. However, while this strategy allowed firms to conquer vast areas of geographic space, it simultaneously created a great deal of new institutional space, often resulting in administrative bloat. Vertically integrated industrial firms responded by implementing yet more intrusive forms of managerial oversight, which usually relied on the creative use of paperwork, especially account books.[36]

Carnegie was an especially committed champion of the account book as a managerial tool to keep tabs on the internal workings of a large and complex organization. Because his Bessemer steel mills required a steady supply of high-quality ore that contained almost no phosphorous, a fluctuation in supply or a drop in quality could bring his entire operation to a standstill. Unwilling to stand by and watch costs mount as expensive machinery sat idle, Carnegie simply began to purchase his own mines. Before long, he also acquired railroads to ship raw materials from the point of extraction to his processing plant.

As his onetime assistant James H. Bridge described it, Carnegie turned his manufacturing enterprise into a "mammoth body" that "owned its own mines, dug its ore with machines of amazing power, loaded it into its own steamers, landed it at its own ports, transported it on its own railroads, distributed it among its many blast-furnaces, and smelted it with coke similarly bought from its own coal-mines and ovens, and with limestone from its own quarries." In this way, Carnegie could make sure that "from the moment these crude stuffs were dug out of the earth until they flowed in a stream of liquid steel in the ladles, there was never a price, profit, or royalty paid to an outsider."[37] By internalizing every step of the production process, Carnegie cut costs and increased profits.

It was not long before political economists began to theorize the business practices that Carnegie and other late nineteenth-century industrialists developed. Among the most intriguing examples was Alfred Marshall, whose 1890 book *Principles of Economics* arguably ranked as the most widely read text on political economy at the time. According to Marshall, the modern industrial firm resembled a social organism whose developmental trajectory "always involves an increasing subdivision of functions between its separate parts on the one hand, and on the other a more intimate connection between them." Just as biological evolution led to a physiological division of labor, industrial evolution pushed firms toward greater and greater "integration," by which he meant "a growing intimacy and firmness of the connections between the separate parts of the industrial organism."[38] That said, Marshall did not agree with contemporaries like J. A. Hobson who wished to see the industrial organism continue along this evolutionary trajectory until all of society's productive capacities were integrated within a single, socialist state.[39] That is because Marshall believed integration came at a cost. Whereas large, integrated firms could benefit from economies of scale, there were advantages that accrued to small firms as well, not least of which was the ease with which its employees could be held to account. In a small workshop or manufactory, Marshall explained, "the master's eye is everywhere," he has "no divided responsibility," and he "saves much of the book-keeping."[40] As industrial firms integrated more parts of the production and distribution process under one roof, however, they became more unwieldy, cumbersome, and difficult to control.[41]

In hopes of combatting the problem of institutional bloat, late nineteenth-century firms turned to the power of bureaucratic authority, implementing procedures to control the flow of information inside a company's ranks. Many adopted a militaristic line-and-staff form of organization, creating a clear institutional hierarchy that sought to eliminate ambiguities in the chain of command. Employees were expected to follow orders from their superiors without question or argument. They were also required to communicate complaints, insights, and innovations directly to their immediate supervisor. But this was only one among many related efforts to exercise what the historian JoAnne Yates aptly describes as "control through communication," all of which sought to channel the flow of information to aid management.[42] With one eye always trained on efficiency, corporate managers especially prized numerical or quantitative information, which could be easily summarized, analyzed, manipulated, and recombined into memoranda and other written reports. As a result, bookkeeping and other accounting practices, which had long been employed to regulate a firm's relationship to the outside world, increasingly came to be used as tools for managerial decision making.[43]

As account books were repurposed as tools to coordinate the internal workings of businesses, one kind of accounting emerged as an especially informative guide: cost management. Rather than calculate profits and revenue by balancing debits and credits, cost accounting was explicitly directed toward reducing expenditures. Watch the costs, the famous saying went, and profits will take care of themselves. Again, Carnegie offers an especially telling example. Having learned the value of careful bookkeeping early in life, he became so fixated on cost management that he made a point of personally scrutinizing his company's books on a regular basis, insisting that cost sheets be sent to him at least once a month (sometimes every week), even when he was traveling overseas. By keeping a watchful eye on costs, Carnegie could ensure the profitability of his steel business from afar without hardly ever having to set foot in Pittsburgh. "There goes that ____ bookkeeper," one of his factory workers is reported to have said, complaining, "If I use a dozen more bricks than I did last month, he knows it and comes around asking why!"[44]

Wealthy capitalists like Carnegie brought these business practices to their philanthropic endeavors as well. According to his biographer, Carnegie devised a "scientific, corporate system of giving, one that guarded against sentiment and made decisions based on hard data." Carnegie managed his philanthropic institutions much like his industrial factories, which resulted in a "highly efficient bureaucracy for giving away his money."[45] Other philanthropic institutions were run along strict business lines too. An internal history of New York's Metropolitan Museum of Art, for example, proudly reported that it "became a modern institution in its business methods around 1905," precisely the year that its trustees elected J. P. Morgan to be the museum's next president. In addition to acquiring the museum's first Remington typewriter and installing a sophisticated "intercommunicating system" throughout the building, Morgan's administration created a centralized registrar's office that developed a "thorough and all-informing" system for keeping track of the museum's collections.[46] As the Metropolitan Museum's assistant secretary reasoned, neither "red tape nor slipshod unbusiness like methods of administration should be tolerated in a museum any more than in a factory."[47]

A similar transformation took place at the New York natural history museum, whose internal operations had to be rendered legible to its parsimonious donors. In a letter that was written before the museum had even been granted its corporate charter, one of its founders urged the "consideration of economy" on the new venture to ensure that it would "bring about the greatest results with the smallest means."[48] As a consequence, the museum soon began implementing the kinds of cost-management practices used by industrial corporations. Indeed, so keen was the museum to prove that it spent its funds wisely that, in 1906, it even contracted for an independent audit of its books twice every year, and by 1941 an outside counsel was able to praise the museum for the way that its financial documents "exactly" conformed to "the manner in which corporations with many subsidiaries prepare their consolidated statements of income."[49]

The controversy that flared up over the "Chicago Idea" notwithstanding, museum curators often proved themselves willing participants in the effort to make their institutions more accountable. A particularly good example in this regard was the vertebrate paleontologist Henry Fairfield Osborn. As a nephew of Morgan and the son of a railroad manager who helped the Illinois Central avert bankruptcy during the 1850s, Osborn was a born bureaucrat.[50] Tellingly, one of his first orders of business upon being named as the museum's president in 1908 was to take stock of the institution's assets and liabilities. To that end, he demanded that Hermann Carey Bumpus, the museum's operating director, immediately draw up a systematic account showing the "total gifts, including expenditures from income, since the time of the foundation of the Museum."[51] Anxious to root out inefficiencies wherever they lurked, Osborn further tasked Bumpus with instructing "the Curators that each department should be mutually self-sustaining," adding "that the drafts of one department upon another for services, specimens or material should not be made without the full authority of the Director."[52] Despite the uprisings that took place in Chicago, then, scientists such as Osborn actively participated in the implementation of administrative structures and accounting techniques designed to keep close tabs on their institutions' financial assets.

The desire to counter institutional bloat through the implementation of bureaucratic management practices extended beyond purely administrative matters, however, exerting a profound influence on the practice of science as well. As new methods to streamline the production of knowledge and systematize the collection and storage of valuable specimens were implemented, personal recollection and oral modes of communication gave way to more formal regimes of written correspondence modeled on systems of "upward reporting" from a corporate context. The practice of collecting, storing, mounting, and studying vertebrate fossils was therefore bureaucratized too. As a result, museum curators increasingly found themselves having to justify the work they were doing in numerical terms.[53]

Because it was founded as a permanent memorial to the 1893 World's Columbian Exposition, the Field Museum of Natural History devised an especially elaborate system for accessioning new objects to its permanent collection. As its first order of business, the museum had to select, acquire, transport, and arrange an impossibly large number of objects from the world's fair in the Palace of Arts Building, the new institution's first permanent home. From the moment the Field Museum was founded, its collections already constituted some of the most extensive in the United States. This included a vast number of specimens initially assembled for exhibition by Ward's Natural Science Establishment and purchased by the museum for nearly $100,000. The Ward Collection included entire cabinets of "Mineralogy and Meteorites," "invertebrate animals," "stuffed animals," "mounted skeletons,"

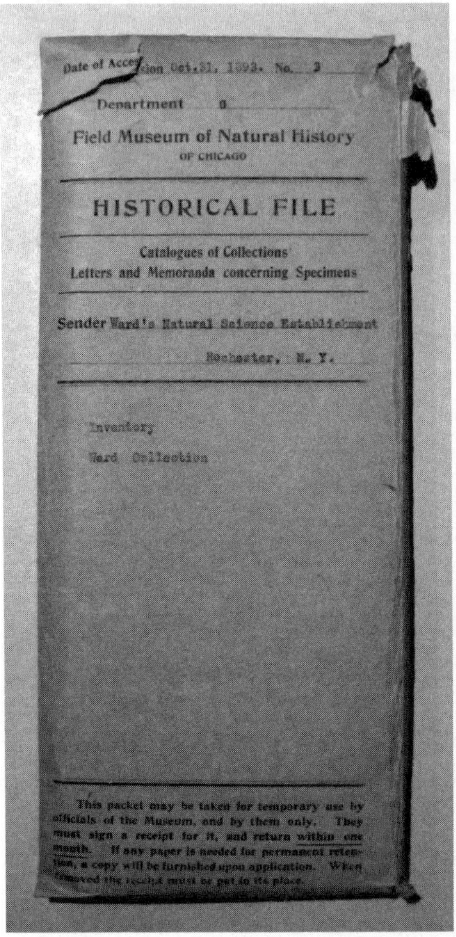

Date of Accession Oct. 31, 1893. No. 3

Department G

Field Museum of Natural History
OF CHICAGO

HISTORICAL FILE

Catalogues of Collections
Letters and Memoranda concerning Specimens

Sender Ward's Natural Science Establishment
Rochester, N. Y.

Inventory
Ward Collection

This packet may be taken for temporary use by officials of the Museum, and by them only. They must sign a receipt for it, and return within one month. If any paper is needed for permanent retention, a copy will be furnished upon application. When removed the receipt must be put in its place.

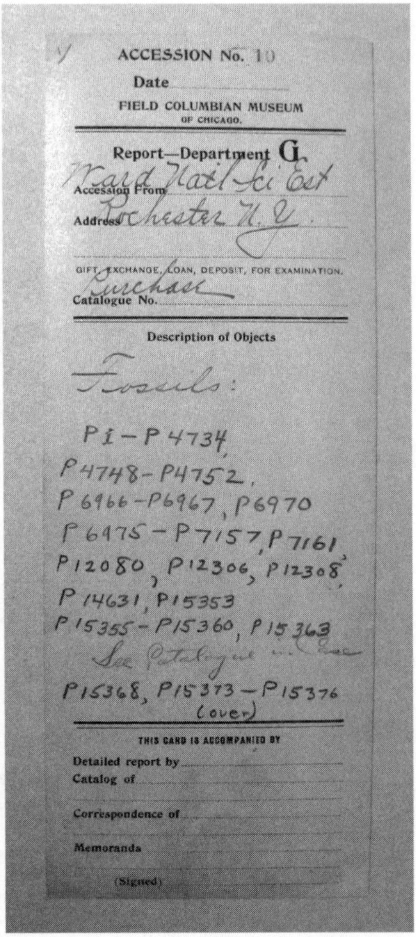

ACCESSION No. 10
Date
FIELD COLUMBIAN MUSEUM
OF CHICAGO.

Report—Department G.
Accession From Ward Natl Sci Est
Address Rochester N. Y.
GIFT, EXCHANGE, LOAN, DEPOSIT, FOR EXAMINATION. Purchase
Catalogue No.

Description of Objects

Fossils:
P1–P4734,
P4748–P4752,
P6966–P6967, P6970
P6975–P7157, P7161,
P12080, P12306, P12308,
P14631, P15353
P15355–P15360, P15363
See Catalogue in Case
P15368, P15373–P15376
(over)

THIS CARD IS ACCOMPANIED BY
Detailed report by
Catalog of
Correspondence of
Memoranda
(Signed)

Figures 4.3 and 4.4. Accession record for fossils acquired from Ward's Natural Science Establishment by the Field Museum of Natural History in 1893.

and "Anthropology specimens." The geological and paleontological specimens alone included 1,126 Silurian fossils, 118 Triassic fossils, and 800 Tertiary fossils, as well as a mastodon, an Irish Elk, a hadrosaur, a *Dinoceras,* a *Glyptodont,* a *Megatherium,* an ichthyosaur, and a plesiosaur.[54] To deal with the logistical nightmare of accessioning all of these objects at once, the museum put a sophisticated administrative structure into place, complete with a centralized recorder's office. The latter ordered that a museum curator had to document the provenance of every new object entering the institution, which was accomplished by filling out a printed card that was kept in a jacketed sleeve and placed on permanent file (Figures 4.3 and 4.4), along with any correspondence and other written material that pertained to the collection in question.

At the Field Museum, accession records primarily served a legal purpose, establishing proof of ownership. To keep track of individual specimens, the Recorder's Office also maintained an accession catalog for each academic department, a duplicate copy of which was kept in the department itself. These consisted of large, bound volumes in which entries were arranged chronologically, reflecting the order in which they had been added to the collection. In addition to requiring a short description of each specimen, the Geology Department's catalog also prompted curators to note when it was received, when and where it was found, the name of the collector, and whether it constituted a gift, a loan, or an exchange. To supplement these bound volumes, the Geology Department kept a card catalog to locate particular specimens by the temporal horizon in which they had been found rather than by their accession number.[55]

Neither the practice of maintaining multiple sets of books—one in each department and another at a centralized location—nor the creation of a card catalog to supplement bound ledgers was unique to the Field Museum. The Smithsonian Museum implemented a vertical filing system as early as 1896, a device that the paleontologist Charles Schuchert described as "the most important agency in the possession of the curator, and its management is the highest test of his capability."[56] Part of the impetus behind the widespread adoption of these information technologies was practical necessity. As collections and scientific departments grew larger in size, it made sense to devise more elaborate means to manage and organize operational affairs. At the same time, rational, systematic, and bureaucratic management practices came to be seen as a kind of good in themselves. It is telling, for example, that a whole section of the Field Museum's annual report was dedicated to progress in collections management, which it described as the "most essential and basic feature of the routine work of the Museum."[57] Catalogs and accession ledgers were increasingly valued by trustees as a way to keep tabs on their institution, and curators were frequently asked to compile detailed statistics that summarized their department's output during a given year. By poring over their copious files, curators at the Field Museum could report that, in

1898 alone, their institution had accessioned 4,900 gifts, 74 loans, 5,831 exchanges, 3,469 purchases, and 5,032 specimens that were collected on expedition. That same year, the museum also boasted that no fewer than ten record books had already been filled with some 21,925 catalog entries by the Geology Department, whereas Botany had managed to fill an even more staggering twenty-nine books with nearly 70,000 entries. Anticipating that a "summarized account" of the botanical collections would "doubtless be of considerable interest" to "the trustees of the Museum," the annual report also included a table showing how many herbarium specimens had been amassed from each of the world's main geographic zones.[58]

Following the corporate lead by implementing bureaucratic documentation regimes, philanthropic museums transformed catalog records from a finding aid into yet another technique of management. Just like account books, specimen catalogs evolved into a way for trustees to keep track of their philanthropic investments. To facilitate this process, curators were sometimes explicitly instructed to record the cash value of every object that had been placed in their care. In 1887, for example, the trustees of the American Museum of Natural History in New York passed a resolution calling on curators to submit to the executive committee "a Capital account consisting of an inventory of all the property of the museum and representing its commercial value."[59] At the outset, they balked at the task. The curator of ornithology, Joel A. Allen, explained that although it would be extremely time consuming and labor intensive to do so, it might be possible to draw up "a general statement of the number of specimens contained in my departments, together with their aggregate cost." But anything more would be asking too much. In particular, he strongly objected to the idea of assigning a "commercial value" to specimens, which he defined as the sum they could be expected to "bring at forced sale." This figure "would be impossible to estimate even approximately," and, in any case, it "would certainly be very far below the actual cost" of acquiring and curating the collection.[60] Less than a decade later, however, the Field Museum's curator of geology, Oliver Farrington, expressed no qualms about assigning cash values to specimens. In a "valuation of the collections" that he submitted to the museum's director, Skiff, during the summer of 1894, Farrington appraised the museum's prized mammoth at $3,000, its mastodon at $2,000, its Irish elk at $1,200, and various fossil mammals and reptiles at between $150 and $700 each.[61] In similar fashion, the Carnegie Museum in Pittsburgh not only listed every item accessioned each year in its annual report (which had been standard practice among many older museums as well), but, by 1900, it was also assigning a definitive cash value to each of its assets, including its exhibition cases, furniture, and library books, as well as the specimens themselves.[62]

Under Osborn's leadership, the American Museum of Natural History in New York pushed the practice of managing by numbers to the extreme, and its archives are teeming with remnants of paperwork documenting the cash value to dinosaur bones. The printed forms constituting its accession records, for example, prompted curators to note not only a specimen's collector and the locality and condition in which it was found but also its "estimated value," helpfully supplying a printed dollar sign to foreclose any misunderstanding about what the museum registrar wanted to know.[63] Similarly, the Department of Vertebrate Paleontology maintained a detailed exchange ledger to record every fossil that was sent to or received from a peer institution, again assigning a precise dollar value to each. However, this does not mean anyone in the Department of Vertebrate Paleontology would have expected, or even tolerated, money to play a role in such transactions. Rather, fossils were assigned a cash value to keep track of who was indebted to whom in the museological gift economy, and museums expected to make up a deficit on either side through the exchange of more specimens, not money. Museums accounted for dinosaurs in order to manage the production of knowledge, not to make a profit.[64]

At the museum, account books functioned as a surveillance technology, a means to increase managerial oversight and control over the practice of science. When Osborn was promoted from his role as the lead paleontologist to museum director in 1908, he instructed his successor, William Diller Matthew, to begin tracking expenditures on each fossil. Once every year, the results of these efforts were consolidated into a single table and sent on to the museum's board of trustees. Using this document, it was possible for them to discern at a glance exactly how much had been spent on the collection and preparation of each specimen in the museum's exhibition halls. For example, they could learn that a paleontological exhibit that consisted of two duckbilled (or "*Trachodon*") dinosaurs had cost the museum exactly $11,024 to collect, mount, and prepare for display (Figure 4.5).[65]

TABLE I, continued,

Description of Specimen	Mus. No. of Spec.	Cost Value of Spec. as Rec'd	Cost of Preparation and Mounting (Labor Cost, not including supervision, materials, or rental charges) Before 1905	1905	1906	1907	1908	1909	1910	Total	Total Cost of Mounted Specimen
CRETACEOUS											
Trachodon group / Cope	5730	$6000		...	....	1982	452	[illegible]	...	3924	$11,024 11024
Trachodon group / Sensiba	5886	1100	.....	...	....	1490	.....	...	...		
Tyrannosaurus group / 1st	973	2200		550	1883	1301	.....	...	... 665	Unfinished	
Tyrannosaurus group / 2nd	5027	2200	.....	...	....	....	126	1227			
Triceratops skel.	5033	1650	.....	...	....	....	.....	...	97	Unfinished	
mount	5116	1000	.....	...	....	....	.....	...	247		

Figure 4.5. Detail from a cost-management account drawn up by William Diller Matthew for the Department of Vertebrate Paleontology, assessing the value of several dinosaur fossils on display at the American Museum of Natural History in New York, 1910.

Matthew's report makes it clear that he compiled these data to aid museum trustees in making high-level decisions. Even a cursory perusal of his table immediately revealed a great deal about the efficiency with which the museum's laboratory staff was doing its job. In later years, Matthew also began drawing up documents called "Tabulated Cost-Sheets of Scientific and Clerical Staff" so they could determine exactly how much had been spent in pursuit of the museum's research goals. Similarly, he calculated what curators were paid to do "routine work" and how much was spent on the installation of exhibits, the storage of specimens, and, in a flourish of recursive glee, on "cataloguing work."[66] His calculations revealed insights that might not have been obvious without such a record. For example, he reported that, as a rule, "it costs less to prepare specimens obtained by our expeditions and properly collected, than to prepare purchased specimens."[67] Matthew's cost sheets therefore allowed the museum's wealthy trustees to rest assured that their philanthropic bequests were being put to good use without ever having to enter the museum's preparation lab, let alone accompany one of its expeditions into the field.

A particularly revealing feature of Matthew's account is the way that he calculated a specimen's raw value. As the first entries in Figure 4.5 clearly show, one of two duckbilled dinosaurs recently mounted for public display had a much higher "cost value" than the other one did. Whereas the "Cope" specimen was assigned an initial cost value of $6,000, the "Sensiba" specimen was only valued at $1,100. This discrepancy was not due to one specimen's having been more complete, scientifically significant, or unique. Rather, it reflected a difference in how the two were acquired. Whereas the first had been purchased as part of a larger collection from the estate of Edward Drinker Cope in the late 1890s, the second was collected by the paleontologist Barnum Brown on a ranch that belonged to Alfred Sensiba outside Laramie, Wyoming, during the early twentieth century. In each case, the value of the individual fossil was calculated by dividing the cost of the entire acquisition by the number of specimens thus obtained (with adjustments to account for a specimen's importance for research and exhibition purposes).[68] In other words, whereas one number represented the *price* of a specimen purchased on the open market, the other represented the *cost* of internalizing the Wyoming fossil fields by integrating backward into the acquisition of raw materials. To Matthew, these numbers indicated that, in general, it was "much cheaper to get specimens through museum expeditions than by purchase." And to avoid giving the impression that the museum had "paid too much for specimens," Matthew used similar accounting techniques to demonstrate that the museum's "field expeditions have been very efficiently conducted, both as to direction and personnel," and had therefore proved to be "very economical."[69]

Conclusion

Although contemporary observers such as Max Weber often equated bureaucracy with modernity, philanthropic museums were hardly the first spaces in which the production and dissemination of knowledge was bureaucratized.[70] Over a century earlier, European empires deployed analogous techniques to govern the large and sometimes unruly territories to which they lay claim. In hopes of gaining mastery over their far-flung possessions, colonial administrators sought to gather a wide range of useful intelligence. This led to a proliferation of statistical surveys, topographical maps, and encyclopedic compendia, which all attest to the importance of information technologies for the functioning of a bureaucracy. If bureaucracy constitutes "rule from the office," then bureaucrats must be able to act at a distance, projecting their authority across space and time. Often this was accomplished by exercising control over the creation and circulation of paperwork, and colonial administrations relied upon identity documents, property registers, and tax rolls make populations and territories legible from afar. But it was not only paper documents that colonial bureaucrats craved. They valued physical specimens too, and the museum functioned as an important element of colonial rule, providing an organized inventory of all the productive resources that were available for extraction in a particular region.[71]

As new transportation and communication technologies made it possible for industrial firms to acquire immense market shares during the late nineteenth century, bureaucratic management practices migrated from imperial governance to corporate administration. The familiar challenge of governing from afar was especially acute among firms such as Carnegie Steel, which sought to internalize all parts of their supply chain by integrating backward into the production of raw materials and forward into mass distribution. Because it promised to reduce the risks of conducting business with unknown and untrustworthy third parties, vertical integration spread like wildfire. But it also posed a substantial logistical challenge in that it required corporate managers to maintain oversight over all parts of a complex and unwieldy production process while keeping close tabs on a far-flung distribution network. Once again, archival techniques that centered on the collection, storage, and analysis of paperwork emerged as the preferred solution. During the Long Gilded Age, scholars and administrators therefore converged on a shared enthusiasm for information technologies such as the printed form, the standardized index card, and the vertical filing system, which could equally well be used to keep track of a company's payroll or a museum's collections. But the same management practices did radically different work in the two institutional contexts. Rather than seek to maximize profits, philanthropic museums modeled themselves on the vertically integrated industrial corporation to help distance their wealthy benefactors from the world of commerce.[72]

Eager to maximize the return on their investment in cultural capital, philanthropists built large and ambitious museums with a complex, hierarchical management structure. Taking their cues from some of the latest corporate governance techniques, these museums quickly evolved into sprawling, multiunit organizations that were run by a small army of salaried managers, employed some of the latest accounting techniques to cut costs and keep track of specimens, integrated backward into the acquisition of raw materials, and integrated forward into mass distribution. As this happened, dinosaur paleontology changed from an individual enterprise dominated by a small number of charismatic (if also querulous) naturalists into a bureaucratically managed team effort overseen by a centralized administration and governed by copious amounts of paperwork. At the same time, philanthropic museums also sought to decommodify dinosaurs by controlling the circulation of specimens. But rather than seek to convert vertebrate fossils from economic capital into cultural capital by *eliminating* the market for dinosaur bones, philanthropic museums deployed information technologies such as the account book to *internalize* the commercial specimen trade.[73]

Bureaucratic forms of authority proved a good fit for philanthropic museums for a variety of reasons. First, they offered a way for wealthy capitalists such as Morgan and Carnegie to maintain oversight of their monetary bequests. Insofar as bureaucracy offered a reliable means to reduce costs, it helped reassure donors who wanted to get the most out of their philanthropic investments. But as the controversy over Skiff's autocratic and overbearing management style revealed, curators sometimes pushed back. Still, the Chicago controversy did not primarily stem from a worry about the commercialization of science. If anything, natural history was further removed from the world of commerce by the early twentieth century than it was during the 1870s and 1880s. Thus, when curators complained about the Chicago Idea in the late 1890s, it was not so much the commercial dimensions of business administration against which they rebelled. Rather, it was the strict hierarchies, the avalanche of printed forms, and the inflexible reliance on formal procedures that raised their ire. In short, when curators attacked the role of businessmen at the museum, they primarily had corporate bureaucrats in their sights.

The widespread obsession with keeping down costs points to a second reason why bureaucratic authority was so widely adopted by philanthropic museums. Because philanthropists viewed these museums as engines for the creation of cultural capital, they did not seek a financial return on their investment. To do so would have been anathema to the project in which they engaged. This helped make bureaucratic management practice that centered on keeping down costs an especially good fit for these institutions. An alternative system, one that was designed to increase revenues, would have been out of place at a museum devoted to the ethos of pure science. Ironically, because they had been designed to reduce costs rather than maximize profits, accounting techniques developed to manage vertically integrated industrial firms were especially well suited to oversee the production of pure science.

The ease with which managerial practices migrated from an industrial to a nonprofit context reveals a broader truth about American culture around the beginning of the twentieth century. Historians often characterize the Long Gilded Age as a period of mass commodification in which the value of everything came to be measured in the terse language of dollars and cents. Recounting the development of modern consumer culture during the late nineteenth century, for example, William Leach argues that "American capitalism began to produce . . . a secular business and market-oriented culture, with the exchange and circulation of money at the foundation of its aesthetic life and its moral sensibility." As this happened, Leach continues, "money value became the predominant measure of all value in society."[74] This chapter has told a different story. While there is no doubt that many aspects of life came to be subsumed by the logic of monetization at this time, not all values were measured through commercial exchange. On the contrary, economic elites often sought to legitimize their status and wealth by actively *removing* things from the market. The emergence of a legal distinction between the for-profit and nonprofit corporation during this period is a clear case in point. So too are the broad range of things that were stripped of their commodity status. Perhaps most notable were objects of conservation, such as wilderness preserves and what eventually came to be called endangered species, which resembled dinosaurs in that both were carefully excised from the cash nexus. This demonstrates that while the Long Gilded Age is often considered a high-water mark of American capitalism, this period also gave rise to profound anxieties about the need to protect the country's most valuable resources—including dinosaur fossils—from the demands of the marketplace, lest these be degraded by the onslaught of commercial development.[75]

Finally, this chapter has argued that museums did much more than make specimens more widely accessible. Philanthropic museums were as much about the accumulation of specimens as they were about the circulation of knowledge, just as modern capitalism is about both the accumulation of wealth and the circulation of commodities. By accessioning objects to their permanent collection, philanthropic museums effectively disentangled specimens from circuits of economic exchange. Because museums functioned as public repositories, this did not make dinosaurs inaccessible. However, museums did leverage bureaucratic authority to restrict or control the economies—moral, commercial, and otherwise—in which dinosaurs traveled.[76]

5

Exhibiting Extinction

Less than a year after the American Museum of Natural History completed its *Brontosaurus* display in 1905, paleontologists from New York announced an even more sensational discovery: *Tyrannosaurus rex.* From the moment it was unveiled, *T. rex* was consistently described as a particularly unrelenting predator. Barnum Brown, who had uncovered the creature's fossil remains in the Badlands of South Dakota, told the *New York Times* that it was "so formidable a fighting machine that he easily preyed upon herbivorous neighbors twice his own size." Describing this "newly discovered monster" as "the absolute war lord of the earth in his day," Brown likened *T. rex* to the folkloric griffin. Even its name, which translated to "tyrant lizard king," was coined by Brown's boss at the museum, Henry Fairfield Osborn, to indicate this creature's identity as a ferocious predator from the deep past.[1]

Before long, the museum set to work on a new exhibit that would emphasize the dinosaur's proclivity for extreme violence. Just a few years after Brown was led to make the discovery of *T. rex* by a tip from the noted conservationist William Temple Hornaday, he found another fossil that contained many of the pieces missing from the first. Much as the museum had done with *Brontosaurus,* this opened the possibility of combining bits and pieces of different specimens into a composite mount. But this time curators wanted to go further and create a paleontological group display. By casting the missing parts of one specimen from the bones of the other, they hoped to combine elements of both to create a chimerical assemblage that showcased two specimens interacting with each other. To make a systematic study of the idea,

Osborn had an assistant produce a scale model that could be adjusted into a variety of poses. After several weeks of experimentation, the museum settled on a vivid scene featuring two *T. rex* skeletons fighting over the remains of a duckbilled dinosaur (Figure 5.1). In a popular article that he penned to promote the new exhibition, Brown invited readers to imagine themselves into the Age of Reptiles, where they could witness the scene's bloody brutality in gory detail. "As this monster crouches over the carcass, busily dismembering it," Brown explained, another *T. rex* arrives "to grapple the more fortunate hunter and dispute the prey." In response, the first *T. rex* "stops eating and accepts the challenge, partly rising to spring on its adversary." In its planned *T. rex* display, the museum sought to illustrate a "psychological moment of tense inertia before the combat," just as the two reptiles were about to "spring together, seizing a vital spot, . . . and hold on till one or the other yields."[2]

This gruesome scene, which was intended to form the centerpiece of the museum's dinosaur hall, was never completed. With all of the fossils already on exhibition, there simply wasn't enough space for such a large and ambitious display. As a result, when *T. rex* was unveiled to the public in 1915, it was in the form of a single skeleton that stood upright with its jaws gaping wide open and using its powerful tail as a tripod for balance (see Figure C.1). Curators hoped the specimen would eventually be incorporated into the proposed dinosaur group, but these plans were abandoned after its second *T. rex* fossil was sold to the Carnegie Museum in an effort to protect it from German submarines during the war.[3] Still, despite the fact that it never came to fruition, the planned *Tyrannosaurus* display reveals much about the museum's overall exhibition strategy. First, curators were keen to create a paleontological analogue to the habitat diorama by mounting multiple skeletons interacting with one another and their surroundings. And, second, these paleontological groups always depicted the Age of Reptiles as an especially nasty and brutish period in the history of life on earth.[4]

While the New York museum never completed its ambitious *T. rex* exhibit, it did mount a number of paleontological group displays. In 1907, for example, an *Allosaurus* was assembled in the act of predating on a section of *Brontosaurus* tail. As a visitors' guide explained, the mounted specimen "gives to the imaginative observer a most vivid picture of a characteristic scene in that bygone

Figure 5.1. A model of *Tyrannosaurus rex* engaged in fierce combat, at the American Museum of Natural History in New York.

Figure 5.2. *Leaping Laelaps,* a painting of two carnivorous dinosaurs attacking each other. Commissioned by J. P. Morgan and executed by Charles R. Knight in consultation with Edward Drinker Cope, 1897.

age, . . . when reptiles were the lords of creation, and 'Nature red in tooth and claw' had lost none of her primitive savagery." A few years later, a pair of duckbilled dinosaurs were mounted to reinforce this message. (See Figure 6.7 for both the duckbill and *Allosaurus* exhibits.) The latter group featured two plant-eating dinosaurs, but the museum guide pointed out that one of them stood upright after it had been "startled by the approach of . . . Tyrannosaurus, their enemy."[5] Additional examples could be listed almost indefinitely, including an early painting commissioned by J. P. Morgan that showed two carnivorous dinosaurs brandishing their oversize claws, with one of them pouncing as the other lay on its back and opened its maw to reveal a row of razor-sharp teeth (Figure 5.2).[6]

Because dinosaurs were consistently depicted as ferocious tyrants engaged in a bloody struggle for survival, historians often charge paleontologists with helping to naturalize the competitive ethos of modern capitalism. At the museum, the argument goes, dinosaurs functioned as object lessons in the evolutionary ethics of social Darwinism. Ronald Rainger, for example, argued that dinosaur "displays meshed with Osborn's interests in glorifying the struggle for existence," whereas Hilde Hein observed that "museum exhibition in the era between 1890 and 1930 had . . . a tone of robust social Darwinism."[7] The literary scholar W. J. Thomas Mitchell was even more forthright, maintaining that America's Long Gilded Age, "so often portrayed as the era of 'social Darwinism,' economic 'survival of the fittest,' [and] ruthless competition . . . is aptly summarized by the Darwinian icon of giant reptiles in a fight to the death."[8] According to this line of argument, dinosaurs were not just materially bound up with the history of modern capitalism. They also projected its ideology backward in time by showing that fierce competition was a fact of nature that long predated the evolution of human society.

This chapter offers a different interpretation. While it is true that dinosaurs were depicted as fearsome predators engaged in a ruthless fight for survival, that did not make them a fitting emblem for modern capitalism. After all, dinosaurs famously suffered a mass extinction event at the end of the Cretaceous period. They did not, therefore, offer a very durable symbol for a political economy that aspired to considerable longevity. Moreover, Gilded Age capitalists such as J. P. Morgan and Andrew Carnegie generally did not embrace a vision of markets dominated by cutthroat competition. Quite the opposite. As we saw in the last chapter, they overwhelmingly preferred the stability, rationality, and organizational complexity that was supposed to characterize bureaucratically managed corporations such as the museum itself. Finally, it was widely believed at the time that political economies were subject to the same evolutionary dynamics as biological organisms. Because they hailed from a bygone world in which the Darwinian struggle prevailed, dinosaurs were most often compared to an earlier and more "primitive" stage in the history of American capitalism. In other words, dinosaurs stood in for the past, not the present. Thus, instead of helping to naturalize the primordial brutality of the marketplace, dinosaurs were used to represent the poverty of an older, laissez-faire model of social organization that much of the economic elite had already come to regard as obsolete.[9]

Philanthropic museums used dinosaurs as part of a broader exhibition strategy to help bolster a progressive reform movement that flourished around the turn of the twentieth century. As nonprofit institutions designed to expose popular audiences to the highest achievements of modern civilization, philanthropic museums were keen to advance a narrative in which the Darwinian struggle for existence had become a thing of the past. The science of vertebrate paleontology played a decisive role in shaping that narrative. Whereas paleontologists characterized reptiles as primitive creatures who were ruled by their instincts and bound to the laws of biology, they described humans as rational agents who were endowed with a unique capacity to control their own destinies. Blessed with superior intelligence, humans could transcend nature by actively shaping the social, cultural, and material context in which they lived. But progressives did not judge all humans to be equally intelligent, and they did not trust most people to administer their own affairs. This led them to champion a range of highly prescriptive reforms that were predicated upon deeply held assumptions about the inherent superiority of existing elites. Progressives did not seek to overturn the social hierarchy. Rather, they sought to empower those at the top to shape the world after their own image. This, too, mirrored the theories of vertebrate paleontologists such as Henry Fairfield Osborn, who went so far as to brand his own version of evolution "aristogenesis." By joining the Greek words "aristos" and "genesis," Osborn created a portmanteau term for the idea that evolutionary progress would only result if the "best in its kind" passed on their genes to the next generation.[10]

Progressive reformers put the evolutionary theories developed by paleontologists such as Osborn into practice, creating a diverse range of measures and institutions to administer human affairs from the top down. In addition to setting up countless government bureaucracies, such as the Food and Drug Administration and the Federal Reserve System, they also championed conservation policies to manage the sustainable exploitation of nature. Perhaps even more telling is that many progressives enthusiastically embraced eugenic programs of controlled breeding, which sought to direct the course of human evolution and maintain the purity of what Madison Grant, Osborn's friend and the director of the Bronx Zoo, described as the "master race."[11] Finally, several progressive reformers also served as trustees for museums of natural history, which they charged with exposing the public to the teleological theory of evolution developed by paleontologists and used to underwrite their paternalistic social policies. Far from naturalizing a political economy characterized by fierce competition, then, the museum was tasked with inculcating the notion that elite experts had discovered a means by which human beings could transcend biology and actively manage their destiny.[12]

In sum, this chapter argues that dinosaurs were hardly a straightforward symbol of American capitalism during the Long Gilded Age. Instead, they played a key role in advancing a powerful narrative of evolutionary progress that was often invoked to explain the course of economic development. Just as proprietary or free-market capitalism was believed to have given way to a more organized marketplace dominated by bureaucratically managed and corporately organized firms, the disappearance of primitive reptiles made way for the evolution of more advanced and intelligent mammals. The contrast between these two eras was made especially vivid by the example of *T. rex*. More so than any other creature from prehistory, the tyrant king served as an illustration of primitive competition. But the museum always reminded visitors that its terrible reign had come to an end, and the extinction of dinosaurs opened the ecological space for more agile, adaptable, and intelligent mammals to evolve. As a museum guidebook put it, the dinosaur's "era of brute force" was eventually tempered by the "gradual amelioration" that came "to pass in future ages through the predominance of superior intelligence."[13] Rather than symbolically undermine the longevity of American capitalism, the extinction of dinosaurs was therefore portrayed as a moment of progress that helped to usher in a more rational world.[14]

Progress and Degeneration

To appreciate how and why dinosaurs played such an outsized role in progressivist theories of evolution requires delving into the history of extinction. As we saw in the first chapter, a French savant named Georges Cuvier is widely credited with convincing the scientific community that most of the animals who once inhabited our globe have since disappeared. But Cuvier wasn't the only one at the Muséum d'histoire naturelle in Paris who was thinking deeply about the history of life on earth; the botanist and invertebrate zoologist Jean-Baptiste Lamarck developed some very different ideas about the way organisms have changed over time. Whereas Cuvier imagined a series of catastrophic revolutions that caused radical breaks to appear in the fossil record, Lamarck envisioned a far more regular and orderly process of developmental change that constituted one of the first modern theories of evolution. And although Lamarck expressed doubts about whether whole species of organisms could truly die out, his ideas had an especially strong impact on the way many late nineteenth- and early twentieth-century paleontologists in the United States would go on to understand the process of extinction.[15]

In a sweeping and ambitious treatise first published in 1809, Lamarck argued that complex organisms had developed out of simpler ones over time, giving rise to an ascending series or natural hierarchy. According to Lamarck, the most organized or highly developed life-forms (Lamarck himself often used the word "perfect") were also the oldest, whereas the bottom rungs of creation were continually replenished through a process of spontaneous generation. Lamarck also postulated that a suite of what he called "subtle" or "imponderable fluids" were responsible for this remarkable change over time. As these fluids (which included electricity and caloric, a substance that was thought to produce heat) came into contact with solid parts of an organism, they naturally induced it to grow more complex. Indeed, the power that inhered in these subtle fluids was so great they could even breathe life into inert matter, thereby creating simple organisms from inorganic materials. In addition, Lamarck also observed the way living things changed as they interacted with each other and their surroundings, especially through the use or disuse of particular body parts. In his most famous example, he described how the neck of a giraffe would gradually elongate as it continually strained to eat leaves that were just out of reach, much like the arms of a blacksmith became more muscular as he practiced his trade. Taken together, the organizing power of subtle fluids and the inherited effects of use and disuse were believed to explain the organic diversity of all living things.[16]

Lamarck's ideas often strike modern ears as foreign if not naive, but during the nineteenth century, a natural tendency toward organized complexity was often seen as a defining feature of life itself.[17] After all, what could be more distinctive of living things than their ability to reproduce themselves through the predictable yet mysterious process by which a complex new organism gradually emerges from a simple, fertilized egg? But this process did not only explain the development of individual organisms. A similar pattern appeared to characterize the entire plan of creation as well, and many nineteenth-century biologists were struck by the degree to which the embryonic stages of highly organized animals seemed to resemble the adult forms of simpler ones. This led many biologists at the time to stipulate that an innate drive toward increasingly organized complexity constituted a basic law of nature, placing embryogenesis within a sweeping trajectory that encompassed the ascending hierarchy of all living things, from monad to man. As the French anatomist Antoine Étienne Serres explained, the "entire animal kingdom can, in some measure, be considered ideally, as a single animal, which, in the course of formation and metamorphosis of its diverse organisms, stops in its development, here earlier and there later."[18]

The idea that embryological development mirrored the order of nature was hardly unique to Lamarck. Indeed, it was not long before naturalists began to argue that different stages of embryological growth actually recapitulated the successive appearance of new organisms in the fossil record. Louis Agassiz, for example, held that a "three-fold parallelism" obtained between the hierarchical order of nature, embryological growth, and the appearance of new creatures over geological time. For Agassiz, the "remarkable agreement" between "the embryonic growth of animals" and "the succession of organized beings in past ages" spoke to "the working of the same creative Mind, through all times, and upon the whole surface of the globe."[19] But others drew precisely the opposite conclusion from the striking convergence between embryological development and species evolution, causing them to abandon the religiously motivated theory of special creation. The anonymously published but wildly successful *Vestiges of the Natural History of Creation,* for example, hypothesized that a "mere modification of the embryotic process" could generate an "advance" from "the simplest forms of being, to the next more complicated."[20] Even more radical was the German biologist Ernst Haeckel, who articulated what he called the "biogenetic law," which held that "ontogeny, or the development of an individual, is a truncated and accelerated repetition of phylogeny, or evolution of the species." According to Haeckel, phylogeny actually served as the efficient cause of ontogeny, meaning that evolution effectively explained embryogenesis.[21]

Toward the close of the nineteenth century, a number of American paleontologists invoked the oft-cited parallelism between ontogeny and phylogeny to advance a controversial theory of evolution called orthogenesis. While they were clearly responding to the work of Haeckel and other developmental morphologists, the theory of orthogenesis was primarily based on their detailed and painstaking observations of fossils. Derived from the Greek word for "straight, upright, or rectilinear," orthogenesis sought to explain why, judging from the fossil record, biological lines of descent were regular and predictable, forming linear sequences of anatomical change that appeared to exhibit a kind of directionality or inertia. This was exactly the opposite of what the Darwinian theory of evolution by natural selection predicted, and Darwin himself famously viewed the paucity of fits and starts in the fossil record as one of the chief difficulties that his theory had to overcome.[22]

Among the most ardent champions of orthogenetic evolution in the United States were Edward Drinker Cope, Alpheus Hyatt, and Alpheus Spring Packard. Although all three worked independently and studied different empirical systems—Hyatt primarily concentrated on mollusks and Packard on insects, whereas Cope largely focused on vertebrate fossils, including dinosaurs—they reached all but identical conclusions. The most important of these was that organisms evolve by speeding up or slowing down the developmental process, which Cope designated the law of acceleration and retardation. As Cope explained the idea, progressive evolution consisted of accelerating the ontogenetic process, which allowed the developing embryo to reach a higher state of perfection during gestation. Regressive or degenerative evolution was just the opposite, involving a form of arrested development in which adult organisms resembled the immature stages of their phylogenetic ancestors.

Although the theory of orthogenesis was compatible with Darwinian evolution, most paleontologists in the United States left little room for debate about which had a more powerful impact on the history of life on earth. Cope was particularly adamant on this point, insisting that "nothing has ever originated by natural selection." Although Darwin's theory could explain the disappearance of maladapted forms, Cope did not see how it could account for evolutionary novelty. As Cope never grew tired of pointing out, whereas Darwinian evolution only explained the *survival* of the fittest, orthogenesis sought to account for their *origin* as well. Since natural selection was a purely destructive principle, not a creative one, Cope judged that it only amounted to "half the question, and indeed the lesser half." Much more significant, he felt, was the positive "growth force"—a kind of "organic energy" that bore a striking resemblance to Lamarck's "subtle fluids"—which induced cells to divide, organs to grow, and new physiological systems to come into being. Cope further echoed Lamarck when he postulated that vigorous and sustained use of a particular organ led to a concentration of growth force in that anatomical region, whereas prolonged inaction led to a local reduction of growth force. Hence, Cope did not rest content in his theory's capacity to explain how and why organisms tended to grow more complex over time. His ambition was to account for their exquisite adaptation to one another and their surroundings too, much as Lamarck had done before.[23]

Orthogenesis sought to explain the disappearance of old life-forms from the fossil record as well as the emergence of new ones. After all, biological development did not cease with adulthood: all organisms eventually grow old and die. As Haeckel wrote, each life commenced with a period of "adolescence" that was "characterized by growth," followed by an "adult" stage "characterized by differentiation," only to end with a "geriatric" stage that saw "the individual's degeneration." As before, Haeckel argued that the same pattern also held true for species and higher orders of biological classification. Early on in their history, he observed, phylogenetic lineages usually "blossomed" by generating new branches and expanding their range. Next, they "bloomed" via a process of divergence and morphological diversification as different groups adapted to their local context and particular circumstances. Finally, they "withered" away as the "physiological assets" of various organisms gradually diminished until the whole lineage had gone "completely extinct."[24] In a striking parallel, the orthogenetic theory that was so popular among paleontologists from the United States was invoked to make sense of degeneration in much the same way. In his study of marine cephalopods, for example, Hyatt noticed that after evolving an increasingly complex morphology, these organisms usually reverted to a much simpler structure before finally disappearing from the fossil record entirely. Just as an "old man returns to a second childhood in mind and body," Hyatt concluded, the final or "geriatric" stage of a species' life cycle resembled the juvenile form of a previous generation, meaning that it bore a "very close resemblance to its own young."[25] For Cope, these observations amounted to further evidence of the way growth force controlled ontogeny and phylogeny alike, as the "senile" or "gerontic" condition in both could be explained by a kind of physiological entropy in which the organizing powers of vital energy tended to dissipate over time.[26]

Cope also developed a mechanistic theory of extinction that he designated the "Law of the Unspecialized." Successful new lineages usually evolved from relatively humble beginnings, he observed, noting that, most often, "the highly developed, or specialized types of one geological period have not been the parents of the types of succeeding periods, but that the descent has been derived from the less specialized of the preceding ages." Cope reasoned that this pattern could be explained by the fact that organisms that had evolved a suite of specific traits in response to particular conditions usually found it difficult to adapt if those conditions suddenly changed. "Degeneracy is a fact of evolution," Cope concluded, "and its character is that of an extreme specialization, which has been, like an overperfection of structure, unfavorable to survival."[27] For Cope, the cyclical life history that orthogenetic theories postulated therefore did not merely result from an internal dissipation of growth force alone. Extinction could be brought on by external circumstances as well. Indeed, internal and external causes often worked in concert, as senescent or degenerative lineages found they could no longer keep up with the struggle to survive in a rapidly changing environment.

A later generation of paleontologists expanded on Cope's and Hyatt's ideas to explain the disappearance of particular creatures, especially dinosaurs. According to Osborn, the fact of extinction followed directly from the theory of orthogenesis. For him, long-necked herbivorous dinosaurs known as sauropods, including *Brontosaurus* and *Diplodocus,* offered a particularly vivid illustration. After an extended period during which these ungainly creatures developed an increasingly ponderous carriage, they "reached a *cul-de-sac* of mechanical evolution from which they could not adaptively emerge," making it impossible to cope with "the new environmental conditions of advancing Cretaceous time."[28] Eventually, these dinosaurs succumbed to their evolutionary inertia, as features that may have been useful in the past continued to develop to such a degree that they represented a morphological burden that simply proved too much to carry, especially if circumstances suddenly changed.[29]

Osborn was hardly the only paleontologist who invoked orthogenetic ideas to explain why dinosaurs went extinct. In a short piece titled "Momentum in

Variation" that appeared in the *American Naturalist* at the beginning of the twentieth century, Frederic Loomis from Amherst College proposed the rule that "*variation started along any line tends to carry that line of development to its ultimate,*" so much so that a character would often evolve "beyond its utility." Loomis illustrated this claim with the oversize canines of the saber-toothed tiger, but he also singled out *Stegosaurus* as a vivid case in point, emphasizing this creature's remarkable "dermal armor," which was made up of large plates that "developed" until they were "over three feet across and several inches thick." "With such an excessive load of boney weight entailing a drain on vitality," Loomis remarked, "it is little wonder the family was short-lived."[30] Similarly, Arthur Smith Woodward from the London Natural History Museum lingered on the anatomical excesses of dinosaurs to illustrate the "infallible marks of old age" that often crept into a lineage over time. Given the "obvious symptoms of old age" they so often exhibited, Woodward concluded that dinosaurs "died a natural death."[31]

By the beginning of the twentieth century, then, dinosaurs had emerged as a preferred model of what was often described as "racial senescence," "senility," or simply "old age."[32] But that did not mean it was all gloom and doom. On the contrary, Cope pointed out that "definite progress has been made, and highly specialized characters have" been passed down "through the vicissitudes of geologic revolutions."[33] Thus, despite the historian Peter Bowler's insistence that "racial senility" illustrated the "pessimistic aspects of orthogenetic Lamarckism" with particular clarity, a cyclical model of evolution did not, in fact, rule out the possibility that history continued to unfold in a definite direction.[34] Indeed, extinction was widely thought to play a generative role in the orthogenetic theory, promoting what modern biologists would describe as the "evolvability" of life over time.[35] As Cope wrote, the most successful lineages were invariably those that "presented a combination of effective structures with plasticity," enabling "them to adapt themselves to changed conditions."[36] Hence, the principal lesson that Cope and his followers drew from the fossil record was the importance of flexibility and adaptability as a basic condition for long-term survival. In this way, the truncated life histories of all those extinct species whose fossil remains littered the American West could be rendered as so many epicycles in a grand narrative of evolutionary progress.

In the eyes of American paleontologists, the mass extinction event that killed off the dinosaurs served as ground clearing, extirpating organisms whose evolutionary potential had become ossified in order to make room for new lineages. From the orthogenetic perspective, extinction was a moment of rebirth and rejuvenation, making way for the proliferation of higher and more advanced types. It was often remarked that just as the annihilation of trilobites set the stage for the evolution of bony fishes, so too did the disappearance of dinosaurs make room for the proliferation of mammals. As Osborn put it, "nature began afresh" with the "extinction of the reptilian dynasties," allowing evolution to "slowly build up out of the mammal stock the great animals which were again to dominate land and sea."[37] Each step in this sequence was taken to represent a definite advance over what came before, manifesting itself in the gradual appearance of consciousness and intelligence. In contrast to the brutal dinosaurs, mammals were far kinder and gentler, often cooperating for the greater good of the family, the community, and even the species. Indeed, paleontologists routinely characterized the evolution of rational thought as the denouement of life's splendid drama, which endowed our own species with the cognitive resources to shape an increasingly complex environment. Rather than merely adapt to external surroundings, then, humans could alter the world to fit their own needs.[38]

Even paleontologists who opposed orthogenesis tended to view evolution in progressive terms, agreeing that successful organisms grew more intelligent over time. For example, although Cope's bitter rival, Othniel Charles Marsh, declared natural selection to be "the most potent" factor in evolution, he nonetheless pronounced Herbert Spencer's "law of progress" to be the "great truth" of "organic evolution."[39] Interpreting the fossil record along thoroughly progressivist lines, Marsh found that the cranial capacity of successful lineages tended to increase, whereas organisms with smaller-than-average brains were destined for extinction.[40] "In the long struggle for existence," Marsh observed, "the big brains won, then as now."[41] Around the beginning of the twentieth century, Marsh's "law of brain growth" attracted a number of prominent supporters, including E. Ray Lankester, a staunch a defender of Darwinian evolution who agreed it was "obvious" that "in successive generations the bigger and more educable brains would survive and mate, and thus bigger and bigger brains would be produced."[42] Indeed, Lankester's faith in progress was so strong he even suspected Darwinian evolution might be surpassed. As organisms grew more intelligent, he reasoned, they also developed a greater capacity to intervene in their world. This made it possible for humankind to begin engineering its own evolution, and Lankester argued that although "Man is held to be a part of Nature," it is "his destiny to understand and control it."[43]

To emphasize the cunning intelligence of modern mammals, paleontologists consistently foregrounded the ferocious stupidity of their saurian predecessors. Osborn, for example, played the intense physicality of carnivorous dinosaurs such as *T. rex*—which he described as "the most destructive life engine which has ever evolved"—off the "excessively small size of [their] brain," concluding that an animal's "mechanical evolution is quite independent of the evolution of their intelligence; in fact, intelligence compensates for the absence of mechanical perfection."[44] An even more forthright assessment was offered by Marsh's successor at Yale, the paleontologist Richard Swann Lull. Christening it the "Age of Brawn," Lull characterized the Age of Reptiles as a time in which "might gave right," marking it out as a "period of dominant brutality, rather than being in any sense a psychic age." Again, *T. rex* was singled out as "the ultimate possible expression of the carnivorous dinosaur, admirably adapted to prey upon his equally dull-witted and slow contemporaries." If it were still alive today, however, "the tyrannosaur would evidently be rather hard pressed to make a living," because "the alert modern mammal would have little difficulty in avoiding them." Lull heaped even more ridicule onto *Stegosaurus,* whose tiny brain he judged to be so incompetent it could not even control the entirety of its own body, necessitating the development of a secondary nerve center near the base of the spine, which Lull took to illustrate "in the most graphic manner the dominance of muscular force over intellect in this armored colossus." If reptiles ruled over a nasty and primitive Earth, the proliferation of mammals ushered in a new age, one that Lull explicitly likened to the Renaissance, rhapsodizing about how, "after a long era, when brute force was dominant, came the close of the dark ages," and with it, "the birth of intelligence."[45]

Paleontologists were so impressed by the cognitive capacities of advanced mammals that many came to suspect that biological evolution had ground to a halt. In its place, there emerged a new process of social and technological evolution. Once a certain level of intelligence had been achieved, organisms acquired the mental powers to exercise increasingly deliberate and effective control over their surroundings and circumstances. As Cope put it, the evolutionary process inevitably "tends to upward progress, in the organic sense; that is, toward the increasing control of the environment by the organism."[46] Eventually, this resulted in the emergence of a species that could begin to control its own destiny, *Homo sapiens*. The paleontologist-turned-sociologist Lester Frank Ward observed that "the power of the human intellect over vital, psychic and social phenomena" meant that "Nature has . . . been made the servant of man."[47] From there, it was but a short step to reach the conclusion that extinction, too, might one day become a thing of the past. This was especially so if the scientific community—whose discoveries were ranked among the highest human achievements—could be enlisted to develop measures that would allow human beings to direct the future of evolutionary progress. Hence, the disappearance of dinosaurs not only set the stage for the emergence of human intelligence. It also held out the possibility that our species, having evolved from its biological ancestors into social and psychological creatures whose technological advances allowed it to conquer nature, might one day be able to outrun its geological fate.

Managing Evolution

The conviction that humans could use their superior intelligence to transcend nature and escape the fate of extinction was not unique to nineteenth- and early twentieth-century paleontologists. It also informed the reform efforts championed by progressives like Theodore Roosevelt. Whereas paleontologists linked the process of evolution to the fact of extinction by modeling the development of entire species on the life history of individual organisms, progressive reformers worried endlessly that advanced civilizations carried within them the seeds of their own destruction. However, just as paleontologists hoped humans might differ from other organism and be able to control their own biological destiny, progressives enacted a number of highly prescriptive reform efforts to stave off extinction, including measures to conserve nature and promote eugenic reproduction. Both nature conservation campaigns and eugenic breeding programs saw progressives embrace science, especially biology, as a preferred means by which human civilization might avoid the fate of the dinosaur. Moreover, in both cases, the museum emerged as a key site where progressive reformers sought to implement social technologies to manage the process of biological evolution. Finally, both examples add an important dimension that has been left out of the story so far, namely, the degree to which widespread assumptions about gender and race informed debates about evolutionary progress. Taken together, the history of nature conservation and eugenics show that anxieties about the extinction of modern civilization were, perhaps more so than anything else, anxieties about the future of white male supremacy.

Progressive conservation campaigns offer a particularly revealing glimpse of the dialectical tension between millenarian optimism and eschatological pessimism, between faith in future redemption and longing for past glory, that so marked the period's thinking about evolution and extinction. Precisely the same decades that saw so much exuberant confidence in the productive capacity of modern capitalism also witnessed the development of a profound anxiety that industrialization would eventually lay waste to the wilderness. This gave rise to a palpable concern that as human beings destroyed nature, they undermined the biological basis of their own species' longevity. As early as 1864, for example, the well-known scholar and politician George Perkins Marsh observed how "the destructive agency of man becomes more and more energetic and unsparing as he advances in civilization." Marsh went on to predict that unless "human cunning" could restore at least some parts of the wilderness, humankind would eventually reduce nature "to such a condition of impoverished productiveness . . . as to threaten the depravation, barbarism, and perhaps even extinction of the species."[48] In the decades that followed, these fears only grew more pronounced, and Marsh's sentiments were soon being echoed by naturalists who had set off to document the location and abundance of valuable resources for economic extraction in the American West. Becoming increasingly convinced that commercial development not only produced wealth but also degraded the land, prominent surveyors such as Josiah Whitney and Clarence King began lobbying for the creation of federally recognized nature preserves, which helped spawn the creation of an ambitious network of national parks such as Yosemite and Yellowstone.[49]

Marsh, Whitney, and King were hardly alone in their fear that industrial capitalism threatened to destroy nature. Around the turn of the twentieth century, a set of concurrent anxieties about the erosion of traditional gender roles prompted a large number of urban professionals to seek out the wilderness in hopes of restoring their masculine vigor. Here, too, the culprit was seen as an excess of civilization. An oft-diagnosed medical condition called "neurasthenia" or "nervous exhaustion," for example, was directly informed by the idea that life in a technologically advanced society tended to erode the strength and virility of male office workers. Neurasthenia presented with a hodgepodge of wildly differing symptoms—ranging from migraine headaches, insomnia, and poor digestion to inebriety, myopia, lethargy, and a loss of sexual appetite—but medical professionals nearly all agreed on its cause. "Civilization is the one constant factor without which there can be little or no nervousness, and under which in its modern form nervousness in its many varieties must arise inevitably," the physician George Miller Beard lectured his patients. In much the same way paleontologists believed the most highly developed lineages would be the first to run out of growth force and succumb to racial senescence, physicians like Beard were convinced the most civilized people were also the most likely to suffer from nervous exhaustion. Whereas less highly developed persons moved through a world of brute force via the expenditure of muscular energy, Beard explained, neurasthenia targeted those with a more sensitive constitution, a "fineness of organization," as he liked to put it. Because the latter was deemed essential for "the development of the civilization of modern times," elite members of technologically advanced societies such as the United States were seen as especially prone to suffer a "bankruptcy" of nervous energy. By contrast, Beard also observed that neurasthenia "scarcely exists among savages or barbarians."[50]

The treatment for nervous exhaustion was as revealing as the diagnosis itself. With characteristic literalness, Beard recommended the therapeutic application of electrical energy, helping to spawn a large market for electro-charged belts and other, sometimes remarkably intimate, devices that promised to replenish one's inner battery. But even more telling still was the way treatments were tailored to particular gender roles. Whereas female patients were usually prescribed a "rest cure" that often involved being confined to the bedroom—the physician Silas Weir Mitchell counseled "a year or more of utter idleness"—male neurasthenics were told to get plenty of exercise and experience nature.[51] "The man who lives an outdoor life," Mitchell wrote, "has a strange sense of elastic strength, may drink if he likes, and may smoke all day long, and feel none the worse for it."[52] If the stresses of modern civilization made people ill, then a direct exposure to wilderness would restore men to health and to vigor, providing the energy to return back to city life and the "brain work" required to promote evolutionary progress. This, in turn, necessitated ready access to wide-open spaces unspoiled by human development, which is one reason why conservation became such a popular cause among the economic elite.[53]

It is no coincidence these fears also dovetailed with a sweeping enthusiasm for big-game hunting among elite conservationists. One of America's earliest and most prominent conservation organizations, the Boone and Crockett Club, was explicitly founded by the progressive reformers Theodore Roosevelt and George Bird Grinnell "to promote manly sport with the rifle."[54] Grinnell, who was born into the family of a New England stockbroker, first became interested in conservation while studying under the Yale University paleontologist Othniel Charles Marsh, and he teamed up with Roosevelt after seeing commercial hunters decimate bison and antelope populations during an expedition to Yellowstone in the 1870s.[55] Upon founding the Boone and Crockett Club, the two men immediately moved to restrict membership in the new organization to those who had killed at least one species of American big game "in a fair chase," further cementing the connection between wilderness conservation and masculine courage.[56] But the club did far more than just police its own membership, which soon grew to read like a who's who of late nineteenth-century high society. It also worked hard to enact legislation that furthered its conservationist goals. In 1894, for example, a club member from Iowa, Senator John F. Lacey, ushered a bill through Congress that imposed a fine of up to $1,000 and a maximum jail sentence of two years for anyone who was convicted of poaching in Yellowstone National Park. Conservation thus emerged as a preferred antidote the excesses of modern civilization, providing elite men such as Roosevelt with an opportunity to enter the wilderness, visibly expose themselves to the rigors of natural selection, and emerge from the experience victorious, pleased to have demonstrated their evolutionary fitness for life outside the feminized urban office environment.[57]

Members of the Boone and Crockett Club also took steps to reverse the process of degradation and restore what had been lost. Perhaps most successful in this regard were the efforts of William Temple Hornaday to bring the American bison back from the brink of extinction. Hornaday was a professional taxidermist whose first foray into ecological restoration was to create an ambitious "monument to the American bison" for the Smithsonian's National Museum that he hoped might render this majestic creature "comparatively immortal."[58] But Hornaday also hatched plans for a captive breeding program that turned into reality when he received a letter from Osborn inviting him to direct the recently created Bronx Zoo. Soon after he had established a stable bison population in New York, Hornaday enlisted Roosevelt to help introduce captive animals back into the wild. With the federal government's official backing, the Bronx Zoo would bring the entire species back to vitality. By 1910, Hornaday had succeeded in establishing four viable "federal bison herds," and he was confident that "the future of the American bison, as a species, is now secure."[59]

Although conservationists such as Hornaday sought to save the wilderness from the onslaught of industrial civilization, they did not fear progress or reject modernity. On the contrary, by taking a census of the remaining bison, selecting individuals to breed in captivity, and exporting their offspring back to federally protected wilderness preserves, Hornaday brought scientific expertise, managerial efficiency, and bureaucratic forms of authority—all elements that were often cited as hallmarks of modern civilization—to bear on the problem of extinction. He even deployed the language of managerial capitalism in defense of his aims, arguing that the "wild things of this earth" have "been given to us *in trust,* and we must account for them to the generations which will come after us and audit our accounts."[60] Other conservationists embraced this language too. For example, Gifford Pinchot, the head of the US Forest Service under Roosevelt, argued that conservation "stands for development" and "the prevention of waste," which gave him confidence that the "outgrowth of conservation, the inevitable result, is national efficiency."[61] Similarly, Grinnell argued that "economic" motives for wilderness preservation were "constantly gaining strength" because conservationists increasingly saw "wild things as assets which possess a tangible value to the community."[62] If industrial capitalism had evolved the capacity to dominate nature to such an extent that it threatened to disappear, conservationists did not advocate turning back time. If anything, they sought to *increase* human control over the wilderness. In their eyes, what was needed was not less civilization but more of it, which led Osborn to remark that conservation itself "marks the advance of a true, as distinguished from a false civilization."[63]

This dynamic was even more clear in the history of eugenics, which involved a deliberate attempt to manage human evolution. In the United States, eugenics was born of the anxiety that modern civilization not only undermined masculine vigor but also threatened to erode the genic basis of biological progress. Having conquered the wilderness, humankind created a world for itself in which the struggle for resources no longer determined survival. In the eyes of progressives, this posed a grave danger indeed. On the one hand, the creation of a social support system to care for the weak, the sick, and the infirm was seen as a hallmark of enlightened modernity. But, on the other hand, these measures also prevented the Darwinian struggle from taking its course. At first sight, these concerns may seem at odds with the orthogenetic theories espoused at the time. But while paleontologists such as Osborn and Cope rejected the creative power of natural selection to produce genuine novelty, the Darwinian process was still seen as essential for maintaining those adaptations whose original evolution had been so hard-won. Absent a mechanism for culling the herd, deleterious traits would proliferate in the population, causing its germplasm to degrade. The paleontologist Lull therefore lamented the widespread availability of glass spectacles, whose ability to correct poor eyesight would invariably lead humans to lose their "keenness of perception" over time, whereas Osborn insisted that the "finest races of man, like the finest races of lower animals, arose when Nature had full control."[64] Indeed, it would be difficult to find a biologist, paleontologist, or social theorist who did not worry that a relaxation of what came to be called "stabilizing" or "purifying" selection posed an existential threat to civil society.[65]

Eugenic anxieties about the cessation of competition prompted geneticists to intervene in the process of evolution directly. In much the same way that conservationists scientifically managed the extraction of scarce resources to ensure a sustainable yield, eugenicists advocated a program of controlled breeding. To that end, in 1910 the biologist Charles B. Davenport set up the Eugenics Record Office near Cold Spring Harbor, New York, to serve as a clearinghouse for the accumulation, analysis, and dissemination of statistical information about all manner of inherited traits. These data were gathered by countless field workers who fanned out across the United States to collect family pedigrees, which they duly recorded on over a million three-by-five-inch index cards whose contents were subsequently cross-referenced using a complex, numerical classification scheme (known as *The Trait Book*). In effect, the Eugenics Record Office sought to amass a complete census of the United States' germplasm, which researchers could use to assess the population dynamics of various traits as they evolved over time. But Davenport's organization served a more practical purpose as well, directing the course of human development by shaping the future gene pool. With that goal in mind, the Eugenics Record Office created an "Index to Germ Plasm" designed to help citizens make eugenic reproductive decisions.[66] No longer would the fate of the species be left to nature; instead, the process of evolution was subjected to managerial oversight and administrative control.

Eugenicists sought to direct human evolution through more coercive means too. In addition to promoting state laws that sought to use forced sterilization to prevent "unfit" individuals from contaminating the gene pool, they also lobbied the federal government to restrict immigration and conserve the United States' traditional "racial stock."[67] These efforts drew inspiration from the work of progressive sociologists such as Edward A. Ross, a student of Ward who argued that a flood of "low-standard" immigrants from southern and eastern Europe threatened to overwhelm resident populations and that a failure to act constituted a form of "race suicide."[68] In response to concerns such as these, Congress appointed Davenport's right-hand man, Harry Laughlin, as the "expert eugenics agent" for the newly formed Committee on Immigration. Laughlin knew how to use the vast trove of statistical data at his fingertips to good effect, arguing, "Inferior individual family stocks are tending to deteriorate our national characteristics."[69] This helped to secure passage of the Johnson-Reed Act in 1924, a law that effectively stemmed the tide of immigrants who were deemed undesirable by creating explicit quotas based on demographic data from the 1890 census.[70]

Eugenics had a number of features in common with nature conservation. Perhaps foremost was a shared faith in the principles of rational management as applied to the natural world. Progressive reformers concluded that if humankind had evolved to dominate nature completely, they were faced with no choice but to take charge and manage their own evolution directly. Convinced that modern civilization not only threatened the wilderness but, in so doing, also upended the power of natural selection, eugenicists responded by enacting a program of artificial selection to stave off degeneration and avoid extinction.[71] Moreover, both movements were highly paternalistic in that elite experts especially sought to control the behavior of the poor, working-class immigrants, and people of color. Whereas eugenic breeding programs and immigration reforms were designed to prevent the "unfit" from spreading their "deleterious" traits, Hornaday explicitly singled out "Italians," "Poor Whites," and "Southern Negroes" as "Guerillas of Destruction" who recognized "none of the rules of civilized warfare" when hunting for food, regularly poaching wildlife without any concern for sustainable yields.[72] Finally, eugenicists and conservationists both cultivated close ties to the museum. Because success in either domain hinged on the ability to enlist public support, both were in need of an institutional base from which to disseminate their ideas. Philanthropic museums of natural history offered an ideal venue for doing just that.

According to Osborn, the principal aim of the museum was to "bring the world of nature within the walls of a great city," and many of its exhibits sought to broadcast a conservationist message among a working class urban audience.[73] Hornaday's hope that his skills as a taxidermist might render the American bison "comparatively immortal" was one case in point, but additional examples abound in the historical record. In 1901, for example, the Carnegie Museum in Pittsburgh unveiled a display featuring two ivory-billed woodpeckers whose explanatory label told visitors these creatures were "on the verge of extinction" because their natural habitats had been "extensively logged for their valuable timber."[74] Two kinds of exhibits were particularly well suited to spreading the conservationist gospel. The first were paleontological mounts made from the fossil remains of prehistoric creatures, which one curator likened to a "passing procession of animals marching into oblivion."[75] The second were habitat dioramas, which were especially designed to exhibit the way different organisms interacted with, and depended upon, one another. As Oliver Farrington from the Field Museum put it, biological group displays illustrated that "nothing in nature is of isolated origin," and Osborn celebrated their capacity to teach "moral lessons" about "cooperation" and "government."[76]

Eugenicists, too, saw the museum as an ideal medium for their message. Inspired by the success of the First International Eugenics Congress in London in 1912, Davenport teamed up with Osborn to host a similar gathering in the United States. Given that Osborn had just become president of the New York natural history museum, it soon emerged as the obvious place to hold the event. When the Second International Eugenics Congress commenced in September 1921, it featured an impressive guest list of international researchers—including the geneticists Herman J. Muller, Robert A. Fisher, and Sewall Wright—whose presence helped to legitimize the scientific basis of eugenic reforms. A popular exhibition prepared by Laughlin was even more consequential, however, introducing thousands of visitors to eugenicists' race-betterment agenda. Laughlin's exhibit consisted of numerous alcoves and booths that sought to explain the principles of Mendelian genetics, teaching visitors how a family pedigree could be used to make scientifically informed reproductive decisions.[77] These museum exhibits were deemed such a success that Laughlin even brought some of the displays into the halls of Congress, where they helped persuade lawmakers of the dysgenic effects that attended human migration, thereby encouraging them to pass the 1924 immigration restriction act.[78]

In a fitting choice of venue, the Second International Eugenics Congress held its scientific proceedings in the museum's Hall of Man, which served as the triumphant climax of its paleontological exhibition. In an elaborate plan of Osborn's devising, the museum's extensive fossil collections were displayed in a series of rooms forming a linear sequence, beginning with the Age of Fishes and continuing through the Age of Reptiles before leading to the Age of Mammals and culminating in the Age of Man. Simply by making their way from one exhibition hall to the next, visitors were invited to follow the path of progressive development with their own two feet. The same teleological narrative governed the sequence of fossils within the Hall of Man. Whereas Neanderthals were depicted as hulking creatures with low foreheads, rounded shoulders, and a stooped posture, anatomically modern humans were shown walking upright, with their heads high in the air. In his official announcement of the new hall's completion, Osborn was particularly keen to extoll the virtues of the "highly evolved Cro-Magnon race," which he described as a "race of warriors, of hunters, of painters and sculptors far superior to any of their predecessors." He especially emphasized the "cultural capacity" of Cro-Magnons, which gave them a "greater artistic sense and ability than have been found among many other uncivilized people."[79] The museum's celebration of humankind's ability to overcome nature was further made visible in a mural painting by Charles R. Knight, which depicted a group of Cro-Magnons helping each other produce a cave painting of the kind that were being uncovered in southern France at the time. By holding the Second International Eugenics Congress inside its Hall of Man, the museum thus made an implicit claim that eugenic reform efforts furthered the sequence of development whose continued vitality required advanced humans to distance themselves from their biological roots by asserting control of the evolutionary process.

During the early twentieth century, the museum emerged as an institutional space in which reformers not only represented their vision of evolutionary progress but sought to implement it as well. As the anthropologist Donna Haraway has described it, "The museum was a medical technology, a hygienic intervention, and the pathology was a potentially fatal organic sickness of the individual and collective body."[80] But while progressive reformers were haunted by the specter of extinction, they also expressed faith in the power of human intelligence to control the destiny of their species. To that end, eugenicists outlined an ambitious biopolitics in which elite members of modern society actively managed their own evolutionary trajectory. Speaking before the Second International Eugenics Congress, Charles Darwin's son Leonard Darwin welcomed the adoption of "rational methods in human affairs" as the only responsible way for advanced humans to avoid the "suffering" that "animals in the wild have to endure because of that struggle for existence to which they must submit." Progressive reform efforts held out a promise for people such as himself to be "continually nobler, happier, and healthier," Darwin argued, whereas Davenport rejoiced that, at long last, our "fate is controllable." Without discounting "the paleontological record," Davenport remained confident that a "willingness to be guided" by "research in eugenics" meant that "the end of our species may long be postponed and the race may be brought to higher levels of racial health, happiness, and effectiveness."[81] In other words, although it was clearly fueled by the general anxieties about decline and degeneration that were so pervasive around the beginning of the twentieth century, eugenics was also an expression of exuberant optimism, leading Osborn to prognosticate that ere long, "the rise of man to Parnassus will again take an upward trend and the future progress of the human race will be secure."[82]

As was the case with so many other social reformers, eugenicists were glad to embrace the paternalism that informed their worldview. Accepting that evolutionary progress came at the price of individual liberty, they did not, as a rule, rue the loss deeply. For example, Madison Grant, an avid eugenicist and the director of the Bronx Zoo, railed against universal suffrage because "democratic theories of government" were based on "dogmas of equality" that failed to account for the profound "influence of genius." In his view, humanity "emerged from savagery and barbarism under the leadership of selected individuals" with the "power to compel obedience," not the average will of the majority.[83] Similarly, Osborn denounced the tendency to confound the "true spirit of American democracy" with the "political sophistry that *all men are born with equal character to govern themselves.*" For that reason, he decried the "rampant individualism" of his time, which he saw manifested "not only in art and literature, but in all our social institutions."[84] In its place, he urged the adoption of what he called "racial values," which entailed rigorous adherence to the following motto: "Care for the race, even if the individual must suffer."[85]

Osborn's eugenic vision involved no less than the creation of a new social order, one in which entire segments of the population were made to sacrifice their reproductive potential. His enthusiasm for a planned society in which the lives of the many would be directed by the intelligence of the few was so keenly felt that, in private, he even flirted with fascism. After the National Socialist Party came to power in 1933, he personally visited Germany, reporting that "one-sided reports in the American press" notwithstanding, he was "greatly impressed by the solidarity of the country and enthusiasm for the rebirth of Germany under the new conditions of the Hindenburg-Hitler regime."[86]

Beyond Laissez-Faire

So far, we have seen that progressive reformers viewed the evolution of civil society as both the root cause of, and the most promising solution to, the threat of extinction. Because modern civilization was understood as the end stage of biological evolution, it immediately raised the specter of extinction. But, paradoxically, it was also seen as a means of salvation. Confident in the power of human intelligence, progressives sought to transcend the state of nature and take charge of their biological destiny. This helps to explain why reform efforts like nature conservation and eugenics so often combined a sense of dread and foreboding with exuberant optimism. In what remains of this chapter, we will explore how this dialectical tension manifested itself in debates about political economy. Here too, reformers embraced a teleological narrative of biological progress that mirrored orthogenetic theories of evolution, arguing that a highly competitive marketplace naturally gave way to a more organized political economy dominated by corporately organized and bureaucratically managed firms. Hence, museum displays that dramatized the transition from ruthlessly violent dinosaurs to intelligent mammals as a step toward enlightened modernity could be used to help naturalize the transition from proprietary to corporate capitalism, framing these economic developments as an instance of evolutionary progress.

The corporate reconstruction of America's political economy at the turn of the century was met with predictable fears and anxieties. Between 1895 and 1904, some 1,800 industrial enterprises merged into just 157 consolidated firms, nearly half of which enjoyed more than 70 percent market share.[87] At the same time, a series of periodic downturns and financial panics gave way to major depressions in 1873, 1893, and 1907. Large corporate firms also used the power conferred by their immense market shares to stifle unions and drive down wages, which often led to violent confrontations with workers such as the battle at Homestead that we encountered in Chapter 3. As labor unrest became increasingly frequent and violent, many progressive reformers grew convinced that American capitalism itself had to be brought under control. But despite the period's well-known enthusiasm for "trust-busting," reformers did not generally advocate for a return to an older, preindustrial society in which unfettered competition reigned supreme. Instead, they were more likely to recommend scaling the bureaucratic management practices used to run large, corporate firms up to help organize a more stable, predictable, and durable political economy.[88]

Professional economists during this period often argued that whereas competition once was the life-blood of a healthy market, it had become so "cutthroat" as to be "destructive" and even "ruinous" in the late stages of industrial capitalism. While a small manufactory could simply reduce production in the face of waning demand, the same was not true of an industrial behemoth like Carnegie Steel, whose operating expenses primarily consisted of overhead such as maintaining heavy machinery or servicing debts. Crucially, such "fixed costs" were incurred regardless of output. This gave industrial firms an incentive to continue expanding their operations beyond the point of incurring a loss in hopes of starving the competition of profits. As the progressive economist Oswald Knauth explained, these market conditions drove industrial firms to inaugurate "a condition of 'severe' or 'destructive' competition" that had "little, if any, reference to productive efficiency." Before long, mass bankruptcies followed, leading to a reduction in competition and an increase in prices. But this only prompted the cycle to repeat anew, once again yielding a period of overproduction in which industrial firms again fought to the death. To dampen these recurrent boom and bust cycles, industrial firms began to form pools and cartels that could manage the output of an entire industrial sector, preventing ruinous competition and ensuring sustainable profits for everyone. However, because they could not punish defectors absent enforcement mechanisms that did not run afoul of the period's emerging antitrust legislation, these informal agreements did not hold up for long. In response, industrial firms turned to the new general incorporation laws being adopted on a state-by-state basis as an alternative means of consolidation. In either case, Knauth observed, "monopoly is the result." For that reason, economists even began to describe industries with especially high fixed costs as "natural monopolies," suggesting that certain sectors of industry would inevitably consolidate into a sole enterprise or single provider.[89]

In the eyes of American capitalists such as Andrew Carnegie, these structural transformations were simply another instance of evolutionary progress. In a piece for the *Century* magazine, he argued that powerful trusts should be welcomed with open arms because "we evolutionists know that in the end [humankind] will hold fast only to that which is good for the organism known as human society." While competition among independent producers prevailed in times past, larger conglomerates now cooperated for the good of the whole. "Everywhere we look we see the inexorable law producing bigger and bigger things," he explained, meaning that the "concentration of capital" was but "another step in the upward path of development." It was not only inevitable but indeed laudable that small, independent producers eventually gave way to much larger, vertically integrated conglomerates. "The day of small concerns" was "over, never to return," Carnegie concluded with discernable glee.[90] Carnegie's erstwhile assistant, James H. Bridge, went even further, analogizing the stupendous growth of America's industrial economy to the development of a biological organism and arguing that whereas the nineteenth century had "marked the growth of the nation to adolescence; the present one promises to witness the perfection of its maturity." This, Bridge continued, entailed a transition from "mutual competition to universal co-operation," which brought the much-maligned "trust" into view for what it really was: the result of "a tendency which, originating in the barbaric past, is giving us the promise of a fuller and more complete national life than the world has ever seen." Viewed in the "broad perspective of history" and "illuminated by the light of evolutionary law," the vertically integrated industrial corporation was therefore revealed to be something far more benign than a "creation of self-seeking capitalists" hell-bent on "undermining the very foundations of society by destroying competition and competitors." On the contrary, it represented "a wholesome, irresistible, natural progression from lower forms of industrial life to higher ones."[91]

But it was not only industrialists and their salaried managers who characterized the development of industrial capitalism as an instance of evolutionary progress. Some of the late nineteenth century's fiercest critics of social inequality did so as well.[92] Henry George's best-selling 1879 book, *Progress and Poverty,* for example, argued that human society resembled a biological organism in that it, too, naturally grew to become more complex over time. Just as a physiological division of labor distinguished highly evolved creatures from their primitive ancestors, an industrial society could produce a wide range of diverse goods more efficiently than its agrarian precursors. In true orthogenetic fashion, however, George also concluded that social progress would eventually degenerate. Because specialization entailed differentiation, it would inevitably also yield inequality. Hence, social progress would invariably give way to class conflict. "The unequal distribution of the power and wealth gained by the integration of men in society tends to check, and finally to counterbalance, the force by which improvements are made and society advances," George reasoned. Industrial civilization thus stood at a crossroads: it could either accept that "petrifaction succeeds progress" in political economy as well as biology, or it could enact policies to ensure a more equal distribution of wealth. The alternative was to invite disaster by nurturing the hordes of "new barbarians" already starting to gather in the most "squalid quarters" of America's "great cities." Echoing the words of naturalists such as George Perkins Marsh, Henry George therefore declared the "civilized world" could either "leap upward" and "open the way to advances yet undreamed of," or it could "plunge downward" and allow itself be carried "back towards barbarism."[93] Whereas one path would see the United States go the way of the dinosaur, the other required the country to abandon its laissez-faire policies, intervene in the market, and begin to manage the evolution of capitalism directly.

George may have been a particularly outspoken proponent of the view that a complex and highly evolved society required constant intervention, tending, and care, but he was hardly alone. Others even suggested that competitive markets were no more than a transitory stage in the evolution of modern capitalism, just as biological evolution by natural selection gave way to controlled breeding programs informed by the science of eugenics. On this view, the consolidation of numerous small businesses into large, multidivisional firms such as U.S. Steel was simply another step in the progressive march toward a planned industrial economy. The American engineer Charles W. Baker, for example, wrote that while a "system of competition" was well "adapted" to "the formative period of civilization," the time had come to abandon "the cruelly terse 'survival of the fittest,'" which "was never meant to control the wondrously intricate relations of the men of the coming centuries."[94] According to optimists such as Baker, the dawn of corporate capitalism thus represented the development of "a vast organism in which each individual, each community, each State, each nation has its prosperity and destiny indissolubly interwoven with the prosperities and destiny of every other one."[95] Similarly, the social reformer Frank Parsons insisted that "monopoly means cooperation instead of conflict, wise management instead of planless labor, economy instead of waste."[96] Or, as the corporate lawyer and early twentieth-century art critic Arthur Eddy put it, "Only savages permit the law of the survival of the fittest to work unchecked; they expose infants, abandon the sick, kill the aged—

they are evolutionists without human compunction, they are biologists without heart."[97]

Of course, not everyone shared these views. Going back to the United States' founding, a tendency for monopolies to concentrate wealth in the hands of the few ensured they were widely regarded as injurious to the common good. But whereas antebellum critics primarily saw corrupt practices and government favoritism as the main cause for concern, those at the turn of the twentieth century were more worried about the effects of ruinous competition.[98] Hence, rather than advocate laissez-faire policies of free competition as the best antidote to monopoly power, as their predecessors had done, progressive reformers were more likely to endorse direct government intervention to ensure the market's smooth functioning. This was especially so after the wave of corporate consolidations that swept through America's political economy in the decade between 1895 and 1905 further propelled the so-called trust question into the national spotlight. As a result, the federal government began actively scrutinizing the behavior of powerful corporations, and Congress held several well-publicized hearings in which numerous business leaders, including John D. Rockefeller, Carnegie, and J. P. Morgan, were called to account for their unfair and collusive business practices.[99]

No doubt the most outspoken critics of corporate consolidation were rural farmers, miners, and others outside the industrial Northeast who worried that small, independent producers could not possibly stand up to the power of large, corporate firms. During the 1890s, rural critics of corporate capitalism coalesced into the People's Party and began advocating a range of interventionist measures to help level the playing field, including the nationalization of railroads and the creation of a public banking system administered through the post office. Although some of their policy proposals were later embraced by more established political parties, progressive reformers often dismissed Populists out of hand, characterizing their criticisms of corporate capitalism as mere yearnings for a simpler, agrarian past among those who had failed to keep up with the times. Theodore Roosevelt, for example, described Populists as "rural Tories," and Woodrow Wilson declared that there was no sense in attempting to "disintegrate what we have been at such pains to piece together in the organization of modern industrial enterprise."[100] Similarly, the progressive historian Frederick Jackson Turner reasoned that those who belonged to a "primitive society can hardly be expected to show the intelligent appreciation of the complexity of business interests in a developed society."[101] Thus, whereas progressive reformers agreed with Populist critics that competition between small, family-owned and operated firms had become a thing of the past, they held fast to the vision of an industrial future ruled by the power of bureaucratic authority.[102]

Among progressive reformers, Roosevelt was especially keen to build an administrative state with sufficient power to manage the corporate economy. Only a few years after assuming the US presidency in 1901, Roosevelt oversaw the creation of a Bureau of Corporations within the executive branch. Not long thereafter, he ordered the Justice Department to bring suit against the Northern Securities Company, before taking steps to dismantle a number of other powerful conglomerates as well, including American Tobacco and Standard Oil. But Roosevelt did not seek to destroy the new corporate order. "We recognize that this is an era of federation and combination," he assured Congress in 1903, promising that his Bureau of Corporations would not seek to "hamper or cramp the industrial development of this country."[103] As he explained in his autobiography, critics of corporate consolidation were correct to point out "the evil done by the big combinations," but they were misguided in seeking to "remedy it by . . . restoring the country to the economic conditions of the middle of the nineteenth century." Rather than listen to "foolish radicals" who sought to "break up all big business," Roosevelt drew a subtle distinction between what he called "good and bad" trusts. Companies who sought "profits through serving the community by stimulating production, lowering prices or improving service" should be afforded "the fullest protection" of the law, he insisted, even if they should become a monopoly. In contrast, anyone who sought to make "profit through injury or oppression of the community" must be "pursued and suppressed by all the power of Government." To help illustrate the distinction, Roosevelt singled out J. P. Morgan's successful attempt to boost confidence by encouraging cooperation among New York investment banks during the financial panic of 1907. "The word 'panic' means fear, unreasoning fear," Roosevelt explained, which is why he gave his explicit consent to have the Morgan-controlled firm U.S. Steel acquire the Tennessee Coal, Iron, and Railroad Company, whose imminent bankruptcy threatened to destabilize financial markets even further. His reputation as a passionate trustbuster notwithstanding, then, Roosevelt remained optimistic about the power of bureaucratic administration to tame the industrial juggernaut.[104]

Although some industrialists—Rockefeller perhaps chief among them—viewed progressive reformers in an antagonistic light, others were more than willing to cooperate with the Roosevelt administration. The steel magnate Elbert H. Gary, for example, welcomed a 1905 decision by the Bureau of Corporations to investigate U.S. Steel with open arms, having correctly surmised doing so could help to legitimize its monopolistic business practices.[105] Similarly, Morgan's right-hand man, George W. Perkins, argued that Roosevelt's administrative state was a product of the same evolutionary dynamic that also drove corporate consolidation. "Competition is no longer the life of trade," he told a reporter in 1911, so the "competitive system must be abolished" and replaced by "a socialism of the highest, best, and most ideal sort," one that forestalled calls for "government ownership" by implementing instead a system of "government control."[106] Indeed, Perkins even joined forces with Roosevelt when the latter decided to form an independent Progressive Party in 1912, even going so far as to become its national chairman.[107]

With surprising regularity, the image of dinosaurs cropped up in debates about the best way to manage the evolution of capitalism. Especially telling is how often financial and industrial elites made reference to extinct creatures in their efforts to characterize the corporate consolidation of America's political economy as an instance of evolutionary progress. "When tadpoles and fish were evolved, there began a mighty gobbling up of the weak by the strong," the publisher Henry Holt speculated in a speech he delivered at Yale's Sheffield School, arguing that primitive competition only grew more intense among "reptiles, big lizards with wings and birds with teeth" who "kept up the game, and made it livelier, perhaps, than ever before or since, even down to the days of Standard Oil." Fortunately, however, "with the evolution of intelligence, there has appeared a new set of factors: sympathy, mercy, justice have begun to restrain and narrow competition, to shape popular opinion, and even to express themselves in law."[108] More outspoken still was the radical journalist Arthur M. Lewis. "Individualism is dead," he proclaimed, declaring it no more than "a reminiscence of a prior stage of social development," a "surviving rudiment" or "legacy that links us with our extinct ancestors of the Silurian age."[109] Indeed, even the noted American economist John Bates Clark likened unrestrained competition to "a monster as completely antiquated as the saurian of which the geologists tell us."[110] In the eyes of political economists such as Clark, dinosaurs represented a Darwinian dystopia in which the struggle for existence prevailed. But their near total annihilation ushered in a new age for the history of life, one in which primordial competition gave way to increased cooperation, rational planning, and efficient administration. According to this perspective, evolution and extinction went hand in hand, part of a larger process in which the "perennial gale" of "creative destruction," as the Harvard economist Joseph Schumpeter would eventually call it, was "incessantly destroying the old" and "creating the new."[111]

Conclusion

When curators at the American Museum of Natural History in New York began to experiment with designs for a *T. rex* display in 1913, they did not consciously seek to legitimize the claim that organized, corporate capitalism naturally evolved from a more primitive and unruly form of proprietary capitalism. But the orthogenetic interpretation of evolution espoused by so many late nineteenth-century paleontologists did cast the history of life on earth as a progressivist narrative in which the extinction of dinosaurs made space for the evolution of more intelligent mammals. Moreover, political economists invoked a strikingly similar narrative when they argued that "ruinous competition" between independent producers was giving way to a more organized marketplace dominated by corporate conglomerates. Paleontologists and political economists therefore drew on a common conception of evolutionary change, one in which the full sweep of earth history, including the development of modern society, could be explained in teleological and progressivist terms. For that reason, while museums such as the one in New York did not explicitly seek to naturalize the corporate reconstruction of America's political economy, exhibits that celebrated the gradual proliferation of cooperation and intelligent planning in geological history nonetheless lent weight to the idea that American capitalism was evolving in the direction of a more enlightened modernity.

At museums like the one in New York, dinosaurs were most often depicted as solitary creatures acting in isolation. In those instances when two or more dinosaurs were exhibited together, their interaction almost exclusively revolved around acts of predation. Whether it was an *Allosaurus* feasting upon a section of *Brontosaurus* tail, a *T. rex* facing off against a *Triceratops,* or a duckbill startled by the sound of its enemy off in the distance, it was violence that bound these creatures together. Theirs was an ecology of intense struggle and ruthless competition most often resulting in death. In stark contrast, early mammals were exhibited as social and intelligent creatures who cooperated to further some common aim. Two famous mural paintings by the artist Charles R. Knight illustrate the contrast especially well. Whereas Figure 5.3 depicts two titans of the prehistoric—the brutal predator *T. rex* and the armored giant *Triceratops*—engaged in an epic standoff against the backdrop of a gloomy, miasmic swamp, Figure 5.4 shows what Osborn explicitly described as a herd of wooly mammoths cooperating with one another by surrounding one of their offspring to protect it against a potential threat from afar. Knight even showed two of these mammoths raising their trunks in the air, suggesting that perhaps they were communicating with some of their conspecifics off in the distance.[112]

The progressivist claim that social cooperation inevitably replaced individual competition was made all the more credible by the exhibition of real, fossil specimens. When curators at the New York museum decided to mount the remains of several large and extinct ground sloths from South America, they produced a veritable celebration of mammalian teamwork. As the paleontologist William Diller Matthew explained in the *American Museum Journal,* the exhibit was "the most realistic that has yet been attempted in the mounting of fossil skeletons," incorporating the insights of numerous luminaries in the field of vertebrate paleontology ranging from Cuvier to Richard Owen, as well as New York's own Osborn. In much the same way as they did with the *T. rex,* curators used scale models to experiment with a number of different poses and configurations, each of which was extensively "criticized and discussed" to produce a display that could be trusted to "represent the most characteristic poses and habits of these animals." Tellingly, the end result consisted of four skeletons procuring a common source of food, working together to pull down a single tree (Figure 5.5). Whereas the largest individual stood on its hind legs, "endeavoring to reach up and drag down" the food source, another was shown "busily digging and tearing at the roots to loosen

Figure 5.3. An epic standoff between two titans of the prehistoric era, *T. rex* doing battle with *Triceratops*. Mural painting by Charles R. Knight at the Field Museum of Natural History in Chicago, 1931.

Figure 5.4. A herd of wooly mammoths crossing the Somme. Mural painting by Charles R. Knight at the American Museum of Natural History in New York, 1916.

and break them," helping "his big friend to uproot and pull the tree down." A third was shown "coming around the base of the tree to assist in the digging operations, while a fourth stands at a short distance, ready to add his weight to drag down the branches when they are brought within reach."[113]

While paleontologists reliably chose to depict dinosaurs as violent creatures who ruled over a world that was red in tooth and claw, then, their exhibits did not celebrate the fiercely competitive social order of modern capitalism. On the contrary, paleontologists constructed a narrative in which a ruthlessly violent world full of primitive reptiles gave way to one ruled by more intelligent mammals. On this view, the extinction of dinosaurs served as an affirmation that evolutionary history progressed in a remarkably straight line. In turn, the orthogenetic theory mirrored a popular narrative about the historical development of American capitalism. Faced with widespread anxieties about corporate consolidation, progressive reformers like Roosevelt and wealthy capitalists such as Morgan—both of whom were trustees of the New York museum—embraced the claim that corporations were engines of cooperation rather than corruption. In their view, the process of corporate consolidation was just another instance of evolutionary progress. Hence, paleontological

Figure 5.5. The ground sloth group on display at the American Museum of Natural History in New York, completed in 1911.

displays at philanthropic museums that consistently emphasized the vicious brutality of dinosaurs engaged in an intense struggle to survive could serve as a powerful means to legitimize the administrative vision of progressive reformers. But for philanthropic museums to fulfill their function of engendering social progress, their paleontological exhibits had to be taken seriously as trustworthy and authoritative accounts of prehistory. As we will see in the next chapter, they primarily did so by assiduously stressing the material link that directly connected modern exhibits with prehistoric creatures who roamed the deep past.

6

Bringing Dinosaurs Back to Life

On June 2, 1922, Sir Arthur Conan Doyle presented a motion picture film featuring live dinosaurs at the Society of American Magicians' annual dinner in New York. Audience members were astonished at what they saw: How could such long-extinct creatures possibly have been captured on film? Doyle was delighted by the incredulity that his performance elicited. Insisting that the pictures would "speak for themselves," Doyle did not offer much help solving the mystery, only venturing to say, "If I brought here in real existence what I show in these pictures, it would be a great catastrophe." In an enigmatic aside, he also allowed that his film had been made by joining the imagination with the "power of materialization." According to a journalist from the *New York Times,* Doyle's images were so "extraordinarily lifelike" that if they were "fakes, they were masterpieces." But the very next day, Doyle admitted his performance had been an elaborate hoax that was "constructed by pure cinema." The footage, Doyle now revealed, was provided by Watterson R. Rothacker, a producer from Chicago who had teamed up with the special-effects wizard Willis O'Brien to make a film version of Doyle's 1912 best-selling novel, *The Lost World.* In exchange for providing the rights to his story, Doyle was given permission to show some of O'Brien's footage before the film was released into wider circulation, and Doyle used the occasion to play a practical joke on his magician friends.[1]

While O'Brien's films were considered extraordinarily convincing and lifelike, nobody was actually fooled by Doyle's stunt. After all, dinosaurs had been extinct for millions of years.[2] Neither did it escape notice that Doyle was an outspoken supporter of spirit photography. At the time, his name was particularly associated with the so-called Cottingley Fairies, about which he had recently penned a widely ridiculed book.[3] Indeed, Doyle's trip to America was primarily undertaken to defend the claims of spiritualism against its many detractors, largely by deploying photographic evidence that ghosts, apparitions, and other elusive entities truly existed. Hence, by the time he exhibited footage of live dinosaurs in New York, Doyle was often mocked as a great believer in the magical, the wonderful, and the occult. This made his presentation of O'Brien's stop-motion dinosaurs all the more enigmatic, especially given that Harry Houdini, Doyle's host at the magician's dinner, was a vehement critic of spirit photography. Audiences therefore had good reason to wonder whether Doyle might have harbored an ulterior motive when he decided to provide "a little mystification to those who have so often and so successfully mystified others," as he told newspaper reporters.[4]

So far, this book has dealt almost exclusively with the way dinosaurs were exhibited in the museum. But they were hardly confined to that cultural context alone. During the early twentieth century, representations of prehistory came to proliferate in all manner of exhibition spaces. In addition to the cinema and the theater, dinosaurs could also be found in the department store and at the fair, as well as in countless newspapers, magazines, and other print publications. However, while dinosaurs moved easily between the museum and more commercial amusement venues, the manner in which they were displayed in various institutional settings differed considerably. Although philanthropic museums did not shy away from a bit of showmanship to attract visitors, they worked hard to be seen as nonprofit institutions that produced and disseminated reliable scientific knowledge, not business ventures that delighted in peddling humbug. This was in contrast to commercial purveyors of public spectacle, which did not hold themselves to the same standards of strict truthfulness and intellectual restraint. As Doyle's playful deception demonstrates, they could adopt a more lighthearted approach when exhibiting dinosaurs. Thus, whereas museums went out of their way to assure visitors that every specimen on display was truly authentic, amusement venues such as the cinema often made light of how difficult it was for viewers to ascertain whether they were seeing the truth or being deceived, which only added an element of intrigue and mystery to the shows they produced.[5]

Commercial exhibits were not only faulted for being deficient in their commitment to strict factual accuracy. Many were also regarded as excessively sensationalist. During the 1920s, a firm specializing in the construction of parade floats and theatrical set designs called Messmore & Damon built a life-size animatronic *Brontosaurus* to advertise the company's engineering skills and mechanical expertise. Controlled from within by a human operator, this creature could open its mouth, roll its eyes, flare its nostrils, and sweep its neck above a crowd of awestruck spectators. When it was first unveiled at a New Jersey department store near the start of the 1924 Christmas season, about four hundred thousand curious shoppers flocked to see *Brontosaurus* perform tricks such as snatching a lit cigarette from an audience member using its mouth. Buoyed by their success in bringing *Brontosaurus* to life, Messmore & Damon went on to produce an amusement park ride called the World a Million Years Ago for the 1933/1934 Century of Progress Exposition in Chicago (Figure 6.1). This fairground attraction invited customers to float down the lost river of time, where they could ogle at animatronic sculptures of prehistoric monsters awkwardly interacting with one another. Before long, however, Messmore & Damon acquired a reputation for lurid and lowbrow spectacles. A decline in ticket sales prompted the firm to introduce a new exhibit about medieval torture chambers that included a soundtrack featuring "the screams and groans of the unfortunate victims" alongside a topless "Crusader's Bride" wearing only an iron chastity belt. Messmore & Damon also hatched a number of controversial publicity stunts to help drum up additional customers, including a well-publicized nudist wedding held under the *Brontosaurus* display. All of these antics came in for intense criticism, and there was even talk of banning them from the fair.[6]

While philanthropic museums often collaborated with commercial sites of public amusement, there were considerable tensions between the two. For example, a celebrity explorer named Roy Chapman Andrews who succeeded Henry Fairfield Osborn as president of the American Museum of Natural History in New York personally agreed to promote the work of Messmore & Damon in a Paramount News short called "Stone Age on Broadway." But the New York museum also balked at Messmore & Damon's proposal to donate some of the firm's earnings to the "furtherance" of vertebrate paleontology in exchange for an official statement from the museum attesting to the "correct-

Figure 6.1. Promotional poster for Messmore & Damon's World a Million Years Ago amusement park ride at the 1933/1934 World's Fair in Chicago.

ness of detail" and "educational value" of their amusement park ride. The museum's reluctance to enter such an arrangement almost certainly stemmed from a fear that doing so would tarnish its reputation as a serious, nonprofit institution devoted to science. Nor did it help matters that a promotional flyer compared Messmore & Damon to a nineteenth-century circus or dime museum, characterizing its animatronic dinosaurs as "Bigger than Barnum's Biggest! Better than Barnum's Best!" As we saw in Chapter 2, the New York museum had actively distanced itself from precisely these institutions since the time of its founding, and it continued to hold commercial spectacles such as Messmore & Damon at arm's length.[7]

Whereas commercial amusement venues mounted spectacular shows that often blurred the distinction between the real and the imaginary, philanthropic museums sought to do exactly the opposite, creating authoritative exhibits whose seriousness of purpose was above reproach and whose accuracy was beyond question. But this was especially difficult to accomplish when dinosaurs were involved. In a very real sense, dinosaurs were objects of the imagination that resembled the ghosts, apparitions, and fairies that spirit photographers captured on film. Because dinosaurs were inaccessible to direct observation, knowledge about them was hard to establish conclusively. As a paleontologist at the American Museum of Natural History in New York lamented, for the casual visitor, the deep past remained "somewhat of a fairy-tale, a fanciful imaginative world peopled with ogres and dragons and belonging to the unreal 'once upon a time.'"[8] Thus, although dinosaurs were especially rewarding display objects that reliably drew a crowd, they also posed a substantial risk to the museum's institutional credibility. This made it imperative to design vivid exhibits that advertised their own authenticity, to ensure that audiences would accept them as fact rather than fiction.

Museums assembled dinosaur fossils into lifelike but realistic exhibits that sought to balance a desire for public spectacle with the requirement to project a strong sense of intellectual authority. They did so by insisting that even the most evocative fossil displays were firmly grounded in the solid bedrock of prehistory. Toward this end they embedded their dinosaurs within a much larger exhibition strategy that effectively gave visitors a fictional account of their own production histories. This exhibition strategy consistently emphasized the direct, material link that tied extant fossils to flesh-and-blood creatures from the depths of time while downplaying the degree to which human inventiveness, imagination, and ingenuity were involved in their creation. Lifelike reconstructions of dinosaurs could therefore be presented to viewers as an authentic trace of the deep past, which allowed philanthropic museums to advance an implicit claim that whereas commercial dinosaur shows such as *The Lost World* and the World a Million Years Ago may indeed have been the unreliable products of an unfettered imagination, their own public galleries offered an unmediated link to deep time.[9]

But fossils were hardly the only representational medium that was understood as a genuine trace of the past. Photographs were endowed with immense epistemic power for strikingly similar reasons. While the reliability of a photograph was judged in much the same way as that of a drawing or an engraving for much of the nineteenth century, by the time Doyle attended the magicians' dinner in 1922, photographs had come to be understood as an especially trustworthy medium.[10] Because photographs were produced by a chemical process, they were widely regarded to stand in a direct causal relationship to what they depicted. Hence, they came to be treated as more than just pictures or likenesses. They also functioned as *evidence*.[11] This not only explains why spirit photography was met with such an immense controversy. It also provides the context required to understand why filmic depictions of extinct dinosaurs elicited such a passionate reaction from audiences.

In large part, what made Doyle's hoax at the magicians' dinner so successful was that his performance undermined the generic expectation of photography as a truth-telling medium. Photographs were seen as especially trustworthy, but filmic depictions of dinosaurs could not possibly have been real. Far from counting as genuine evidence of their former existence, O'Brien's stop-motion animations were an elaborate trick that had more in common with a magician's sleight of hand or a spiritualist's obfuscation than with a scientific experiment. Cinematic depictions of dinosaurs thus used the evidentiary status of their medium to produce a very different effect from that created by museum displays. Rather than serve as a marker of credibility, the direct connection of a photograph to its subject offered a powerful means to emphasize the technical wizardry that was required to film dinosaurs. This only made the spectacle of bringing prehistoric monsters back to life by animating their likenesses all the more stunning. In short, stop-motion animations and other filmic depictions of dinosaurs were seen as especially wonderful because, not in spite of, the fact that everyone understood that they constituted a material trace of something that humans could not possibly have captured on film.[12]

Instead of vouchsafing their authenticity, cinematic depictions of dinosaurs used the evidentiary status of photographs to produce a more engrossing spectacle. In advancing that claim, this chapter draws on a distinction between iconic and indexical representations first proposed by the American pragmatist philosopher Charles Sanders Peirce just a few decades after the invention of photography and before the development of cinematography.[13] According to Peirce, whereas indexical representations stand in a relatively straightforward causal relationship to the original, icons function as likenesses, representing their subjects through the possession of shared formal qualities.[14] For example, whereas a snow angel indexically represents the motions of a child at play, it is an icon of a biblical being. That said, Peirce recognized that paintings and drawings are causally linked to their subjects as well. As Peirce put it, in the case of a portrait, the appearance of a subject "made a certain impression upon the painter's mind and that acted to cause the painter to make such a picture as he did."[15] The crucial difference between the two modes of representation, then, is not just that a causal relationship exists between the representation and the original. Rather, Peirce's distinction turns on whether the original exercises its causal power with or without the direct intervention of human consciousness.

Insofar as they were regarded as material traces, indexical representations conformed to one of the most powerful conceptions of scientific propriety at the time: mechanical objectivity. According to the historians Lorraine Daston and Peter Galison, nineteenth-century scientists became increasingly anxious that human subjectivity distorted their understanding of nature. As a result, they came to prize representations whose production minimized or even eliminated the involvement of human consciousness. "Wary of human mediation between nature and representation," Daston and Galison explain, the scientific community enlisted technologies of the index such as "cameras, wax molds, and a host of other devices in a near-fanatical effort" to create representations of "birds, fossils, snowflakes, bacteria, human bodies, crystals, and flowers" that were certifiably free from the corrosive effects of "human interference."[16] Because they were seen as devoid of personal interests and psychological biases, machines embodied this new conception of epistemic virtue especially well. Consequently, the images that came to be seen as the most trustworthy were those that had been mechanically produced, with little or no direct human involvement. To the extent that both fossils and photographs satisfied the stringent criteria of mechanical objectivity, they were afforded an epistemically privileged status and regarded as especially reliable truth-telling media.

Enthralled by the epistemic virtues of mechanical objectivity, museum curators consistently highlighted the direct causal connection between dinosaur fossils and real prehistoric creatures. They did so in hopes of lending scientific authority to some of their most popular but also highly speculative exhibits. In contrast, the producers of cinematic depictions were more interested in making a captivating spectacle that could turn a large profit. As a result, animators such as O'Brien put a great deal more effort into wowing audiences with their audacity than into convincing them of a film's veracity.

To do so, cinematic depictions of dinosaurs frequently played up their status as an index of the impossible—that is, a material trace of something that everyone understood to be fictional. This allowed producers of trick films to show off their technical wizardry, which further heightened the spectacle of their productions. While both the museum and the cinema emphasized the indexicality of the representational media they deployed, then, they did so for precisely the opposite reasons. Whereas museums stressed the fossil's status as a material trace to vouchsafe its authenticity, early trick films deployed the same feature of cinematic representation to further stun audiences into disbelief. In addition to reflecting the divergent institutional goals of both kinds of performance spaces, this illustrates the ontological indeterminacy of the index, which could be deployed equally well as a guarantor of the truth or as a demonstration of skillful deception and artifice.[17]

Faking Photographs

Much as Doyle's performance at the Society of American Magicians' annual dinner confounded audiences by bringing extinct creatures to life on the screen, the text of *The Lost World* dramatized the difficulty of distinguishing between the real and the imaginary. Initially serialized in the *Strand* magazine from April to November 1912, it was subsequently released as a standalone book. Borrowing heavily from the conventional tropes of adventure stories, *The Lost World* repeatedly signaled its status as a work of serious nonfiction. However, despite the fact that Doyle had recently made a name for himself as a journalist with a graphic exposé of the atrocities committed in the Belgian Congo, savvy readers would have known they were dealing with a playful deception. Thus, while the *Strand* promised that Doyle's story was "guaranteed to give a thrill to the most jaded reader," much of what made *The Lost World* fresh and intriguing was the way that it self-consciously poked fun at the genre's own hackneyed conventions.[18] For example, Doyle's text began with a foreword informing readers that one of the story's protagonists—an irascible naturalist named Professor George Edward Challenger—recently withdrew a libel suit that prevented its publication.[19] Traditionally, adventure writers deployed clever manipulations of nondiegetic elements of the text such as this to add an uncanny touch of realism to their stories. But in Doyle's case, the foreword could also be read on a more ironic register, as an inside joke that flattered the sophistication of knowledgeable readers who recognized Challenger's lawsuit as a formulaic element of adventure fiction more so than a literary device that actually sought to fool them into believing the story was true.[20]

The narrative content of Doyle's novel largely revolved around the difficulty of distinguishing fact from fiction as well. Its plot was set in motion when the romantic advances of a young journalist named E. D. Malone were rebuffed by a woman who pined for "a harder, sterner man" who "could look Death in the face and have no fear of him." Determined to prove himself worthy, Malone ran off to ask his news editor for an assignment with "adventure and danger in it." His editor responded by suggesting, "What about exposing a fraud—a modern Munchausen—and making him ridiculous? You could show him up as the liar he is!" This led Malone to visit Challenger, who recently claimed to have discovered an elevated plateau in the Amazon on which the "ordinary laws of Nature," including the "various checks which influence the struggle for existence" had been suspended. As a result, ancient and strange-looking creatures such as the *Pterodactyl* and *Stegosaurus,* as well as other extinct saurians from the Jurassic, were "artificially conserved" there. To win over the young journalist, Challenger adduced several pieces of documentary evidence, including a number of photographs and a physical specimen of the *Pterodactyl's* leathery wing. Feeling the "cumulative proof" presented by Challenger was both "conclusive" and "overwhelming," Malone exclaimed that he was "a Columbus of science." But when Malone rushed back to report on the great naturalist's discoveries, he was met with a skeptical audience. As Malone's wizened editor reminded the impressionable reporter, "Things don't happen like that in real life" and should be left "to the novelists," prophetically adding, "You can fake a bone as easily as you can a photograph."[21]

Despite the withering ridicule heaped on his discoveries, Challenger remained undeterred. "Truth is truth," he insisted, regardless of "the noise" made by "foolish young men" and "their equally foolish seniors." He therefore proposed a follow-up expedition back to the Amazon, which eventually led Challenger and Malone to find themselves stranded in South America alongside an entomologist named Professor Summerlee and an intrepid adventurer named Lord John Ruxton. It was not long before the party discovered that Challenger's plateau would not disappoint. As they come face to face with several species of flesh-and-blood dinosaurs, even the hard-nosed empiricist Summerlee could not doubt any longer. To avoid being branded an "infernal liar and a scientific charlatan" upon his return home, Challenger set out to capture a live specimen. When the party eventually succeeded in finding their way back to London, a huge crowd gathered at the Queen's Hall on Regent Street to see whether the expedition had failed or succeeded. Almost immediately, a skeptical inquirer rose to question their account of what had transpired in the name of "scientific truth," wondering what could possibly "constitute final proof" of such "revolutionary and incredible" claims? What had been proffered by way of "corroboration of these wondrous tales"? he wanted to know. "Some photographs," perhaps, but, "in this age of ingenious manipulation," how could these possibly "be accepted as evidence?" This prompted Challenger to draw a live *Pterodactyl* out of a large "packing-case," but the creature immediately escaped from Challenger's clutches and began flapping its leathery wings "while a putrid and insidious odor pervaded the room." Before anyone could make sense of what had transpired, the creature escaped through a window. With that, Challenger's sole piece of incontrovertible evidence was lost to the ages, and all that remained to corroborate the group's eyewitness account were a handful of unreliable photographs.[22]

Besides making countless textual allusions to the problem of demarcating truth from untruth, Doyle also introduced a number of metafictional elements that destabilized the boundary between Challenger's world and that of the reader. Tellingly, many of these centered on the veracity of photography. For example, a collage of illustrations by Harry Rountree that appeared as the frontispiece to the story's first installment in the *Strand* included a photograph of Doyle dressed up as Challenger that was visibly touched up (Figure 6.2). These alterations imparted the photograph with a painterly style that matched Roundtree's drawings of ape-men and dinosaurs. Doyle also scrawled a handwritten note under the portrait, which read, "Yours truly (to use the conventional lie), George Edward Challenger." In a letter he sent to his mother, Doyle explained that these "photos were all my idea and carrying out," and he later elaborated that it was in "a rather impish mood" that he "set myself to make the pictures realistic."[23]

Figure 6.2. Frontispiece to Arthur Conan Doyle's "The Lost World," from the *Strand*, 1912.

The central but highly contested role that photography played in Doyle's romance reflected the medium's complex history as an instrument of science and showmanship alike. Since its introduction nearly a century earlier, photography had developed a contradictory reputation as both a magical illusion and an especially trustworthy medium. This was already reflected in the differing attitudes of its two rival inventors. Whereas the British scholar Henry Fox Talbot famously likened it to "the pencil of nature," describing photographs as pictures "impressed by Nature's hand," the French artist Louis Daguerre was more interested in the medium's uncanny ability to create an engrossing illusion.[24] Thus, although photographs were widely perceived to exist on the same ontological plane as their subjects, that did not mean they could be relied on to tell the truth. Indeed, almost immediately after the new medium was invented, artists began to experiment with techniques to generate images that were engineered to deceive. Ranging from multiple exposure and chemical manipulation to various forms of optical illusion, what these shared in common was an uncanny ability to mislead the eye and strain credibility. For that reason, photography quickly acquired a reputation for visual trickery as well as empirical accuracy.[25]

These complexities directly fed into the proliferation of spirit photography, which so captured Doyle's imagination. Given the medium's close association with the natural sciences, it did not seem so farfetched that a photograph might reveal parts of the world that were invisible to the naked eye, not unlike the microscope, the telescope, and countless other optical devices had already done. However, skeptics also cited the ease with which a negative could be exposed more than once (perhaps inadvertently) to explain why spirit photographs so often featured a partially translucent apparition near a more opaque and substantial subject. Moreover, although the techniques used to produce them were new, spirit photographs did not differ so radically from older and more established genres of optical illusion such as the phantasmagoria. In both cases, scientific knowledge was exploited to produce unsettling, mystical, and otherworldly visual effects that astonished the viewer. As a result, photography's status as a mechanical inscription device was hardly taken to guarantee truth.[26]

By the time Doyle published *The Lost World,* the cutting edge of photographic illusion had moved to motion picture technology. Perhaps even more so than still photography, early cinema evinced a clear fascination with the medium's capacity for elaborate tricks and contrivances. The first motion picture films were regarded in much the same manner as fairground attractions. During the 1890s and into the early twentieth century, the cinema was seen as an ingenious mechanism that could pull off the unlikely feat of endowing ordinary photographs with the ability to move. To emphasize this fact, early films shown by the Lumière brothers often began with a still image that was projected onto a screen for some time before whirring to life.[27] At the time of its inception, then, the moving picture itself constituted a kind of special effect. But it was not long before the mere fact that a photograph could be animated was supplemented with additional techniques such as stop-motion to instill audiences with a sense of wonder and awe. Georges Méliès, a French filmmaker who initially rose to prominence as a magician, stumbled on this technique when his camera jammed as he was filming a Paris street scene in 1896. He went on to exploit his fortuitous discovery of stop-motion cinematography to great effect in numerous shorts, including *The Vanishing Lady,* while other filmmakers developed additional special effects, such as manipulating the speed at which moving pictures were shot and feeding a filmstrip through the projector backward.[28]

Before long, countless such "trick films" were being produced throughout Europe and the United States, including *Visit to the Spiritualist,* which an exhibitor's catalog described as "the funniest of all moving magical films." Soon thereafter, filmmakers began animating puppets and miniature sculptures as well, and stop-motion cinematography was increasingly used to represent visually inaccessible events, objects, and circumstances. Tellingly, they often featured the deep past, and by the early twentieth century, dinosaurs had become stars of the genre, showing up everywhere from Buster Keaton's first feature film and a short piece for the Pathé Review by the British sculptor Virginia May to an educational film by Johann Ewald named *Aus der Urzeit der Erde.* As the historian Dan North argues, because the technique "preserves the textural or anthropomorphic attributes of a model or puppet, but renders the resulting figure distinctly 'othered,'" stop-motion was "particularly suited to depictions of the monstrous, the imaginary or the impossible." In addition to being immensely popular, then, dinosaurs were especially well-suited to heighten the uncanny sensation that stop-motion animations engendered among spectators.[29]

The most celebrated stop-motion dinosaur films were unquestionably those of O'Brien. A special-effects master who would gain worldwide acclaim for his work on the 1933 production of *King Kong,* O'Brien enjoyed early success with a one-reel "caveman-comedy" from 1915 called *The Dinosaur and the Missing Link.* Financed by a Nickelodeon proprietor from San Francisco, the film garnered enough attention to convince the Edison Company to purchase it for one dollar per foot as part of its "Conquest Pictures" series. Over the next several months, O'Brien produced many more shorts for Edison that were set in the deep past, including *R.F.D. 10,000 BC* (1916) and *Curious Pets of Our Ancestors* (1917). Then, after Edison terminated O'Brien's contract, a sculptor and army officer who had worked in the automobile industry named Herbert M. Dawley approached O'Brien with the proposition of making another dinosaur film. Dawley had devised an improved model design for stop-motion animations that used a skeleton made of steel, and he suggested teaming up to produce a narrative piece that combined live-action sequences with animated dinosaurs, although not in the same frame. The result was called *The Ghost of Slumber Mountain,* and it earned more than $100,000 after premiering at New York's Strand Theater in 1918. Despite the success of their joint effort, however, O'Brien had already entered negotiations for a more lucrative contract with the producer Watterson R. Rothacker behind Dawley's back. When Dawley found out, the two parted ways, but not before Dawley publicly accused O'Brien of infringing his intellectual property by stealing the improved dinosaur models.[30]

With the commercial backing of First National Pictures, O'Brien and Rothacker secured the rights to turn Doyle's *The Lost World* into a special-effects extravaganza. Directed by Harry O. Hoyt and based on a script written by Marion Fairfax, the film took considerable liberties with Doyle's narrative, including the addition of a love interest to spice up the action. But the real stars of the show were the dinosaurs, whose movements were deemed enormously convincing and lifelike.[31] Of all the amazing special effects that O'Brien pulled off in the film, perhaps most astonishing was his success in combining live action with stop-motion to create the illusion that Challenger and his crew interacted with these prehistoric monsters directly. The trick—a first in the history of cinematography—was accomplished by carefully shielding a part of the frame with a matte (a piece of cardboard made to prevent light from exposing a particular section of the negative). By rewinding the film and masking a different part of the frame, complex scenes could be built up in pieces from multiple sets of exposures. An especially dramatic example appeared near the end of the film, when Challenger revealed to an incredulous audience in London that he had brought conclusive proof of his exploits back from the Amazon in the form of a live *Brontosaurus*. Before he could even unveil the discovery, however, the monstrous reptile escaped from its cage and went on a rampage, stomping pedestrians and smashing buildings before it eventually crushed a bridge and fell into the Thames.

Despite the fact that they prominently featured humans interacting with dinosaurs, O'Brien's films were consistently marketed as scientifically accurate depictions of prehistory. A press release circulated by the Edison Company in 1916, for example, assured potential audiences that the monsters in "O'Brien's 'stone age' stories" were "precisely in accordance" with the "prehistoric animals" on display at the American Museum of Natural History.[32] Dawley deployed a similar tactic, styling himself as "the Discoverer and Photographer of the Land of Mystery" and wryly assuring skeptical audiences that "the camera does not lie." Elsewhere, he promised that the authenticity of the animals was "endorsed by the most eminent scientists in the world," even using a photograph of a curator from the American Museum of Natural History posing next to a series of mounted fossils to back up the claim that every animal in the film was "vouched for by every scientific authority." "The animals Major Dawley has pictured are not creatures of imagination," a press booklet asserted. But the same booklet also added the following coy aside: "How he pictured them is the one mystery that is not disclosed." Dawley even pressed theater owners to offer a reward of $1,000 to any customers who could prove they had actually seen a live dinosaur. Artfully failing to clarify whether a filmic depiction sufficed, these ubiquitous boasts about factual accuracy were clearly meant to be lighthearted and ironic.[33] A similar strategy was used to promote *The Lost World,* and one flyer even insisted that paleontologists were "astonished at the sight of these prehistoric monsters actually living and doing battle on the screen."[34] Thus, in much the same way that Doyle's original narrative playfully insisted that it was a work of nonfiction, O'Brien's fantastical trick films were often marketed as truthful portrayals of events that everyone understood could not have transpired.

While promotional materials teased audiences about the factual accuracy of O'Brien's trick films, journalists primarily focused on his technical virtuosity. Whereas the *Boston Daily Globe* gushed that *The Lost World* was "a truly marvelous revelation of what the art of photography can accomplish," Regina Cannon from the *New York Evening Graphic* claimed to be at a loss for words "to express just how breath-taking is this miracle of photography." "It looks exactly as if the ferocious creatures were actually there, and you wish you had brought a pistol or a club, or something to defend yourself against it, should he take the notion to walk off screen right into the audience," she wrote. Another critic judged the film to have opened a "new realm" for "the vision of the picturegoer," one "not of fantasy so much as of scientific imagining," ensuring it would be "hailed a masterpiece." The *New York Times* even sent a reporter to visit O'Brien's studio, where O'Brien was only too eager to disclose that "everything was a matter of mathematical precision" on his set.[35]

Despite all the tongue-in-cheek marketing that touted its scientific pretensions, then, *The Lost World* was primarily viewed as an elaborate piece of visual trickery. Journalists were eager to join in on the fun, and newspapers engaged in countless attempts to unmask the technical contrivances that made such an unlikely spectacle possible. The *New York Times* article quoted earlier, for example, carried the headline "How Miniature Replicas of Monsters Were Filmed." The article went on to give readers a meticulous account of how each "little change in movement," even those that were only "of trifling consequence," had to be shot separately on a single frame that only amounted to "one sixteenth of a second." The *Times* further reported that many scenes required "not only double and triple exposures" but in some cases even a "quadruple" or a "septuple exposure."[36] More detailed still was a heavily illustrated piece that appeared as the cover story of the May 1925 issue of *Science and Invention* (Figure 6.3). Promising to expose how O'Brien succeeded in "filming the impossible," this article carried the headline "Trick Photography Involving Complicated Hand Moved Miniature Models Explained in Detail." Taking on some of the film's most eye-catching sequences, the article carefully dissected the means by which complex scenes could be

Figure 6.3. Front cover of *Science and Invention* magazine, May 1925.

built out of multiple exposures that combined different set pieces with stop-motion and live-action cinematography. The article also revealed how invisible wires had been used to make a *Pterodactyl* fly and cause stone buildings to crumble to pieces when they were attacked by a *Brontosaurus*. In addition, readers learned that life-size models were used whenever live actors had to

Figure 6.4. Detail of illustration in the *Science and Invention* article explaining the stop-motion technique that O'Brien used to film *The Lost World*.

make physical contact with one of the dinosaurs. But perhaps most impressive of all was that O'Brien had fitted the dinosaur models with inflatable bladders so they would appear to be breathing in close-ups, a detail that made everything seem all the more lifelike (Figure 6.4).

The ease with which journalists acquired behind-the-scenes access to O'Brien's set reveals much about the mode of spectatorship that trick films sought to engender during the first decades of the twentieth century. In stark contrast to the sort of immersive experience that viewers would come to associate with more conventional narrative cinema, trick films invited audiences to adopt what the film historian Colin Williamson has described as an "investigative viewing practice." In this respect, trick films betrayed their deep roots in the history of theatrical magic. Much like a magician performing a sleight of hand, O'Brien induced audiences to engage in a form of detective work, prompting them to inquire into what lay behind his most compelling illusions. *The Lost World* therefore invited audiences to ascertain for themselves precisely how O'Brien had succeeded in "filming the impossible," as the cover of *Science and Invention* put it. In many respects, this was exactly the opposite of what more immersive forms of narrative cinema asked of their viewers. Whereas the latter required audiences to suspend their disbelief and accept that the action depicted on-screen truly took place, films such as O'Brien's induced spectators to hold fast to their skepticism. Enjoying a trick film involved refusing to simply believe one's own eyes and instead make an effort to discern the technical means by which such a compelling illusion had been made possible. In this respect, the experience of viewing a trick film such as *The Lost World* resembled a visit to the dime museums that we encountered in Chapter 2, which also invited audiences to question how fantastical creatures such as the Feejee Mermaid had come into being.[37]

In much the same way that Doyle made the problem of authenticity a core element of *The Lost World,* early twentieth-century trick films engendered an inquisitive mode of spectatorship that was sometimes incorporated into their narrative content as well. A particularly good case in point is *Gertie the Dinosaur,* which was among the first animated films by the well-known cartoonist Windsor McCay. After spending much of his youth drawing quick portraits at mid-western dime museums, McCay eventually settled in New York, where he became famous for a regular newspaper comic called *Little Nemo in Slumberland.* In the years that followed, he also began to perform on the vaudeville circuit, earning especially high praise for a series of "lightning sketches" in which he showed off his skills as a draftsman by transforming simple line drawings while telling a story about what they evolved to depict in real time.[38] It was as part of his stage act that McCay introduced his first animated cartoon, inspired by *Little Nemo,* in April 1911. Aside from being a pioneer of the new genre, this film was especially noteworthy because the narrative arc revolved entirely around its own making. The opening scene showed McCay enter a bet with fellow comic strip artists who ridiculed him for boasting that he could draw and sequentially photograph enough images to animate his cartoons. Next he was shown slaving away as a team of assistants hauled boxes of paper and barrels of ink into his studio, after which audiences saw the process by which the results were transferred to film. The performance then drew to a close with a short sequence in which a cartoon drawing of Little Nemo came to life on the screen, dancing and engaging in various high jinks before returning to a static state on the page.[39]

After producing another short film called *The Story of a Mosquito,* McCay released a third animation that featured a playful dinosaur named Gertie. Initially intended as part of his vaudeville performance, *Gertie* was produced with an eye toward interactivity. Upon Gertie's initial appearance on screen, McCay would face her and command her to dance and perform various physical stunts. After some visual gags in which Gertie swallowed a boulder and drank up a lake, the flesh-and-blood McCay threw a real apple behind the projection screen only to have the dinosaur devour a cartoon representation of it as well. Finally, there was a grand finale in which McCay personally stepped behind the screen while *Gertie* seamlessly scooped his drawn double onto her back. As might be expected, audiences loved Gertie, and when McCay premiered his new act at the Palace Theater in Chicago on February 8, 1914, it was a resounding success. The act was so popular, in fact, that his principal employer—the newspaper publisher William Randolph Hearst—began to regard it as a competitive threat and invoked the exclusivity clause in McCay's contract, forcing him off the stage.

That did not spell the end of *Gertie* the dinosaur, however. McCay soon secured an agreement from the Box Office Attraction Company to distribute his animation as a standalone film. To make up for his physical absence on stage, McCay replaced his vaudeville performance with intertitles and added a prologue that was strikingly similar to the one used in *Nemo*. As before, the film's genesis was explained as the result of a foolhardy bet. Only this time, McCay's wager was precipitated by a chance visit to the natural history museum, where the group of cartoonists encountered an actual *Brontosaurus*. Moreover, whereas McCay boasted that the production of *Nemo* required some four thousand drawings, with *Gertie* he more than doubled that number to ten thousand. But there were other important differences too. While the animated figures in *Nemo* appeared to be floating freely, Gertie was depicted against a simple backdrop that grounded her presence in physical space. The overall framing conceit remained largely unchanged, however, featuring a live-action sequence that showed McCay winning his bet as the dinosaur he had drawn on a large sheet of paper began to move on her own when his easel was imperceptibly switched out for a projection screen. The film version of *Gertie* therefore again foregrounded the materiality of its own production process, taking audiences on a fictional tour behind the scenes and showcasing the method by which cartoon animations were made. As an early reviewer of McCay's trick films remarked, "the camera, that George Washington of mechanism, at last is proved a liar."[40]

Articulating Prehistory

In many respects, a trip to the museum could not have been any more different from a night at the cinema. Eager to safeguard their reputation as nonprofit institutions devoted to the production and dissemination of reliable knowledge, philanthropic museums did not adopt the cavalier attitude about truth that characterized trick films. On the contrary, they made every effort to convince visitors that there was no need to question the veracity of their exhibits. Thus, whereas a trick film such as *The Lost World* actively blurred the line separating fact from fiction, museum displays were carefully designed to uphold that very distinction by reinforcing the difference between what was real and imaginary.

However, while museum curators worked hard to provide visitors with an authentic exposure to nature, their exhibits had to do more than pass muster among knowledgeable experts. They were also designed to attract a large audience and provide an immersive experience, transporting visitors out of the city and into the wilderness. To balance the different and sometimes contradictory elements of their complex institutional goals, curators adopted a flexible approach in their exhibition designs. Frederic Lucas, who served as the director of the American Museum of Natural History in New York for most of the 1910s and the 1920s, put it well when he argued that a successful exhibit should hold a "mirror up to nature," but one that "let it reflect an image of nature as she looks when alive, not as she appears when dead and shriveled."[41] To that end, museums built elaborate displays to deceive the human sensory apparatus and make a visitor "think for a moment that he has stepped five thousand miles across the sea into Africa itself," to quote the celebrated taxidermist Karl Akeley.[42] The most ambitious among these were dioramas that featured a large group of specimens interacting with one another alongside remnants from their natural habitat. Usually placed in a recessed alcove, habitat dioramas also included a realistic backdrop that was painted in linear perspective and lit from above. All of these elements were designed to draw visitors into the scene on display, and they received lavish praise from journalists who were in awe of the museum's ability to generate an engrossing "illusion of nature" with about as much "beauty and sublimity as may be caught by the imitation of art."[43] Thus, while curators may have been keen to ensure that philanthropic museums would be regarded as serious institutions of science, the exhibits they mounted were not born of a slavish devotion to factual accuracy. They were also designed to inspire the imagination.

To help them appeal to a large, urban audience, philanthropic museums often drew inspiration from commercial exhibits, especially the department store window display. Rather than systematically arrange specimens based on their taxonomic position in the Linnaean classification system as older museums tended to do, philanthropic museums increasingly began privileging ecological relationships in their exhibition halls, showing how plants and animals interacted with one another in the wild.[44] This was in keeping with the latest developments in department stores, which gradually stopped piling similar products on top of each other in favor of creating evocative tableaux that depicted an aspirational domestic scene. Instead of arranging similarly sized and shaped goods into an impressive architectural structure, a family of clothed mannequins might be shown socializing with one another, surrounded by a sumptuous set of living-room furniture. The historian Andrew McClellan therefore concludes that "What museums and stores had to 'sell' might have been quite different, but the means of engaging their publics could be the same."[45] Indeed, museums even hired commercial artists who drew on the latest advertising techniques to attract and sustain the attention of visitors.[46] "We must stoop to conquer," Henry Fairfield Osborn wrote, while John Cotton Dana from the Brooklyn Museum boldly proclaimed that a "first class" department store did a far better job of engaging the public than "any of the museums we have yet established" because it "display[ed] its most attractive and interesting objects" in a way that was commensurate with "the knowledge and needs of its patrons."[47]

The striking convergence in exhibition design notwithstanding, museums policed the cultural boundary that separated them from commercial institutions such as the department store with assiduous care. Well into the twentieth century, the New York museum refused to charge a fee for admission, proudly declaring itself to be "Always Free."[48] Similarly, museum shops at the time consisted of no more than a small desk in the lobby that sold a few postcards or books, usually with just a modest markup.[49] Even the notion that exhibits should cater to the tastes of a popular audience sparked controversy among nineteenth-century traditionalists such as Elliot Coues, who felt that visually arresting displays were "entirely out of place in a collection of any scientific pretensions, or designed for popular instruction."[50] Eye-catching displays also rankled twentieth-century curators such as George Dorsey, who excoriated the New York museum for inviting confusion and spreading misinformation in its anthropology halls, while a zoologist from Dresden accused curators in Chicago of having forsaken "the scientific foundation that a scholarly museum must under all circumstances maintain."[51] Engaging and lifelike displays may have been what the public wanted, but many curators continued to insist on educating and uplifting visitors rather than reinforcing what they already knew by catering to their existing desires.[52]

During the winter of 1910, these controversies grew so inflamed that they led to the ouster of Hermon Carey Bumpus from his post as director of the American Museum of Natural History in New York. One curator denounced his exhibit designs as "childishly pictorial" and lacking any "adequate scientific basis," and another described them as "fundamentally wrong" and motivated by a principle of "picturesqueness" more so than by a reverence for the truth.[53] Even Osborn, who praised Bumpus as "ingenious and very clever in devising exhibitions," sniped at his "tendency to cheap and showy methods of illustration," which caused the museum to be "severely criticized by scientific men."[54] As might be expected, newspapers had a field day with the affair, pitching it as a bloody battle between a director who "sought to make the museum appeal to the public at large" and curators who preferred to "coop" themselves up in the office and "write scientific treatises."[55] Whereas "Bumpus was for 'popularizing' the various exhibits," others felt the museum was "first of all a scientific institution, and as such should not be 'popularized' at the expense of scientific usefulness."[56] If we can look past the hyperbole of these newspaper accounts, the "Bumpus affair" clearly spoke to a far-reaching debate about the museum's institutional identity, one that posed foundational questions about to whom it was primarily responsible and how it should allocate limited resources.

Bumpus may have been done in by the impression that he was preparing shows along "department store lines," as yet another newspaper alleged, but his dismissal in 1911 did not mean the New York museum was about to abandon its policy of appealing to people's taste for excitement and drama entirely.[57] Indeed, the ideas about reaching a broad audience that Bumpus so ardently championed would continue to inform much of the work that was done there, even after he had departed. What emerged was a delicate strategy of defending against the charge of frivolity and commercialism by drawing a fine line between an exhibit's content and form, distinguishing what was shown from the way it was put on display. Whereas the department store would serve as an inspiration regarding issues of form, curators pledged to remain faithful to the high standards of science when it came to content. This is what Oliver Farrington from the Field Museum in Chicago sought to express when he explained that, no matter how "desirable" the "introduction of the best art into our natural history museums" may be, "it should not usurp the place of science."[58] Rather than allow either to gain the upper hand, museums drew equally, even seamlessly, on both, producing exhibits whose goal was to delight the eye and enrich the mind all at once.

Museum curators expressed faith that, if they were properly calibrated, science and art need not be at odds. Indeed, they could be made to strengthen and reinforce one another. As Osborn explained, "To express the whole of nature the element of beauty must go hand in hand with the element of truth."[59] This was not only because talented artists were vital in making exhibits that would appeal to the aesthetic faculties of visitors, but also because art was required to communicate higher truths about nature. Lucas even went so far as to allow that some degree of "fabrication is a necessity," for without it, a museum exhibit would always fall short of its educational goal. Wild plants and animals simply did not come into close physical contact often enough to make a compelling display. Hence, it was impossible to help visitors understand the relationships that existed in nature without the freedom that an artistic license bestowed.[60] As a result, museums adopted an exhibition strategy that deliberately sacrificed a measure of realism for the sake of greater comprehension and understanding. But that did not mean anything goes, and Farrington felt compelled to sound a "note of caution" regarding the "tendency to prefer imitation to reality."[61] Illusion and spectacle in the service of truth were certainly warranted, he agreed, but not if they veered into overt misdirection or falsehood. Notably, this did not only hold for compromises made to increase an exhibition's popular appeal. It also prohibited unwarranted speculation and theoretical overreach. "The best books, written by the best scientific men, soon become out of date," Osborn lamented, whereas "a bare fact of nature, simply and clearly displayed . . . will be the same for thousands of years." For that reason, he insisted, the American Museum of Natural History was "scrupulously careful not to present theories or hypotheses, but to present facts, with only a sufficient amount of opinion to make them intelligible to the visitor."[62] In practice, this entailed a rigorous preference for observation over generalization, which consistently manifested itself as an obsession to present visitors with material evidence direct from the field.

These tensions—between illusion and reality, artifice and authenticity, science and spectacle—were particularly acute when it came to the exhibition of fossilized dinosaurs. Dinosaurs were attractive display objects because their great size and outlandish appearance made them especially memorable and sensational. But as we saw in previous chapters, it hardly sufficed to put a jumble of fossil bones on display. To make dinosaurs comprehensible and appealing to visitors, their remains were assembled into freestanding skeletons that resembled live animals. This posed a significant challenge, however, because doing so involved relying on a great deal of conjectural knowledge about the anatomy and behavior of long-extinct creatures to whom paleontologists had no direct observational access. Dinosaurs at the museum embodied a profound speculation, prompting visitors to cast their imaginations across a vast temporal chasm and enter a world that had entirely disappeared. This threatened to undermine the exhibit's status as a scientific display, inviting the conclusion that dinosaurs were no more real than Doyle's Cottingley Fairies or O'Brien's stop-motion animations. To combat this impression, curators consistently emphasized the exhibit's indexical features to assure viewers that such displays offered a direct and unbroken link to prehistory.

When curators in New York unveiled their first freestanding *Brontosaurus* display in 1905, the anxiety of creating an unreliable exhibit ran deep. In part, this was because previous attempts to reconstruct dinosaurs had not aged well, coming to be seen as woefully inadequate. The life-size sculptures that Benjamin Waterhouse Hawkins had fabricated for the Crystal Palace in London during the mid-1850s were especially controversial. A curator of natural history at the British Museum in London named John Edward Gray, for example, wrote to a representative of the Crystal Palace Company during the mid-nineteenth century to denounce the "Crowning humbug" of their exhibition, deriding Hawkins for "out Barnuming Barnum himself" by perpetuating his "gross delusion" on an unsuspecting public. "The models are not what they profess to be," he continued, "but merely enormously magnified representations of the present existing animals presumed to be the most nearly allied to the fossils—and often on very slender and sometimes on what is known to be erroneous grounds."[63] Perhaps even worse was the ridicule that paleontologists heaped upon Hawkins during the decades that followed. With the benefit of hindsight, his reconstructions had come to appear utterly ludicrous in the eyes of many a scientist. As William Berryman Scott remarked about a *Hadrosaurus* skeleton that still stood at the Philadelphia Academy of Natural Sciences, "there is a lot of conjectural restoration and the head, in fact, as put up by the late Waterhouse Hawkins, an English scientific man, not so very scientific, but known more for being artistic—the head is entirely grotesque."[64]

For all their talk of merging beauty with truth, then, museum curators had reason to fear that their paleontology exhibits would be judged as unsound and defective. Writing about the *Brontosaurus* in a leaflet guide intended for visitors, William Diller Matthew was remarkably candid about the risks associated with putting a dinosaur on display. The "proper articulating of the bones and the posing of the limbs" were both "difficult problems," Matthew admitted, given that sauropod dinosaurs "disappeared from the earth" so long ago. But that did not mean the end result was a mere flight of fancy, and Matthew was careful to point out that extraordinary measures were taken to mount the fossils correctly. He especially stressed the decision to assemble the dinosaur's limbs in a provisional pose for experimental purposes (Figure 6.5).

Figure 6.5. Fossil preparators from the American Museum of Natural History in New York posing the forelimbs of *Brontosaurus* in 1904.

That way, Matthew explained, the "action and play" of each muscle could "be studied, and the bones adjusted until the proper and mechanically correct pose had been reached."[65] The message that Matthew sought to project was clear: Having been determined through a prolonged, detailed, and physical engagement with the fossils themselves, the pose of the *Brontosaurus* was much more than mere speculation.

No matter how much care curators took to assemble their dinosaurs correctly, however, intense disagreements about the functional anatomy of these creatures continued to divide the community. The protracted debate over whether sauropod dinosaurs such as *Brontosaurus* walked upright on columnar legs or dragged themselves along on their bellies in the manner of a lizard or crocodile that we encountered in Chapter 3 is just one example, albeit a notable one. A whole host of other controversies erupted as well. These often arose from the fact that even the most complete fossils usually lacked some of the bones required to mount a freestanding display. Unlike art museums, however, which began to exhibit unrestored marble sculptures from classical antiquity at the time, curators in natural history museums felt that their educational mission prevented these institutions from exhibiting partial and incomplete skeletons. Hence, they usually filled in the missing portions with simulacra of various kinds, usually made out of plaster. The *Brontosaurus* again serves as a good case in point, because it required many elements (including a scapula, humerus, radius, and ulna) that had been cast from other specimens to appear complete.[66]

Perhaps most damning of all was that *Brontosaurus* was missing its skull. This forced Osborn's chief preparator, Adam Hermann, to model a replacement in plaster (Figure 6.6). However, since nothing that approached a reasonably complete skull for this creature had ever been found, Hermann resorted to the use of a close relative—*Morosaurus*—as an anatomical model to guide his hand. That decision came into question a few years later, when paleontologists from the Carnegie Museum in Pittsburgh discovered the remains of another *Brontosaurus* that included a reasonably complete skull. Upon examination, it bore a much closer resemblance to the slender *Diplodocus* skull than to the robust head of *Morosaurus,* which prompted the Carnegie Museum's director, William J. Holland, to publish an article that ridiculed the New York specimen. "The problem is naturally perplexing, and in certain aspects amusing," Holland quipped, using the occasion to take a jab at his institutional

Figure 6.6. Adam Hermann modeling the *Brontosaurus* skull, ca. 1904.

rivals in New York. “My good friend, Dr. Osborn, has in a bantering mood ‘dared’ me to mount the head” on the “atlas” of his own specimen, Holland recounted, “which it fits.”[67] However, because the long, tapering necks of sauropod dinosaurs ensured that their skulls would be found at some distance from the rest of the skeleton, it remained difficult to prove the association conclusively. For that reason, it was not until 1978, when David Berman and John McIntosh published an influential analysis in Holland’s favor, that the New York museum finally acquiesced and switched out the skull of its *Brontosaurus* display.[68]

The authenticity of original fossils was called into question as well. Mounted dinosaurs were effectively chimeras, often cobbled together from fragmentary pieces belonging to separate specimens, many of which had been collected in different quarries. For example, no fewer than three different individuals besides the principal Nine Mile Quarry specimen were called on to supply material for the American Museum of Natural History's *Brontosaurus* display. Sometimes, curators even combined fossils from individuals that did not belong to the same taxonomic group, effectively inventing a new kind of organism in the process. In 1907, for example, the New York museum used portions of a *Dimetrodon* fossil to round out the exhibit of *Naosaurus*, a Permian fin-backed lizard. Osborn defended this practice by stressing that although "the assemblage [was] largely composite," it nonetheless served to provide visitors with "an adequate conception of the unique and imposing characters of these great extinct forms."[69] But others were not convinced. Matthew privately described the museum's Permian collection as a "grave disappointment," complaining that "I am extremely reluctant to combine different individuals in this way." In public, however, Matthew stood by the museum, and it is notable that an announcement of the new specimen written for a popular audience did not mention its composite construction.[70]

The official position of the New York museum was to justify these sleights of hand on the grounds that "filling in missing parts is essential if the skeleton is really to look like the framework of a living animal and clearly to convey the impression of a living organism." At the same time, it was recognized that the "public wants to see originals, in natural history as in art."[71] This all but ensured that the scientific community would become bitterly divided over the role restoration should play in the articulation of fossil dinosaurs. Discussion turned on the questions of whether, to what extent, and the purpose for which museum displays could legitimately mislead the public. Initially, the New York museum labeled bones that were modeled in plaster as such, either by outlining the restored portions in red or by marking the entire bone with an X.[72] Before long, however, it adopted a more subtle strategy, and an internal report indicates that despite being encouraged to "differentiate restoration enough so that no doubt arises under close examination," exhibit makers were instructed to use plaster of a "harmonizing color that will not leap to the eye and give a patchy appearance."[73] As might be expected, however, exhibits that were designed to fool the casual visitor came in for withering critique. Incensed at the failure to distinguish clearly between real and fake bone, the paleontologist Oliver Perry Hay insisted that "beauty ought not to be secured at the expense of truth," arguing that any such "mischievous" attempts "to hide the restoring materials" would only result in "suspicion" being "thrown on the whole exhibit." Even more egregious, Hay judged, was the exhibition of composite skeletons, which he described as veritable "monsters." All such attempts to deceive the public would only backfire in the end, Hay concluded, and visitors could hardly be blamed if they came to view paleontological displays as no more than "products of the unchastened scientific imagination."[74]

Ironically, by far the most vociferous attack was directed at Marsh, who had refused to assemble his dinosaur fossils into a freestanding display decades earlier because he feared that doing so would require him to engage in unwarranted speculation. In 1890 Erwin Barbour, one of Marsh's former assistants, accused him of routinely directing his laboratory staff to "imitate the color and texture" of authentic specimens when creating plaster cast replicas. This, he alleged, made it all but impossible for viewers to discern which aspects of a display were authentic and which ones were fake, leaving the public "in doubt as to what is real, [and] what [is] conjectural." Indeed, not even the trained eye of an expert paleontologist could be trusted "to distinguish between the rusty, frost-cracked, weather-beaten, moss and lichen effects, craftily wrought in plaster, and the conditions wrought by time on the specimens themselves." Hence, despite Marsh's own misgivings about the proliferation of erroneous dinosaur mounts, Barbour was sure that his specimens "must sooner or later stand as monuments of reproach to the man who has so far deceived the world."[75]

Although they constituted an extremely rewarding object for public display, then, dinosaurs posed a substantial risk to the museum's credibility as an authoritative institution of science. This dilemma prompted curators such as Osborn to develop a complex and sophisticated exhibition strategy that capitalized on the popular appeal of these creatures while mitigating the risks of mounting an overly speculative display. They did so by incorporating actual pieces of fossilized bone into a dense, mixed-media installation designed to provide visitors with imaginative access to the deep past while simultaneously conveying its status as a source of reliable knowledge. In effect, because mounted dinosaurs were so far removed from the world of the visitor, curators went out of their way to design fossil exhibits that explicitly advertised their indexicality by consistently drawing the attention of audiences to the material connections that united real fossil bones to extinct creatures from prehistory.

Figure 6.7. Duckbilled dinosaur group display at the American Museum of Natural History in New York, with the *Brontosaurus* and *Allosaurus* displays in the background, 1927.

In the years that followed the unveiling of its first *Brontosaurus* mount in 1905, the New York museum followed a policy of adopting the lessons learned from the habitat diorama in its dinosaur hall. As early as 1907, the museum's Department of Vertebrate Paleontology was already boasting that its "methods of installation" were "progressing along the same lines of development as may be seen in recent zoological exhibits," and Lucas praised the museum's efforts to bring the "group idea" of the habitat diorama into "the dim and distant past."[76] Beyond simply articulating a skeleton as a freestanding mount, this involved combining different specimens to create an ecological tableau. Lucas was particularly impressed by two exhibits that we already encountered in Chapter 5. The first was an *Allosaurus* specimen that was shown predating on a section of *Brontosaurus* tail, and the second featured a number of giant ground sloths cooperating with one another to tear down a tree so they could eat its uppermost leaves. Another notable example (Figure 6.7) was a pair of duckbilled dinosaurs that were mounted with remnants of their ecology in the display, incorporating fossil "shells, leaves, and other plants of this period" into the "matrix" near the animal's feet.[77] One of these two was posed with its head low to the ground, giving the impression that it was feeding, whereas the other held its head high in the air. In a guide leaflet, Matthew encouraged visitors to imagine that it sensed danger off in the distance, explaining that it may have been "startled by the approach of a carnivorous dinosaur, *Tyrannosaurus,* their enemy," and lifted its head "to look over the surrounding plants and determine the direction from which it is coming."[78] In addition to reinforcing the conventional image of dinosaurs as ruthless and bloody tyrants, Matthew's description hinted at the existence of a whole world beyond what was depicted in the actual mount, creating a more engrossing illusion than a mere skeleton could achieve on its own. The museum's annual report therefore praised the improved style of mounting as a significant "step forward in methods of exhibiting fossil vertebrates," primarily because it was seen to add "greatly to the realistic effect."[79]

One major difference between the traditional habitat diorama and the paleontological group was that the former usually featured a painted backdrop, whereas the latter did not. But the museum did hire artists to create elaborate murals and paintings that heightened the immersive experience. Many of these were executed by Charles R. Knight, the period's foremost illustrator of prehistory, and they usually mirrored in detail the poses and arrangement of the skeletons alongside which they were shown. But Knight also gave flesh and blood to his dinosaurs, and he represented them in full, splendid color. According to Knight, this allowed visitors to "look from one to the other and thereby glean perhaps some idea of the life which flourished so vigorously on our planet in ages past," meaning his illustrations functioned as "an explanation of the skeletons themselves."[80] Much as the leaflet guide quoted earlier allowed curators such as Matthew to speak directly to visitors, Knight's paintings thus offered a way for paleontologists to project their speculative theories about the anatomy and behavior of long extinct creatures onto their fossil remains.

Knight's work excelled at its intended function, which, in his own words, was "to put *life* into the dead bones."[81] But his professional identity as a visual artist meant that Knight suffered from a deficit in scientific authority. To overcome this problem, the museum consistently characterized his paintings as the result of an intense collaboration between art and science. Newspaper and magazine reproductions of his paintings, for example, always included a caption that indicated the original had been executed under Osborn's direct supervision. Knight made sure to convey this message as well, especially singling out Matthew as an important "consultant and adviser on matters of pose and difficult bone structure."[82] And when giving a radio interview, he declared adamantly that "no hokus-pokus" was required to bring long-extinct reptiles back to life. "It is all done along strictly scientific lines and only after long consultation with . . . men who are experts in this particular field," he said.[83] For his part, Osborn never trusted Knight to produce illustrations that could pass scientific muster alone, arguing, "It is impossible for him to paint extinct animals without guidance by myself or some other lifelong student of the habits and structure of extinct animals."[84] Osborn therefore developed an efficient workflow designed to ensure that curators would retain absolute control over Knight's creative output, even insisting that Knight sign a contract that required him to defer to Osborn on all matters of technical interpretation.[85]

Like the habitat diorama, dinosaur displays evolved into elaborate mixed-media installations designed to transport the visitor out of the museum and into the depths of prehistory. Unlike a zoological specimen, however, mounted dinosaurs could not stand on their own. Because dinosaur bones are too heavy and fragile to support themselves under their own weight, a metal scaffold was required to create a freestanding display. According to Hermann, the usual practice was to fashion a metal rod as a brace to support the animal's backbone, with pins extending outward that could either be bored into the center of each vertebra or split into an exterior clamp. Limbs and other body parts, such as the hip, shoulder, and skull, could then be attached individually to the central backbone support. In every case, the primary criterion for success was that any and all supporting materials—which included clay, plaster, and wire mesh, as well as the main backbone support—be as inconspicuous as possible, lest they draw attention to themselves and away from the fossils. As Hermann explained, the duckbill group discussed earlier was exemplary because it had "the least noticeable support of any we have had so far."[86] Similarly, after the Carnegie Museum of Natural History mounted a *Diplodocus* specimen for display, Holland praised his staff for making an

Figure 6.8. A fossil preparator welding the backbone support for *Diplodocus* at the Carnegie Museum of Natural History, ca. 1904.

exhibit that combined "absolute safety together with a lifelike pose." He singled out his chief preparator, Arthur S. Coggeshall, for the "great ingenuity" with which he had "devised a system of steel supports conforming themselves to all the sinuosities and rugosities of the under surface of the vertebrae" (Figure 6.8). This made it possible to fashion a support system that was "so reduced in size [as not to] offend the eye by a display of 'open plumbing work,' which has heretofore been very much in evidence in the mounting of dinosaurian remains."[87] Hermann concurred, writing that he considered the Carnegie Museum's technique a "neat style of mounting" that succeeded admirably in its goal of hiding the metal support structure whose presence threatened to distract from the exhibit's capacity to convey the lifelike impression of a real animal.[88] Museum staff also worried that any visible nonfossil material in the display would divert attention from its indexical features, thus highlighting the degree of human creativity required to bring dinosaurs back to life.

Knight's characterization of his paintings as an explanation of fossil exhibits points to another, more figurative sense in which dinosaurs could not stand on their own, one that derived from the fact that a mounted skeleton's meaning was hardly self-evident. The fossilized skull of a dinosaur simply did not have the expressive capacity to indicate that it was on the lookout for predators, as Matthew suggested the duckbill exhibit was intended to show. Thus, whereas the painted backdrop of a traditional habitat diorama primarily served to create the illusion of distance and space, Knight's explanatory paintings were intended to help visitors understand the exhibit by providing additional details about the ecology and behavior of these alien creatures. But whereas the flexibility of Knight's medium allowed his paintings to excel at this task, it also meant they were only as trustworthy as the person responsible for their execution, which is why the museum took so much care to advertise the involvement of trained scientists in his creative process. It is also why Knight's paintings so closely resembled the pose of assembled skeletons. In effect, whereas his artworks were intended to help explain the fossils on exhibition, the skeletons served to ground his artworks by offering a material link to the actual past. For this reason, too, guide leaflets functioned as a crucial element of the exhibit. Because they brought curators out from behind the scenes and into the public gallery, these documents lent further authority to the museum's elaborate paleontological group displays. As we have seen, though, controversy about even relatively straightforward aspects of dinosaur biology existed within the curatorial ranks. An exhibit's credibility could not, therefore, rest entirely with scientists and their speculative theories. The mark of truth had to reside in the fossils themselves.

Guide leaflets and other didactic materials were enlisted to fulfill a second function besides broadcasting the voices of respected scientists into the exhibition hall. Whenever possible, they also directed the attention of visitors to the indexical features of a display. Describing the alert, upright pose of a duckbill, for example, one guide instructed viewers to imagine that "the erect member of the . . . group had already had unpleasant experiences with hostile beasts, for a bone of its left foot bears three sharp gashes which were made by the teeth of some carnivorous dinosaur."[89] Similarly, a descriptive label that was made to accompany an *Allosaurus* predating on a section of *Brontosaurus* tail explicitly instructed visitors to "note that the spines of the tail vertebra are scored and bitten off and somewhat torn apart—as they were found in the quarry." The label went on to inform them that "several broken teeth of *Allosaurus*" were discovered right next to the *Brontosaurus* tail vertebrae out in the field. Finally, it also drew their attention to how well the "marks on the bones correspond with the spacing of the teeth in the jaws of this animal," all of which served as material proof that "this *Brontosaurus* was devoured by *Allosaurus*."[90]

To endow highly imaginative paleontological group displays with a sense of tactile reality, museums consistently emphasized their indexical qualities. Fossils were prized as a direct, physical link that united the world of museum visitors with a distant past populated by strange and extinct reptiles. But as we have seen, a whole series of curatorial interventions were required to assemble a dinosaur, which threatened to undermine its claim of indexicality. Thus, despite Osborn's insistence that his museum was "scrupulously careful not to present theories or hypotheses," its most prized paleontological displays could hardly be said to be free of conjecture. If anything, mounted dinosaurs better illustrated Doyle's claim that "you can fake a bone as easily as you can a photograph."[91] Indeed, as the history of photography and cinematography showed, the fact that a representation constituted a physical trace hardly guaranteed that it was accurate. Given the immense popularity of trick films and other genres of optical illusion, visitors to the museum would have been acutely aware of the ease with which even highly indexical representations could be made to deceive the mind and mislead the eye. For that reason, it did not suffice for the museum to emphasize the material connection between a fossil display and the distant past. Curators had to go further and restore some of the missing links in the causal chain that tied their exhibits to flesh-and-blood creatures from deep time.

The American Museum of Natural History sought to establish a seamless set of material connections between its elaborate dinosaur displays and the animals they depicted. To do so, several additional elements were introduced into its dinosaur hall, each of which further anchored the imaginative experience of visitors in the solid bedrock of prehistory. Two are especially worth singling out. The first was a series of large-format transparencies made from photographic negatives that were taken out in the field. These were then hand colored and hung in the windows that lined both sides of the dinosaur hall, not unlike the use of stained glass in a European cathedral. These photographs offered visitors a means to visualize an early stage in the elaborate scientific process that eventually resulted in a fully assembled specimen. A good example is Figure 2.3, which shows the excavation of large sauropod dinosaur bones at Bone Cabin Quarry in Wyoming during the late 1890s. Although the specimen depicted therein was fragmentary at best, visitors could be expected to notice its partial resemblance to the complete *Brontosaurus* display. In addition, museums also displayed so-called panel mounts, which were constructed by embedding an assembled dinosaur skeleton within a plaster backing. Because the plaster was dyed in a color that simulated the appearance of rock matrix, panel mounts formed a kind of relief sculpture that depicted a half-excavated skeleton. This impression was reinforced further by panel mounts such as the one shown in Figure 6.9, wherein curators articulated a specimen in its characteristic "death pose." For many dinosaurs, as for birds, this involved posing the specimen with its tail fully extended and its head curved back toward the spine, which simulated the effect of strong ligaments in the neck contracting as they desiccated prior to fossilization.[92]

Whereas paleontological group displays were designed to invoke the imaginative experience of seeing a live dinosaur in the flesh, panel mounts such as the one pictured in Figure 6.9 mimicked the discovery of their fossil remains in the field. Similarly, while Knight's murals helped paleontological groups come to life in the imaginations of visitors, photographic transparencies depicting an actual dig site helped them to understand the meaning of a panel display by giving them access to the fossil fields of the American West. In effect, museum exhibits offered a narrative account of their own production history, bringing the moment of fossilization and the point of discovery into the exhibition hall. But just as a freestanding group embodied a great

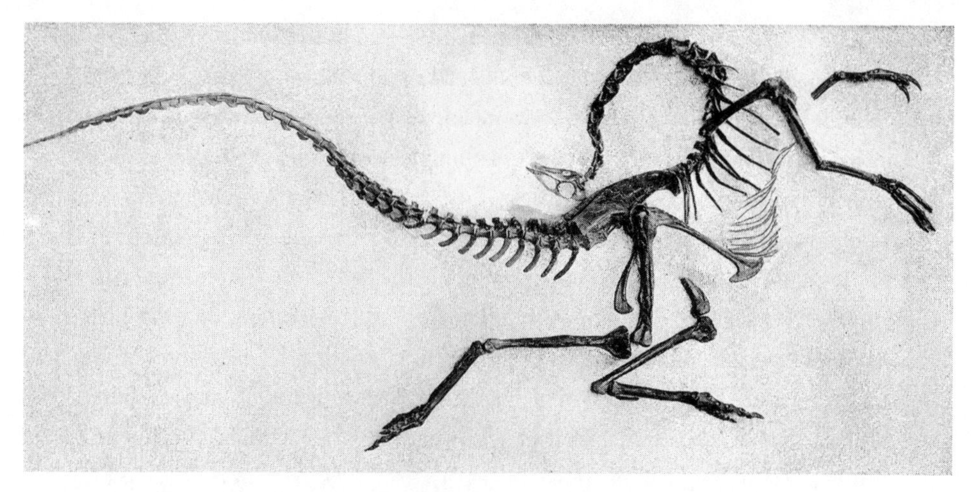

Figure 6.9. A panel mount of an *Ornithomimus* dinosaur shown in its characteristic "death pose," photographed by H. S. Rice and Irving Dutcher in July 1927.

deal of conjecture, panel mounts were also highly idealized. As anyone who has spent time collecting fossils will know, the likelihood of unearthing a dinosaur that was so complete, so beautifully preserved, and so exquisitely articulated as the one pictured in Figure 6.9 is vanishingly small. Still, the overall effect was to endow the paleontology hall as a whole with the credibility of an indexical representation, making a material argument that dinosaurs in the field really looked like their mounted skeletons on display, even before they were collected and reassembled. Thus, whereas a freestanding skeleton relied on the indexicality of each fossil to make the whole seem convincing and real, expedition photographs and panel mounts made an implicit, visual claim that the relationship *between* individual bones—that is, the articulation of a fossilized skeleton—was indexical too.

Conclusion

The obsession with establishing a material link between the past and the present that characterized the American Museum of Natural History's dinosaur hall did not wane in the decades that followed the initial decision of curators to begin mounting paleontological group displays. If anything, it only became more pronounced over time. By the mid-twentieth century, the museum had

Figure 6.10. A young visitor examining *Stegosaurus* at the American Museum of Natural History in New York, with the *Brontosaurus* skull on the upper left and tools of the paleontological trade below, 1959.

incorporated its *Brontosaurus* skeleton into an even larger assemblage that also included a freestanding *Stegosaurus*. In addition, there were two dinosaur limb bones displayed in mid-excavation. Not only were these made to look as though they remained half-buried, but they were flanked on either side by pickaxes, shovels, burlap, shellac, and other tools of the paleontologist's trade. The *Brontosaurus* and *Stegosaurus* specimens were therefore incorporated into a new kind of group display, one that combined lifelike skeletons with an imaginative rendering of the dig site in which they were found (Figure 6.10). This made the new exhibition even more self-referential than previous groups, suggesting a narrative account of its own production history that once again stressed the material links that united a highly imaginative exhibit with real flesh and blood creatures that no longer lived.

Insofar as it promiscuously crisscrossed the temporal boundary that separated live dinosaurs from the people who excavated their fossil remains, paleontological group displays resembled a trick film that combined live action with stop-motion animation. But there was a crucial difference between them as well. Whereas the visual spectacle of *The Lost World* depended on the audience's awareness that it was being playfully manipulated, the dinosaur displays that we have encountered here sought to convince visitors of exactly the opposite. Museums deployed the quality of indexicality to vouchsafe the trustworthiness of their own exhibits, while the cinema cleverly manipulated that very same quality to engage the investigative faculties of viewers, heightening rather than suspending their disbelief. Differently put, while the indexicality of a photograph made the cunning deception that was required to film dinosaurs all the more wonderful, the very same quality served to downplay the fact that a museum display necessarily constituted an immersive illusion. Perhaps most remarkable of all is that, in both cases, viewers were granted strategic access behind the scenes. To achieve the desired effect, museums began to develop an exhibition strategy in which dinosaur displays advanced a fictional narrative about how they had come into being, much like Windsor McCay had done in his *Gertie* cartoon. However, whereas *Gertie* narrated its own production history to highlight the skill and the creativity that were required to create such an amazing illusion, paleontological group displays did so to convince viewers that an unbroken chain of causal connections linked the museum's exhibits to real dinosaurs as they lived and they breathed.

The stark contrast between the way dinosaurs were displayed in these two exhibition spaces demonstrates their divergent institutional goals. Whereas the cinema was a commercial amusement venue that sought to attract a large number of paying customers, the museum was a nonprofit institution dedicated to the production and dissemination of reliable knowledge. In addition, the different exhibition strategies that obtained in these spaces also help to lay bare what might be described as the ontological indeterminacy of the index. As cinematic depictions of dinosaurs show, the photograph's indexical qualities could be used to produce a variety of experiences beyond epistemic assent. Because it was obvious that a stop-motion animation of dinosaurs constituted an index of the impossible, the medium's direct causal connection to its subject could just as well serve to show off the filmmaker's technical wizardry. Finally, the assemblage of dinosaur fossils into a visually arresting and physically imposing, three-dimensional museum display demonstrates that indexicality need not be a feature of individual objects alone. To scale up the indexicality of an individual fossil so that it might cover the entire dinosaur hall, the New York museum constructed an array of exhibits that sought to fill in the missing links of a causal chain connecting real dinosaurs from the deep past to a museum display in the present. In this way, the evidentiary status of individual fossils was buttressed by their incorporation into a much larger and denser mixed-media exhibition. Not only were dinosaurs built out of numerous bits and pieces, but their claim to indexicality was laboriously assembled as well.[93]

Conclusion

Feathered Dragons

The stock market crash that touched off the Great Depression in the autumn of 1929 effectively brought America's Long Gilded Age to a close. Around the same time, enthusiasm for dinosaur paleontology also began to recede among professional paleontologists and the broader public alike. In more recent decades, however, the scientific study of and popular fascination with dinosaurs have experienced a resurgence, and we are often said to be witnessing a dinosaur renaissance. Much of the recent excitement stems from new theories about the evolutionary relationship between extinct dinosaurs and modern birds. Many paleontologists now believe that birds are deeply embedded within the dinosaur family tree. In much the same way that humans are primates, this would mean that birds are modern-day dinosaurs. It also means that dinosaurs did not go extinct. In a remarkable case of backtracking causation, so-called nonavian dinosaurs have therefore been almost entirely reimagined. Contemporary paleontologists now envision them not as drab, sluggish, and solitary creatures who roamed the deep past but as active and social animals that were often covered with colorful feathers. Moreover, whereas the overwhelming majority of dinosaurs that fueled both the public and scientific imagination during the Long Gilded Age hailed from the American West, the most spectacular fossils today come from northeastern China. Still, despite all of these differences, some of the most striking features that characterized dinosaur paleontology in the late nineteenth- and early twentieth-century United States continue to manifest themselves in today's globalized world.

By the end of the Long Gilded Age, dinosaur paleontology had reached a high-water mark, having become a pervasive feature of American popular culture. To cite just a single example, the 1933/1934 World's Fair in Chicago featured not one but two attractions dedicated entirely to these creatures: in addition to the Messmore & Damon exhibit we already encountered in Chapter 6, the Sinclair Oil Company also sponsored an impressive dinosaur display. Much like Messmore & Damon's World a Million Years Ago, it too contained animatronic sculptures, including a larger-than-life *Brontosaurus* that still features as Sinclair's corporate logo. However, whereas the World a Million Years Ago was often derided as excessively commercial and lowbrow, Sinclair Oil was careful to consult with paleontologists from the American Museum of Natural History in New York to ensure that its exhibit could not be so easily dismissed. To help cultivate a visibly close working relationship with vertebrate paleontologists, the petroleum company went so far as to sponsor a number of field expeditions. At one point, Sinclair even donated an airplane that curators from the New York museum used to conduct aerial surveys and spot promising outcrops for closer examination. In exchange, paleontologists agreed to alert the petroleum company if they happened to see a promising site for a test well. They also consulted on the production of promotional materials, such as an advertising pamphlet that explained, "Sinclair uses dinosaurs to symbolize the vast age of the crude oils which are refined into Sinclair Motor Oil" because "it was during the lifetime of such prehistoric creatures that nature was mellowing and filtering *under the earth* the crude oils which are refined into Sinclair Motor Oils."[1]

The period saw important new scientific discoveries, too. Perhaps most notable was a large and ambitious expedition the American Museum of Natural History sent into the Gobi Desert during the 1920s. Led by Roy Chapman Andrews—a handsome and charismatic, if also conceited and arrogant, adventurer who is often likened to Indiana Jones—the Central Asiatic Expedition was organized to test a racially motivated theory promulgated by Henry Fairfield Osborn, who argued that a large number of the world's most charismatic megafauna, including humans, first evolved in Asia rather than Africa. After searching for nearly a decade, however, Andrews was forced to abandon his quest to find physical evidence to prove Osborn's theory. Still, the expedition made headlines, most notably after a fossil preparator named George

Olsen found a nest of beautifully preserved dinosaur eggs on July 13, 1923. Along with most of his crew, Andrews initially doubted the authenticity of Olsen's discovery, suspecting that the curious objects he had unearthed were no more than geological accretions whose shape resembled an egg. But as more and more of these specimens were found, it became harder to doubt their legitimacy. Olsen's discovery turned out to be a momentous occasion, offering material proof that dinosaurs did not bear live young.[2]

If dinosaur paleontology reached a high-water mark during the 1920s, it entered a period of decline in the decades that followed. The economic hardships touched off by the 1929 stock market crash took a heavy toll on the sort of conspicuous consumption that had become commonplace during the Long Gilded Age, as did the passage of a constitutional amendment that led to the creation of a permanent graduated income tax. Even among those who continued to prosper, displays of extravagant spending seemed out of place during a time of severe economic downturn. The money that was needed to sustain large, philanthropically funded museums of natural history thus became harder to find. By 1941 things had gotten so bad that the New York museum commissioned an external audit in hopes of identifying redundancies, streamlining operations, and reducing costs. The audit concluded that "despite its brilliant past, there are good reasons for the conclusion that all is not well with this institution."[3] Without a community of wealthy philanthropists trying to outdo one another, the resources to maintain an ambitious program in dinosaur paleontology simply dried up. Something similar happened to more popular amusements as well. In the age of the cinema and other forms of mass popular spectacle, natural history came to be seen as a quaint and old-fashioned leisure pursuit. Even Messmore & Damon was forced to admit that, after two years at the fair, the World a Million Years Ago had not made enough money to justify the expense of continued operations.[4]

At the same time, an institutional transformation was taking place. The life sciences increasingly moved into the laboratory as experimental practices that stressed the importance of rigorous hypothesis testing came to overshadow exploratory research based on the collection, description, and comparison of specimens. While the growing importance of ecology during this period ensured that the outdoor field site would not cease to be a legitimate place to conduct scientific research, museum-based natural history came to seem like a throwback to older and less sophisticated modes of knowledge production.[5] As a result, the public persona of the vertebrate paleontologist was increasingly regarded as something of a historical oddity or museum curiosity in its own right. This was most famously illustrated by the character David Huxley, a mild-mannered scientist played by Cary Grant in the 1938 screwball comedy *Bringing Up Baby*. Usually the butt of somebody else's joke, Huxley was perennially frustrated in his attempts to track down enough philanthropic support to fulfill his dream of building a dinosaur. It was not until the last decades of the twentieth century that it again became fashionable to make comparative claims across large specimen collections. But now specimens were usually reframed as tissue samples and collections as databases, and the comparison was most likely performed by a computer.[6]

The theory of evolution underwent a dramatic change also. Often described as the "modern synthesis," this intellectual transformation further compounded the declining fortunes of vertebrate paleontology. During the mid-twentieth century, biologists largely abandoned the orthogenetic ideas championed by Osborn and Edward Drinker Cope. Instead, they embraced natural selection as the primary mechanism of evolutionary change. Supplemented by Mendelian genetics and enriched by new theoretical models, the process of evolution was recast as a numbers game, with chance variation playing a far more important role than was previously believed. Consequently, evolution came to be seen as something that happens to populations rather than individuals, the result of a complex interplay of many contributing factors (including not only selection but also migration, assortative mating, and genetic drift, among others) whose relevance could only be assessed statistically. But the infamous poverty of the fossil record made it exceedingly difficult for vertebrate paleontologists to collect enough specimens to engage in these types of analyses. Even George Gaylord Simpson's well-regarded 1944 book *Tempo and Mode in Evolution* primarily set out to demonstrate that paleontology was consistent with the latest findings in population genetics, not to offer novel insights or new theories.[7] Thus, rather than being in the vanguard of evolutionary theorizing, as Cope and Osborn had been, paleontologists now found themselves dutifully supplying material evidence to bolster the theory of common descent. As the evolutionary biologist John Maynard Smith quipped, "The attitude of population geneticists to any paleontologist rash enough to offer a contribution to evolutionary theory has been to tell him to go away and find another fossil, and not to bother the grownups."[8]

Despite these far-reaching changes, paleontologists did not find themselves out of the spotlight for long, and even Smith felt compelled to welcome them back to the "high table" of evolutionary theorizing after Stephen Jay Gould was invited to deliver the 1984 Tanner Lectures at Harvard.[9] Some of the late twentieth century's most ambitious challenges to the modern synthesis emerged from the study of fossils. These included the discovery of periodic mass extinction events as well as the controversial theory of punctuated equilibria, which held that entire species and perhaps even higher taxa might be subject to selection. These unorthodox ideas aroused heated debate, and they helped reinvigorate the paleontological community, many of whose members rebranded their field as "paleobiology" to distance themselves from their intellectual and institutional forebears.[10]

In addition to the posing of big questions, what primarily united the new generation of paleobiologists was their enthusiasm for sophisticated quantitative methods. In effect, Gould and his colleagues recaptured the theoretical confidence of their late nineteenth-century predecessors by embracing the mathematical modeling techniques that were a hallmark of population biology and employing them to study the deep past. However, this meant that the most groundbreaking work done by paleobiologists featured invertebrate fossils—not dinosaurs—because they were preserved in sufficiently large numbers to admit of the statistical treatment that had become an essential component of respectable biological practice. Still, dinosaurs were not entirely left out of the paleobiological revolution. Here too, much of the excitement was due to a renewed willingness to engage in bold theorizing and overt speculation, especially the explosive idea that dinosaurs might be the warm-blooded ancestors of modern birds.[11]

Darwin's well-known disciple, Thomas Henry Huxley, had articulated the idea that dinosaurs might constitute a "missing link" between reptiles and birds as early as 1868.[12] But it did not gain mainstream traction until John Ostrom undertook a systematic examination of early avian fossils and concluded that primitive birds such as *Archaeopteryx* had so much in common with meat-eating dinosaurs such as *Deinonychus* (a close relative of *Velociraptor*) that they must have evolved from them directly.[13] Ostrom's proposal touched off an immense controversy, and it inspired one of his undergraduate students—Robert Bakker—to embark on a crusade to reshape the image of dinosaurs, going so far as to insist that they were most likely warm-blooded. If birds had evolved directly from dinosaurs, and birds were endothermic, he asked, why wouldn't we expect the same of their ancestors? Bakker speculated that dinosaurs must have been far more active than modern reptiles, which helped to explain how they managed to occupy so many ecological niches now populated by mammals.[14] Taken together, these two ideas had enormous consequences. If dinosaurs were warm-blooded and birdlike, then almost everything about the way they appeared in the public imagination was probably wrong. As Bakker put it, the traditional image of dinosaurs "as symbols of obsolescence and hulking inefficiency," as "know-nothing conservatives that plod through miasmic swamps to inevitable extinction," could not have been further from the truth. Right up to the mass-extinction event that demarcated the Tertiary from the Cretaceous period, dinosaurs "were not a senile, moribund group that had played out its evolutionary options," he insisted. "Rather, they were vigorous, still diversifying into new orders and producing a variety of big-brained carnivores with the highest grade of intelligence yet present on land."[15] The late twentieth century thus witnessed these creatures receive a dramatic makeover. Formerly thought to be hulking, stupid, and slow-moving reptiles, they were coming to be regarded as quick-witted social creatures who often hunted in packs and carefully tended their young.

New theories about avian evolution had an enormous impact on the way modern audiences are taught to visualize dinosaurs. Having become something of a celebrity, Bakker was invited to work as a consultant on Steven Spielberg's 1993 film adaptation of Michael Crichton's best-selling novel *Jurassic Park.* Together with the paleontologist Jack Horner, Bakker advised Spielberg on how to make the dinosaurs in his movie as striking as possible without contradicting the latest paleontological findings. The results were widely heralded as a breakthrough in show business spectacle, and the movie showcased the latest theories about dinosaur physiology and behavior in especially dramatic fashion. In one scene, an entire herd of ornithomimid dinosaurs was shown being chased down by a ferocious *T. rex,* and in another, a number of (oversize) *Velociraptors* cooperated with one another to outflank their (human) prey. In addition to the sophisticated computer-generated imaging technology that Spielberg adopted in place of traditional stop-motion techniques, what primarily made *Jurassic Park* so different from predecessors like Willis O'Brian's *Lost World* was the image of active, fast-moving, and birdlike dinosaurs that it introduced to a mass popular audience.

Around the same time, the new image of dinosaurs entered museums as well. During the mid-1980s, several young paleontologists leveraged the power of recently developed computer programs to test Ostrom's phylogenetic theory using much larger data sets than could be analyzed by hand. Comparing dozens if not hundreds of anatomical characteristics across multiple lineages, they repeatedly found modern birds to be nested firmly within the dinosaur family tree.[16] This boosted the community's confidence in Ostrom's hypothesis, with some members even making the argument that, strictly speaking, dinosaurs had not gone extinct.[17] Partially in response to these findings, the American Museum of Natural History embarked on an ambitious construction project to overhaul its paleontology halls that lasted six years and cost nearly $50 million. Completed during the mid-1990s, the revamped exhibit features a complex branching pattern inscribed on the gallery floor, depicting a massive tree diagram that leads all the way from primitive archosaurs (the extinct ancestors of modern crocodiles) to modern birds. The exhibit invites visitors to follow the evolutionary trajectory of these creatures with their own two feet, tracing the phylogenetic history of birds back to dinosaurs and beyond.[18]

The museum also remounted many of its aging displays to better reflect modern theories about dinosaur anatomy and behavior. A particularly good case in point is the museum's *Tyrannosaurus* specimen. Initially posed in the classic Godzilla style, the *T. rex* formerly resembled an oversize kangaroo, with its head high in the air and its tail dragging behind to act as a stabilizing tripod (Figure C.1). Now, the formidable monster is shown with its tail held up off the ground to counterbalance the massive skull far out in front. Other museums did much the same. As Samuel Taylor, director of the Carnegie Museum in Pittsburgh, told the *New York Times,* "We realized our dinosaur hall had become, well, a dinosaur."[19] When the Field Museum of Natural History in Chicago purchased a famous *T. rex* skeleton popularly known as Sue for nearly $9 million during the late 1990s, curators made sure their newest star specimen was mounted to reflect the emerging consensus among paleontol-

Figure C.1. The original *Tyrannosaurus rex* mount at the American Museum of Natural History in New York, photographed by H. S. Rice and Irving Dutcher in July 1927.

ogists (Figure C.2). Laboratory technicians assembled Sue with its massive skull slightly turned to one side, to give the impression of having been caught in mid-stride, skillfully tracking down its unfortunate prey.[20]

The theory that modern birds descended directly from meat-eating dinosaurs became a great deal less theoretical after a series of feathered dinosaurs were discovered in China. The first was found by a rural farmer from western Liaoning Province named Li Yinfang during the mid-1990s. Although he was not a trained paleontologist, Li immediately realized that he was dealing with something special. "I was very surprised because it was very different from the small fish and insect fossils we normally find here," he recalled several years later.[21] Making the most of his good fortune, Li sold one half of the fossil to the Geological Museum of China in Beijing and the other half to the Nanjing

Figure C.2. The *Tyrannosaurus rex* popularly known as Sue on display at the Field Museum of Natural History in Chicago.

Institute of Geology and Paleontology for the equivalent of several hundred US dollars. Although not a large sum for such a significant discovery, this was a windfall for Li, equivalent to more than a year's wages. The specimen had many unique features, but paleontologists were especially excited about the wispy, feathery fringe that ran all along the animal's back, down to the tip of its tail. This was the first time that anyone had found a dinosaur fossil with intact feathers, and it led paleontologists from Beijing to declare that *Archaeopteryx,* a specimen that had been found over one hundred years earlier in the lithographic limestone of Solnhofen, Germany, was now superseded as the oldest and most significant ancestor of modern birds.[22]

Li's dinosaur was christened *Sinosauropteryx prima* (meaning "first Chinese lizard wing"), and it caused a scientific sensation, both inside and outside China. After the discovery had been announced in a Chinese-language geology journal, *Sinosauropteryx* was unveiled to Western paleontologists in a series of carefully choreographed events. First it was shown to Philip J. Currie, a Canadian paleontologist who had been cultivating a close working relationship with Chinese scientists for over a decade. Currie happened to be in Asia that summer, and he was invited to see the specimen at the Geological Museum of China. "The detail of the fossil was fantastic," he later recalled, so much so that "within milliseconds I believed in feathered dinosaurs."[23] Several weeks later, Chen Peiji from the Nanjing Institute of Geology and Paleontology brought several photographs of the specimen to the Society of Vertebrate Paleontology's annual meeting, which took place at the American Museum of Natural History in New York that year. As Chen showed these images to his awestruck colleagues in hallways and corridors between formal lectures and technical presentations, word of the fossil spread like wildfire. Before the meeting had even concluded, the find was prominently featured in the *New York Times*. "Rarely are scientific findings of this possible importance presented so casually," the *Times* reported, quoting Ostrom, who said that seeing Li's fossil left him "in a state of shock."[24] Because *Sinosauropteryx* bore a close family resemblance to therapod dinosaurs, it offered the single best piece of material evidence for Ostrom's theory of avian evolution. One Chinese paleontologist even described feathered dinosaurs from Liaoning as a "smoking gun."[25] As the name *Sinosauropteryx* implied, however, the value of Li's specimen extended beyond its evidentiary status. It had clear cultural and political implications as well, helping to secure China's place in the history of vertebrate paleontology.

In much the same way that the United States came to be seen as a paleontological wonderland during the late nineteenth century because of its abundant and well-preserved fossils, the recent discovery of feathered dinosaurs has led China to inherit that role. Since *Sinosauropteryx prima* made its debut in the mid-1990s, many more feathered dinosaur specimens have been unearthed in Liaoning. Some, such as *Microraptor gui* and *Beipiaosaurus inexpectatus,* are relatively diminutive, considerably smaller than modern birds of prey such as eagles. But others are much more substantial. A close relative of *T. rex* named *Yutyrannus huali,* whose discovery was announced in 2012, for example, measures about thirty feet in length, making it one of the largest meat-eating dinosaurs (Figure C.3).[26] Moreover, large numbers of other fossilized creatures have been found in the Cretaceous deposits of Liaoning as well. These include a diverse array of insects and plants, as well as early mammals and birds (including the earliest known bird with a beak, *Confuciusornis sanctus*). Besides offering insights into the evolution of feathers and documenting the origin of modern birds, Liaoning fossils shed light on the early diversification of angiosperms and the radiation of placental mammals. As a result, present-day paleontologists have come to regard China in much the same way that their late nineteenth-century predecessors thought of the United States. Peter Dodson from the University of Pennsylvania, for example, recently stated that China has "surpassed the United States . . . to become the greatest country on earth for dinosaur paleontology," while Dong Zhiming from Beijing predicts that "the day when [the] number of Asian dinosaurs surpasses that of the North American ones is just to come."[27]

The fossil deposits of Liaoning Province provide evidence of an entire terrestrial ecosystem, offering a rare glimpse of the many diverse species that flourished during the late Jurassic and early Cretaceous periods. To commemorate the fact that some of Liaoning's most productive quarries are located in a region that was formerly known as Jehol Province, researchers refer to this terrestrial ecosystem as the Jehol Biota.[28] Apart from their sheer number and diversity, what makes fossils from the Jehol Biota so significant is their exceptional preservation. Paleontologists believe the region that is now western Liaoning once featured a large number of active volcanoes and extremely deep lakes. Since violent eruptions were common during the late Jurassic and early Cretaceous periods, organisms from the Jehol Biota often died in large numbers and sank to the bottoms of these lakes, where they were covered by a layer of volcanic ash. This process resulted in the production of Liaoning's characteristic paper shales, extremely thin layers of volcanic rock that yield fossils of exceptional quality, often featuring high-fidelity traces of soft body parts, including integumentary structures such as feathers and skin. Because the fossil-bearing strata of western Liaoning provide such a vivid

Figure C.3. Restoration of *Yutyrannus huali* and *Beipiaosaurus inexpectatus* patrolling the Jehol Biota in what is now Liaoning Province, China, by Brian Choo.

and detailed snapshot of life in the deep past, at least two Chinese geologists have likened the area to a Mesozoic Pompeii.[29]

Some of Liaoning's fossils are so well preserved that scientists have even used them to infer the color of dinosaurs. During the first decade of the twenty-first century, a graduate student at Yale University named Jakob Vinther began testing the theory that subcellular organelles are preserved in the fossil record.[30] Using a high-powered microscope, Vinther succeeded in detecting trace evidence of melanosomes in a fossilized feather. Melanosomes are organelles that synthesize a light-absorbing pigment called melanin, which gives color to the skin, fur, and feathers of mammals and birds. Notably, melanosomes come in two varieties. Whereas elongate, or sausage-shaped, eumelanonsomes synthesize pigments that range from black to dark brown, submicrometric, or meatball-shaped, phaemelanosomes produce colors ranging from rufous red to buff yellow. Once it was shown that both could be recovered from fossils, the race was on to color a dinosaur, and in 2010, two separate research teams announced within days of each other that they had succeed. One, led by scientists from the Institute of Vertebrate Paleontology and Paleoanthropology in Beijing, found that the tail of *Sinosauropteryx* was covered by reddish and chestnut-brown stripes.[31] The other group, which included Vinther, examined the fossil remains of a Jurassic therapod, *Anchiornis huxleyi.* With the aid of a scanning electron microscope, they measured the shape, size, and density of melanosomes in both fossil and extant feathers. Next, they correlated melanosome morphology and distribution with observed plumage color in modern birds, which gave them a statistical baseline from which to infer the color of dinosaur feathers. What emerged was a picture of *A. huxleyi* that is often compared with a fancy chicken or spangled hen, having predominantly black feathers, a vibrant red crest on the back of its head, and brilliant white stripes on its forewings and legs.[32] Although Vinther's methods have not escaped criticism, microfossils have since been used to reconstruct the color of a wide range of other extinct organisms, including *Archaeopteryx, Confuciusornis,* and *Microraptor*.[33]

In sum, Jehol fossils from Liaoning Province in northeastern China have completely upended the modern conception of dinosaurs. In 2011 the paleontologist Mark Norell marveled that "now, instead of scaly animals portrayed as usually drab creatures, we have solid evidence for a fluffy colored past."[34]

Fossilized feathers have even led paleontologists to revise their ideas about dinosaur behavior. Because many therapods such as *Sinosauropteryx* had colorful but relatively simple feathers that did not impart the aerodynamic properties necessary for flight, it is now widely believed that complex or "pinnate" feathers were only secondarily adapted for that purpose.[35] Initially, they most likely evolved for thermoregulation, for camouflage, and as a signaling mechanism to communicate and attract mates. This further cements the once revolutionary idea that, far from having been solitary brutes, at least some dinosaurs were intensely social creatures that may have developed complex family structures.[36] But the political implications of these recent discoveries are arguably just as important as their scientific significance, as they have completely reshaped the intellectual geography of vertebrate paleontology. Not only do the most important new dinosaur fossils now hail from China, but even a cursory perusal of the recent scholarly literature reveals that Asia has emerged as a powerhouse in the world of vertebrate paleontology.

The current conception of dinosaurs could hardly be any more different than it was during America's Long Gilded Age. But as the saying goes, the more things change, the more they stay the same. In particular, these remarkable creatures continue to provide a compelling case study of how the practice of science is bound up with the culture of commerce. For that reason, the recent history of paleontology, especially its migration to China, offers an opportunity to revisit some of the core themes and principal arguments of this book.

In the introduction to this book, I argued that dinosaurs as we know them came into being through the complex interplay of geological and cultural history. Crisscrossing multiple temporal boundaries and ontological registers, they are a product of biological evolution and human imagination alike. As a result, dinosaurs tell us a great deal not only about the history of life on earth but also about the preoccupations of those who assembled their fossil remains into large and spectacular museum displays. Hence, rather than offer a straightforward history of paleontology, this book has used dinosaurs to examine much broader changes that took place during the late nineteenth and early twentieth centuries. In this respect, my approach resembles the study of so-called model organisms among modern biologists. These include the fruit fly, the house mouse, and mustard weed, among many others, and they are responsible for the vast majority of what we know about genetics, development, physiology, and evolution. As we have seen, the colossal size and extreme rarity of dinosaurs made them a poor choice of animal models for paleobiologists. But as I have tried to show in this book, they can serve as an excellent model for the social, cultural, and economic historian. By following dinosaurs from the dig site to the museum, and thence into a much broader, popular context, it becomes possible to assemble a remarkably rich and informative snapshot of the commercial culture of America's Long Gilded Age.[37]

My strategy in this book has been to exploit the way dinosaurs circulated across social, geographic, and epistemic space to tell a much bigger story about the way science and capitalism were entangled with one another during the Long Gilded Age. But dinosaurs are hardly unique in their propensity to travel. Indeed, several historians have gone so far as to insist that the circulation of ideas, objects, and utterances is a constitutive feature of all knowledge production, especially scientific knowledge, which so often aspires to universality. While a new claim to knowledge may originate in a locally circumscribed space such as the laboratory, it cannot remain there. Instead, such claims must find a way to travel beyond their point of origination and into a broader cultural context. This is because it is only after a claim has become widely accepted among members of the scientific community that it comes to be regarded as knowledge within that community. Differently put, if knowledge only comes to be recognized as such once it is held in common, it must always already be shared. This is especially so in the case of a popular science like vertebrate paleontology, whose ambition it is to see elite knowledge gain currency among a much broader and more socially diverse group of people than the scientific community.[38]

In contrast, this book has argued that not only must truth claims be put into motion for them to become knowledge, but the reverse also holds true. To describe knowledge as scientific is to make a normative claim about its quality and reliability. As a result, the practice of science always involves an element of exclusion and demarcation. While a claim must be widely shared among the members of a community in order to be seen as legitimate, the constitution of that very community is often achieved by erecting barriers to exclude those who are not regarded as having a proper place in the group.

Thus, while communication certainly plays an indispensable role in the history of knowledge, efforts to promote circulation always operate in tandem with attempts to control, manage, and, at times, explicitly arrest the movement of objects, ideas, and people. Decisions about which knowledge claims are put into motion, and in which contexts they circulate, are therefore an exercise of epistemic power, a means of exerting control over what can be known and who knows it. Moreover, strategic acts of exclusion are just as constitutive of scientific knowledge production as are the communicative practices that allow new theories to transcend their local context and become universal laws. Again, this is especially so in the case of popular science, whose visibility often makes it a target of critique among purists who view its appeal to nonscientists as a threat to its seriousness and credibility. The fact that many biologists routinely dismiss paleontologists such as Robert Bakker who make it their business to consult for Hollywood movies like Jurassic Park as celebrity cowboys is an especially clear case in point. Hence, dinosaurs are boundary objects in more ways than one. They not only promote collaboration and contact between different communities. They are also sites of contestation and conflict, used to erect barriers of entry and uphold intellectual hierarchies. This too is constitutive of the scientific community, which has long sought to differentiate itself from other, more popular forums of knowledge that are not seen to have the same status and thus do not command the same cultural authority.[39]

The language of circulation not only informs debates on how local knowledge acquires its claim to universality. It has also been embraced by historians who are eager to displace a parochial but powerful narrative about the global history of science, which holds that reliable knowledge about nature was first generated in Europe before "diffusing" to other parts of the world.[40] Several more recent accounts tell a different story by following material objects, such as spices and specimens, that played a crucial role in the history of European imperialism as they crisscrossed the globe. Instead of simply traveling from colonial centers out to the periphery, such "matters of exchange" moved in every direction at once. What results is a history of knowledge production that emphasizes cross-cultural encounters and stresses the important contributions of local informants, indigenous expertise, and cultural brokers or "go-betweens." Attending to the circulation of ideas, commodities, and people therefore effectively displaces Europe from its central role in the history of science and capitalism alike.[41]

Despite their success in producing a more cosmopolitan history of science, accounts of how knowledge circulates are not immune to critique. A particular source of concern is that, in their zeal to show how the practice of science transcends local geographies and individual people, historians who foreground transnational circuits of exchange threaten to enshrine a liberal account of epistemic value. In much the same way that many economists elevate the marketplace to a central place in society by arguing that a commodity only acquires its value through the interaction of supply and demand at the moment of exchange, global histories of science that focus on intercultural communication implicitly (and perhaps inadvertently) deny the status of indigenous knowledge as knowledge before it comes into contact with European science. In both cases, value—epistemic or economic—is seen as a product of exchange rather than as residing, locally or intrinsically, at the site of production. Hence, in spite of their efforts to displace older narratives that celebrated the unique contributions of Europe to the history of knowledge, such accounts nonetheless continue to privilege the centrality of modern science and capitalism over alternative modes of knowledge production and value creation.[42]

Dinosaur fossils from China exemplify the fraught, complex, and controversial role of movement, communication, and exchange in the global history of knowledge with particular force and clarity. The Jehol Biota is only the last in a long line of important paleontological discoveries made in Asia. In fact, one of the oldest known written descriptions of vertebrate fossils from China appeared almost two thousand years ago. In *The Chronicles of Huayang,* a history of the Yangtze River valley composed around 350 CE, the scholar Qu Chang described a region near Chendu in modern-day Sichuan Province that was known for its abundant "dragon bones," which were highly sought after for their medicinal properties: "In Wucheng County," Qu tells us, "there is a mountain called Somber Warrior Mountain, also called Three Corner Mountain, that has six bends and six rises. Dragon bones are taken from it. It is said that dragons flew up from these mountains, but when they found heaven's gates closed, they could not enter, and thus fell dead in that place, and later sank into the earth. That is why one can dig out dragon bones."[43]

Because Mesozoic exposures that are rich in dinosaurs abound in this part of Sichuan, modern paleontologists have concluded that Qu must have been referring to their fossil remains when he wrote about dragon bones.[44] But Qu could not have been writing about dinosaurs in 350 CE, because the word dinosaur refers to a biological category that only came into being once the anatomist Sir Richard Owen successfully argued for the formal inauguration of a new taxonomic unit—the Dinosauria—during a meeting of the British Association for the Advancement of Science that took place in 1841.[45] While there is no doubt that Qu was a skilled observer, he did not understand himself to be practicing paleontology, and it would be misleading to describe *The Chronicles of Huayang* as a scientific treatise. To say this is not to question the reliability or significance of Qu's text. Rather, it is to respect and take seriously the specific tradition out of which it emerged. It is also to insist that a science like vertebrate paleontology is not fundamentally different from other products of human culture. Social practices, including those that pertain to the generation of knowledge and the collection of specimens, occupy a particular location in time and space. Just as Chinese medicines such as dragon bones did not exist in Europe during classical antiquity, so too did European science only enter China during a much later period in its history. If the history of paleontology is to become truly global, it must first accept the provincial origins of its subject matter.[46]

The first people to describe Chinese dinosaur bones as such were Western explorers and adventurers in the early decades of the twentieth century, most of whom traveled to Asia on behalf of imperial states. This included a colonel in the Russian military who alerted Soviet paleontologists to the existence of large fossil bones near the Armur River in Heilongjiang, which led to the formal description of a new species of hadrosaur. This was the first Chinese dinosaur known to Western science, and its remains were eventually assembled for exhibition in Saint Petersburg.[47] Similarly, although the Central Asiatic Expedition that we encountered near the start of this chapter was sponsored by the New York natural history museum rather than a European state, it too was regarded as an imperial enterprise among Chinese scholars who resented the notion that fossils from Asia could only become truly known if they were studied abroad. Thus, when dinosaurs from China first began circulating outside Asia during the early twentieth century, the relationship between Chinese scholars and Western scientists was not one characterized by cross-cultural communication or mutual exchange. Rather, Western explorers traveled to Asia to locate specimens for export, without offering much in return.[48]

The pervasive asymmetries that characterized the relationship between Western scientists and Chinese scholars began to break down after Sun Yat-sen's revolutionary Guomindang Party succeeded in forming a Nationalist government in Nanjing during the 1920s. It was only once the Guomindang general Chiang Kai-shek's forces asserted military control over large parts of China that local geologists acquired the institutional power required to gain access to specimens held by European and American scientists. Embarrassed by the Qing dynasty's century-long capitulation to European imperialism following its defeat during the Opium Wars, the Guomindang sought to rejuvenate China's stature on the world stage. One way that it did so was by asserting its sovereignty in a commercial and epistemic, as well as a military, context. This included an effort to begin stemming the unidirectional flow of knowledge across its hotly contested borders. To that end, a group of prominent scholars in Beijing formed the Chinese Association of Learned Societies during the spring of 1927. In a blustering manifesto, the association demanded an end to all foreign expeditions that "infringe our sovereignty, plunder our research materials, and cause great loss to the future of Chinese academic development."[49] Their first order of business was to prevent Sven Hedin, a Swedish explorer who had been sent by Germany to map out a new trade route between Europe and Asia, from embarking for the Tarim Basin in Xinjiang, a remote region north of Tibet. After an acrimonious negotiation, Hedin finally agreed to allow all of the specimens collected by his expedition to remain in China. Hedin also accepted Shu Beihuei from Peking University as a co-equal leader of the recently renamed Sino-Swedish Northwestern Science Expedition. This outcome was viewed as a significant political victory by members of the Chinese Association of Learned Societies, one of whom described the agreement as the first "reversed unequal treaty" with a significant foreign power.[50]

Not long after concluding its negotiations with Hedin, the Chinese Association of Learned Societies was replaced by the politically more well-connected National Commission for the Preservation of Antiquities. Armed with broad powers to prevent the plunder of cultural artifacts and other national treasures, the commission ordered municipal governments all over China to stop foreign explorers from exporting collections abroad. This led to the confiscation of about eighty-five boxes of specimens amassed by the Central Asiatic Expedition in the Gobi Desert. To justify this action, the commission accused Andrews of prospecting for oil and conducting espionage in addition to digging up fossils. Both turned out to be true. Although Andrews consistently denied it, internal documents reveal that Standard Oil had agreed to underwrite some of the expedition's costs in exchange for "specialized information" about the Gobi's geology, with Andrews giving his personal assurance that his "geologists are among the most eminent in America and no oil bearing strata would escape their attention."[51] In addition, Andrews engaged in strategic espionage for the US military, offering "detailed information" and "special knowledge" about Chinese infrastructure, geography, and politics to the Office of Naval Intelligence.[52] But it was the claim that Andrews engaged in looting China's national patrimony that gained the most traction. A strongly worded newspaper editorial, for example, argued that in failing to recognize China as "the sole owner" of its epistemic resources, Andrews had effectively violated its hard-won territorial sovereignty. Chinese intellectuals viewed Andrews and his team of explorers not as cultural brokers but as an invading force that sought to undermine its international standing.[53]

North American paleontologists responded by accusing their counterparts in China of failing to share the cosmopolitan ideals of the scientific community, whose ability to produce universal knowledge required its members to transcend national politics and share discoveries freely. Andrews, for example, chalked all of the "trouble" up to the "anti-foreignism" of Republican China's Guomindang government, while the expedition's lead paleontologist, Walter Granger, took it upon himself to instruct the Antiquities Commission on the difference between "purely scientific" endeavors and those undertaken to promote "political or commercial interests."[54] Similarly, Osborn lectured the Chinese ambassador in Washington, DC, about the importance of international cooperation in science, arguing that "in every country the distinction is drawn between the archeological and artistic works of man, which should rightly be retained within the country which created them . . . , and the natural works of nature, which are freely interchanged."[55] According to the New York museum, whereas the immense cultural value of archeological artifacts derived from their antiquity, fossils were so old they only had scientific value. In effect, because dinosaurs were older than all of humanity, they belonged to no one in particular.

Despite the museum's best efforts, the National Commission for the Preservation of Antiquities refused to allow Chinese specimens to circulate outside the country's borders. Having convinced himself that the Chinese were unwilling to "cooperate in a scientific spirit," Andrews decided the only hope that remained was "unfavorable world publicity" to convince the commission that "they can't adopt this attitude of indifference to the progress of world knowledge without great detriment to themselves." The New York museum therefore began waging a public relations campaign to humiliate Chinese scholars before the international scientific community. An article in the English-language *China Weekly Review,* for example, claimed that Chinese scholars showed a complete "lack of understanding" of "the nature of the work which the Central Asiatic Expedition is doing," while the *Seattle Times* argued that "fossils of prehistoric animals" have "little if any interest for anybody except scientists engaged in piecing out the history of life on earth." Similarly, the *Peking Leader* attacked the preference of Chinese scholars to "save face" even at the expense of "world science." But most damming of all was an editorial penned by Osborn and Andrews for the September 1929 issue of *Science,* wherein the two men accused China's Guomindang government of misrepresenting the "purely scientific and educational aims of other countries" by charging the Central Asiatic Expedition with "'stealing China's priceless treasures,' 'infringing her sovereign rights,' 'seeking for oil and minerals,' 'being spies against the government,' etc., etc." Although "it has been one of the dreams of our ever-hopeful and optimistic country to see China realize the full benefits of the 'advancement of science,'" they insisted, it had become clear that the "spirit of anti-foreignism" spreading through China meant that "all foreign scientific work in the country must cease."[56]

By accusing the Chinese of a failure to embrace the scientific community's cosmopolitan norms, the New York museum hoped to pressure the National Commission for the Preservation of Antiquities into allowing the Central Asiatic Expedition to proceed unencumbered. For a time, it appeared the ploy might prove successful. During the summer of 1930, Andrews was given a special dispensation to reenter the Gobi so long as he promised to collect only paleontological specimens and return any duplicate material to China after it had been cleaned and examined in New York. Andrews also agreed to allow three Chinese scholars who were appointed by the commission to join his expedition. Unbeknownst to the New York museum, however, the 1930 field season would prove to be Andrews's last in the Gobi. When he asked for permission for yet another excursion the following summer, his request was summarily denied. Refusing to give up on his life's work, Andrews entered into negotiations with the Japanese puppet regime in Manchuria—the Manchukuo—in hopes of accessing the Gobi from the northeast rather than from Beijing. But this plan was soon scuttled, and it was not until the last decade of the twentieth century that North American paleontologists returned to the red cliffs of the Gobi.[57]

Meanwhile, a community of resident paleontologists developed in China. In a striking coincidence, just as negotiations between the New York museum and the National Commission for the Preservation of Antiquities broke down, a young scientist named Yang Zhongjian returned home from abroad. Having studied geology at Peking University, Yang completed his training at the University of Munich in 1927 and adopted the Westernized name C. C. Young. Upon his return to Asia, Yang took up a position at the Geological Survey of China's Cenozoic Research Laboratory. He then became one of the three Chinese scientists Andrews was forced to include in the Central Asiatic Expedition's final excursion into the Gobi. Between 1931 and 1949, Yang undertook a large number of his own field expeditions, which led, among other things, to the discovery of the Lufeng Dinosaur Valley, a large and extremely rich quarry of prosauropods outside Kunming. In 1941, as a result of Yang's efforts, *Lufengosaurus huenei* became the first Chinese dinosaur to be mounted for public display within China. After the Communist People's Republic was formed under Mao Zedong in 1949, Yang helped to found the Institute of Vertebrate Paleontology and Paleoanthropology and became the director of the Beijing Museum of Natural History. He remained in Beijing until his death in 1979, authoring numerous scientific papers while he trained a whole generation of Chinese paleontologists.[58]

In spite of the fact that he had studied in Europe and adopted a Westernized name, Yang does not conform to the figure of the cultural broker or go-between. While Yang proved adept at cultivating intellectual relationships that spanned across much of the globe, he primarily directed his efforts to building the social structures and institutional frameworks required to develop a robust paleontological community within China. In effect, Yang and the Antiquities Commission agreed with the editorial that Osborn and Andrews wrote for *Science,* in which they argued that "the Chinese themselves cannot do the work" of the New York museum "for they have neither adequately trained men nor the money to conduct investigations, and will not have [them] for many years."[59] However, they disagreed about what should be done in response. Yang and his colleagues felt that the development of Chinese science could only be furthered by protecting its sovereignty, not by opening its borders so foreign researchers could plunder its resources. As the precursor to the Antiquities Commission put it, because foreign expeditions such as the one led by Andrews "infringe our sovereignty" and "plunder our research materials," they caused "great loss to the future of Chinese academic development."[60] This is why, looking back on the experience of having accompanied Andrews into the Gobi, Yang described "the so-called 'Sino-American collaboration'" as a lesson in "how they take advantage of us, and how we take advantage of them."[61]

These protectionist policies largely remained in place until the last third of the twentieth century. Around the same time that Deng Xiaoping's market reforms led to significant trade liberalization in China's economic policy, Chinese paleontologists embarked on a robust program of international collaboration with scientists from North America and Western Europe.[62] By that point, Chinese scientists had amassed the power to insist that any joint ventures take place on a more or less equal footing. For example, in 1986 a series of cooperative expeditions was inaugurated between the Institute of Vertebrate Paleontology and Paleoanthropology in Beijing and two Canadian institutions, the Royal Tyrrell Museum and the Canadian Museum of Nature.[63] More recently, the New York museum has teamed up with Chinese and Mongolian paleontologists to reenter the Gobi. Crucially, all these collaborative projects stipulate that any specimens collected in China or Mongolia must remain there and cannot be exported abroad. Ironically, then, knowledge about feathered dinosaurs has circulated so far and wide because of, not in spite of, the fact that vertebrate fossils from the Jehol Biota stay put in China. This reveals that movement and circulation are always, and perhaps foremost, predicated on power. When the New York museum extolled the free trade in specimens during the 1920s, it did so because the export of fossils would benefit Western science—North American paleontology in particular. Similarly, the insistence that Chinese fossils remain in Asia has helped to promote the formation of a robust paleontological community there, which, in turn, has contributed to the discoveries that have so completely upended our conception of dinosaurs.

In science as well as in capitalism, circulation is only half of the story. Processes of accumulation are just as important. It was by arresting the movement of specimens across international borders that Chinese paleontologists accumulated the research materials required to build up a resident community of vertebrate paleontologists who could extract meaningful cooperation as a concession from their counterparts in the West. Nor are such dynamics confined to the past. Paleontologists from China continue to use fossils from sought-after localities such as the Jehol Biota as a strategic resource, trading access to specimens for authorial credit in international journals like *Science* and *Nature* that still primarily publish the research of European and American scientists. Thus, while the intellectual geography of vertebrate paleontology has undergone a dramatic reorganization over the past several decades, it remains profoundly uneven. These global asymmetries are a product of inertial tendencies that stem from unequal access to capital (in all of its manifestations) that has accumulated over the course of the past several decades if not centuries. To borrow Thomas Piketty's evocative phrase, the complex and contested history of vertebrate paleontology in twentieth-century China reveals how the past devours the future, in a science and capitalism alike.[64]

Whereas early twentieth-century explorers such as Andrews chastised China for failing to embrace the cosmopolitan ideals of the international scientific community by refusing to allow Chinese fossils to circulate globally, contemporary paleontologists are more worried that dinosaurs are too readily available on the commercial market. These fears do not only extend to fossils from Asia, however. Concerns about the corrosive effects of the specimen trade grew especially acute after the Field Museum of Natural History purchased Sue, an especially large and complete *T. rex* from the Badlands of South Dakota during the 1990s (Figure C.2). This fossil was named after Susan Hendrickson, who found the specimen while working for a commercial dealer, the Black Hills

Institute of Geological Research. Hendrickson's discovery was located on land owned by a member of the Cheyenne River Sioux tribe, Maurice Williams, to whom the institute paid $5,000. Due to an ownership dispute between the institute, Williams, and the Cheyenne River Sioux, the FBI confiscated Sue's fossil remains. After a drawn-out custody battle, a court ruled that Sue should be sold off to the highest bidder on Williams's behalf by Sotheby's in New York. Having secured the financial backing of Disney and McDonald's, the Field Museum acquired the specimen for close to $9 million.[65] Although Sue turned out to be a major attraction, the museum came in for intense criticism among vertebrate paleontologists, who worried that headlines about the sale would encourage more people to enter the commercial specimen trade and drive up the price of fossils, making it difficult for museums to compete for the most sought-after specimens with wealthy collectors who covet dinosaurs fossils as a mark of their social distinction. As David Krause, the president of the Society of Vertebrate Paleontology, told the *Chicago Sun-Times* in 1996, "The American public wants to see fossils in museums, not lost to private collections."[66] For that reason, many American paleontologists lobbied for legal restrictions to prohibit the sale of vertebrate fossils collected on federally owned land. In addition, the Society of Vertebrate Paleontology's scientific journal explicitly prohibits the publication of studies based on fossils that have not been deposited in a recognized nonprofit collection, such as a public museum. Because specimens that feature in prominent publications often increase in value, the ban on publishing studies that rely on privately owned fossils not only seeks to ensure that new theories about the deep past can be independently tested and verified. It is also designed to help dampen the commercial specimen trade.[67]

Sue's auction may have been an especially high-profile and controversial event, but it was hardly the last of its kind. The past several decades have seen countless spectacular dinosaur fossils come up for sale, many of which were illicitly collected in Asia. These controversial sales often made headlines in major newspapers and magazines. An especially noteworthy case involved Eric Prokopi, who was arrested for smuggling a nearly complete *Tarbosaurus bataar* out of Mongolia, but the revelation that the American film actor Nicholas Cage has purchased a skull from the same dealer has caused a large uproar as well.[68] A far larger number of illegally collected specimens have flown under the radar, however, and been purchased largely unnoticed. There is now a thriving black market in dinosaur fossils, many of which are traded illegally because their export remains forbidden by Chinese and Mongolian law. And while the United States does not prohibit the commercial exchange of dinosaur fossils per se, their collection on federal land is prohibited by the 2009 Omnibus Public Land Management Act. The purchase and sale of vertebrate fossils thus remains exceedingly controversial, much as it was during the Long Gilded Age. This is not only because the commercial specimen trade confounds the near-sacred distinction between monetary and scientific value. Paleontologists also complain that it puts the most valuable fossils out of their reach, arguing that commercial specimen dealers actually impede rather than promote the circulation of knowledge about dinosaurs.[69]

Finally, paleontologists often charge the commercial circulation of dinosaur fossils with threatening to erode public trust in science.[70] The risk of deception and fraud continues to structure the commercial exchange of vertebrate fossils, just as it did in nineteenth-century America. As anyone with access to eBay can easily verify, this is especially so in the case of fossils from Asia, whose often dubious legal status makes them difficult to authenticate.[71] Fake, composite, or otherwise adulterated fossils are routinely sold to unwitting consumers, a practice that has induced several eBay users to post extensive instructions purporting to teach neophytes the art of spotting forgeries and other fraudulent specimens. "Caveat emptor," declares one such guide, authored by the dealer "triassica," adding that "knowledge is power."[72] The website paleodirect.com even includes a page on "fake Chinese fossils," which cautions that "each year, thousands of trusting buyers are duped by both inexpensive and very expensive, highly realistic fakes." This online specimen dealer, which offers "investment grade" fossils complete with a "certificate of authenticity," goes on to explain that common methods of forgery range from "assembling genuine fossils of unrelated specimens together to make a complete, impressive fossil" to "crushing genuine fossil bone and mixing [it] with glue to fabricate body parts, skulls and skeletons that appear genuine with the proper color."[73] Online retailers are far from the only ones who have to work hard to distinguish real dinosaur fossils from fake specimens, however. Forged fossils pose a major problem for professional paleontologists too. For example, a spectacular *Darwinius* fossil (an early primate similar to a lemur) popularly known as Ida stirred enormous controversy in 2009 when it was sold to the Norwegian paleontologist Jørn Hurum. Hurum went on to publish a paper on Ida that received much fanfare despite speculations that his specimen might have been doctored by its original owner.[74] Cases like these have led a number of museum paleontologists to publish detailed guides that purport to teach even inexperienced curators how to use CT scans, chemical analysis, and ultraviolet light to help them spot faked or forged specimens.[75]

Paleontologists were especially troubled by the revelation that a much-heralded "missing link" between dinosaurs and modern birds—*Archaeoraptor liaoningensis* from northeastern China—constituted a fictitious creature that was constructed by combining the bones of a primitive bird with those of a nonflying dromaeosaurid dinosaur. The *Archaeoraptor* affair began in the fall of 1998, when rumors that a spectacular new fossil was available for purchase began to circulate during the Society of Vertebrate Paleontology's annual meeting. When the fossil was offered for sale at the Tucson Gem, Mineral, and Fossil Showcase the following spring, an amateur paleontologist and dinosaur artist named Stephen Czerkas convinced patrons of the Dinosaur Museum in Blanding, Utah, to supply the $80,000 required to make the purchase. Czerkas then approached his friend Philip J. Currie from the Royal Tyrrell Museum in Drumheller, Alberta, proposing to publish a scientific article on the specimen. Currie agreed and alerted Christopher Sloan, an editor at *National Geographic,* who jumped at the chance of publishing a major story about the new find. But as soon as Currie began to investigate *Archaeoraptor* closely, problems began to emerge. To begin with, Sloan and Currie both insisted the specimen be returned to China, from which they believed it must have been exported illegally. More troubling still was that Timothy Rowe from the University of Texas, who had been recruited to make a CT scan of the fossil, concluded that it was a composite specimen assembled from numerous distinct slabs of stone. From there, things only got worse, as both *Nature* and *Science* refused to publish a scholarly study of *Archaeoraptor* because of concerns about its provenance, legal status, and authenticity raised by external reviewers. With its publication deadline fast approaching, *National Geographic* nonetheless went ahead and printed its article, disregarding the fact that *Archaeoraptor* had been purchased on the commercial market for a private museum and was not yet officially described in a peer-reviewed journal. Not long thereafter the Chinese paleontologist

Xu Xing discovered conclusive evidence of the forgery, and an analysis published the next year by Rowe, Xu, and Currie, among others, concluded that the specimen had been built out of eighty-eight pieces that could have come from as many as five separate specimens.[76]

The case of *Archaeoraptor* and others like it have generated a huge amount of attention around the issue of paleontological forgeries, especially pertaining to fossils from China. At least some of the uproar can be attributed to a long history of Western hysteria over the threat posed by Chinese bootleggers, copycats, and counterfeits more broadly.[77] Consumers in Europe and North America have been primed to distrust Chinese goods and commodities as low-quality imitations, so why should dinosaur fossils be any different? But the reality is both more complex and more interesting. The regions that yield some of China's most spectacular fossils tend to be rich in natural resources—including large vertebrate fossils—yet poor in most other respects, including in cultural and political capital. As a result, farmers from areas such as Liaoning often supplement their meager incomes by collecting fossils and selling them on the commercial market. Knowing that more complete, more spectacular, and more novel specimens usually fetch a premium, small-time collectors and independent fossil hunters have an incentive to augment their discoveries in various ways. Sometimes, as in the case of *Archaeoraptor,* they do so by combining two or more specimens collected in separate quarries to create a composite. Other times, they restore missing pieces by artfully sculpting them out of nonfossil material, a process that requires exacting care and a great deal of skill. "The Chinese are excellent craftsmen," Currie has said, and Xu likens them to artists for the way they "carve the bone from the rock."[78] Indeed, these practices have become so widespread that Li Chun from the Institute of Vertebrate Paleontology and Paleoanthropology in Beijing estimates that over 80 percent of all marine reptile fossils on display in Chinese museums have been "altered or artificially combined to varying degrees."[79]

Despite all the anxiety about doctored and fraudulent specimens, the past several decades have seen China cement its reputation as the best place to find new and spectacular dinosaurs. It is thus fitting that a large number of new museums and exhibitions focused on dinosaurs are being constructed in Asia, much like they were in late nineteenth- and early twentieth-century America. The Lufeng Dinosaur Valley Museum in Yunnan Province, for example,

Figure C.4. A herd of Jurassic dinosaur skeletons on display at the Lufeng Dinosaur Valley Museum, Yunnan Province, China.

resembles Dinosaur National Monument in Ogden, Utah, in that it gives visitors a chance to see an entire dinosaur quarry and to examine specimens that have only been partially taken out of the ground. A pavilion next to the excavation site also features dozens of mounted dinosaur fossils, some of which have been positioned to give the impression of a running dinosaur herd (Figure C.4). However, in contrast to American natural history museums, which worked hard to distance themselves from commercial purveyors of public spectacle, the Lufeng museum also features a dinosaur theme park complete with three-dimensional models of these extinct creatures and several amusement park rides (Figure C.5).

It has become almost a cliché to compare modern-day China to America during the Long Gilded Age. Much as the United States did in the late nineteenth century, China is undergoing a period of rapid industrialization. Moreover, China's explosive economic expansion resembles that of the United States in that it has been fueled by an abundance of natural resources, including huge tracts of arable land and large stores of mineral wealth. Finally, economic expansion in both countries has come at the price of widespread corruption,

Figure C.5. Sculptural reconstructions of *Lufengosaurus* on display at the Lufeng Dinosaur Valley Museum, Yunnan Province, China.

escalating labor unrest, and concerns over environmental degradation, as well as a precipitous increase in economic inequality. Perhaps most striking for our purposes, however, is that as well-to-do segments of both societies developed a taste for conspicuous consumption, dinosaur fossils joined artworks as some of the most sought-after means by which new members of the wealthy elite seek to demonstrate their class status and social distinction. Not unlike Andrew Carnegie did during the 1890s, for example, the industrialist Zheng Xiaoting recently drew on the vast fortune he made in gold mining to amass a spectacular collection of paleontological specimens. Zheng has used these to found the Shandong Tianyu Museum of Nature, which holds the Guinness World Record for the largest collection of dinosaurs.

The comparison between late nineteenth-century America and present-day China can easily be taken too far, however, and the global economy has changed dramatically in the past hundred years. Whereas the Long Gilded Age was a period of increasing consolidation and market integration, we are now on a very different trajectory, due in no small part to the emergence of Asia as a dynamic industrial center. Whereas the late nineteenth and early twentieth centuries saw the rise of vertically integrated corporations, the past several decades have seen a new mantra of efficiency taking hold, one that celebrates small, agile, and adaptable startups whose so-called disruptive innovations have eroded the power and profitability of large and heavily bureaucratized industrial firms. Thus, while Standard Oil and General Electric were on the cutting edge of economic development during the Long Gilded Age, the most celebrated companies today are online service providers like Google and biotechnology startups such as Genentech, not to mention high-tech manufacturing concerns such as Foxconn and Huawei that are headquartered in Asia. Moreover, while late nineteenth-century industrial behemoths like Carnegie Steel sought to internalize the entirety of their production and distribution networks, the world's most profitable manufacturing company during the early twenty-first century—Apple Computers—publicly celebrates that whereas its most iconic products have been designed in Cupertino, California, they are manufactured by third parties in Shenzhen, China. Finally, rather than seek to leverage the power of bureaucratic authority to reduce transaction costs, modern manufacturers are more interested in building a global supply chain that allows them to reduce labor costs and evade expensive health, safety, and environmental regulations.[80]

Given all of these changes in the global economy, is it any surprise that our understanding of dinosaurs has undergone such a dramatic transformation? The past several decades have seen business and political leaders alike embrace Joseph Schumpeter's notion that a process of "creative destruction" underlies economic development, whereas biologists have revolutionized evolutionary theory with the introduction of game-theoretic models of rational decision-making. Gone are Osborn's and Cope's faith that an orthogenetic process would inexorably produce higher and higher levels of evolutionary complexity. Instead, bottom-up explanations in which the self-interested behavior of autonomous agents yields adaptive outcomes in the population at large have proliferated. Indeed, many biologists now believe that even the most exquisitely complex social assemblages have been produced through natural selection acting on a molecular level.[81] Moreover, while evolutionary biologists have reinterpreted altruistic acts of self-sacrifice as selectively advantageous from a genic perspective, political economists have renewed their commitment to classical liberalism, even as the financial panic of 2008 has caused many of them to shy away from the excesses of the perfect market hypothesis.[82] No wonder, then, that dinosaurs have been transformed in our imaginations from lumbering behemoths of the prehistoric into agile, intelligent, and intensely social creatures covered by colorful feathers, many of which hail from Asia rather than North America.

Notes

Abbreviations

AHC	American Heritage Center, University of Wyoming, Laramie
AMNH	American Museum of Natural History, New York, NY
Big Bone Room Archive	Big Bone Room Archive, Department of Vertebrate Paleontology, CMNH (largely available online through the Digital Collections of the Carnegie Mellon University Archives)
Carnegie Collection	Andrew Carnegie Correspondence Collection, Archives and Special Collections, Carnegie Free Library of Pittsburgh, PA (available online through the Digital Collections of the Carnegie Mellon University Archives)
Century of Progress records	Century of Progress records, Special Collections and University Archives, University of Illinois at Chicago
CMNH	Carnegie Museum of Natural History, Pittsburgh, PA
DVP	Department of Vertebrate Paleontology, AMNH
FMNH	Field Museum of Natural History, Chicago, IL
Holland Papers	W. J. Holland Papers, Archives of the CMNH
Marsh Papers	O. C. Marsh Papers, Sterling Library, Yale University, New Haven, CT (available online through the Digital Collections of the Yale Peabody Museum of Natural History)
MCZ	Museum of Comparative Zoology, Harvard University, Cambridge, MA
NARA	National Archives and Records Administration in Bethesda, MD
NHM London	Natural History Museum, London
NYHS	New-York Historical Society, New York, NY
NYPL	New York Public Library, New York, NY
Osborn Papers, AMNH	H. F. Osborn Papers, Archives of the AMNH
Osborn Papers, NYHS	H. F. Osborn Papers, NYHS
S. H. Knight Papers	Samuel H. Knight Papers, AHC
SIA	Smithsonian Institution Archives, Washington, DC

Introduction

1. For more on the *Barosaurus* mount, see Lowell Dingus, *Next of Kin: Great Fossils at the American Museum of Natural History* (New York: Rizzoli, 1996), 21–28. On the construction of dinosaur displays in the early twentieth century, see Lukas Rieppel, "Bringing Dinosaurs Back to Life: Exhibiting Prehistory at the American Museum of Natural History," *Isis* 103, no. 3 (2012): 460–490.

2. For more on dinosaurs as unobservable entities, see Derek Turner, *Making Prehistory: Historical Science and the Scientific Realism Debate* (Cambridge: Cambridge University Press, 2007). For a different perspective, see Adrian Currie, *Rock, Bone, and Ruin: An Optimist's Guide to the Historical Sciences* (Cambridge, MA: MIT Press, 2018). See also Claudine Cohen, *La Méthode de Zadig: La Trace, Le Fossile, La Preuve* (Paris: Éditions du Seuil, 2011).

3. The scholarly literature on dinosaurs is vast. See especially Paul Brinkman, *The Second Jurassic Dinosaur Rush: Museums and Paleontology in America at the Turn of the Twentieth Century* (Chicago: University of Chicago Press, 2010); Richard Conniff, *House of Lost Worlds: Dinosaurs, Dynasties, and the Story of Life on Earth* (New Haven, CT: Yale University Press, 2016); Mark Jaffe, *The Gilded Dinosaur: The Fossil War between E. D. Cope and O. C. Marsh and the Rise of American Science* (New York: Crown, 2000); W. J. Thomas Mitchell, *The Last Dinosaur Book: The Life and Times of a Cultural Icon* (Chicago: University of Chicago Press, 1998); Ilja Nieuwland, *American Dinosaur Abroad: A Cultural History of Carnegie's Plaster Diplodocus* (Pittsburgh, PA: University of Pittsburgh Press, 2019); Brian Noble, *Articulating Dinosaurs: A Political Anthropology* (Toronto: University of Toronto Press, 2016); Ronald Rainger, *An Agenda for Antiquity: Henry Fairfield Osborn and Vertebrate Paleontology at the American Museum of Natural History, 1890–1935* (Tuscaloosa: University of Alabama Press, 1991); Tom Rea, *Bone Wars: The Excavation and Celebrity of Andrew Carnegie's Dinosaur* (Pittsburgh, PA: University of Pittsburgh Press, 2001).

4. For an introduction to the "new" history of capitalism, see Sven Beckert and Christine Desan, *American Capitalism: New Histories* (New York: Columbia University Press, 2018). Additional works in the field include Sven Beckert, *Empire of Cotton: A Global History* (New York: Knopf, 2014); Christine Desan, *Making Money: Coin, Currency, and the Coming of Capitalism* (Oxford: Oxford University Press, 2014); Jonathan Levy, *Freaks of Fortune: The Emerging World of Capitalism and Risk in America* (Cambridge, MA: Harvard University Press, 2012); Seth Rockman, *Scraping By: Wage Labor, Slavery, and Survival in Early Baltimore* (Baltimore, MD: Johns Hopkins University Press, 2009); Caitlin Rosenthal, *Accounting for Slavery: Masters and Management* (Cambridge, MA: Harvard University Press, 2018); Amy Dru Stanley, *From Bondage to Contract: Wage Labor, Marriage, and the Market in the Age of Slave Emancipation* (Cambridge: Cambridge University Press, 2011); Michael Zakim, *Accounting for Capitalism* (Chicago: University of Chicago Press, 2018).

5. Attempts to connect the history of science to political economy have been a mainstay in science studies. For an overview, see Lukas Rieppel, Eugenia Lean, and William Deringer, "Introduction: The Entangled Histories of Science and Capitalism," *Osiris* 33 (2018): 1–24. Recent examples include Dan Bouk, *How Our Days Became Numbered* (Chicago: University of Chicago Press, 2015); Courtney Fullilove, *The Profit of the Earth* (Chicago: University of Chicago Press, 2017); Michelle Murphy, *The Economization of Life* (Durham: Duke University Press, 2017); Jamie L. Pietruska, *Looking Forward: Prediction and Uncertainty in Modern America* (Chicago: University of Chicago Press, 2017); Harold J. Cook, *Matters of Exchange: Commerce, Medicine, and Science in the Dutch Golden Age* (New Haven, CT: Yale University Press, 2007).

6. Historians continue to debate how best to describe a period that usually goes by the unwieldy name of the Gilded Age and Progressive Era. The *Journal of the Gilded Age and Progressive Era* has even published two forums on this subject. See Richard Schneirov, "Thoughts on Periodizing the Gilded Age:

Capital Accumulation, Society, and Politics, 1873–1898," with comments by James L. Huston and Rebecca Edwards, in *Journal of the Gilded Age and Progressive Era* 5, no. 3 (2006): 189–240; "Forum: Should We Abolish the 'Gilded Age'?" with contributions by Rebecca Edwards and Richard R. John, in *Journal of the Gilded Age and Progressive Era* 8, no. 4 (October 2009): 461–485. For a spirited argument that both the Gilded Age and the Progressive Era ought to be understood as part of a Long Reconstruction, see Heather Cox Richardson, "Reconstructing the Gilded Age and Progressive Era," in *A Companion to the Gilded Age and Progressive Era,* ed. Christopher McKnight Nichols and Nancy C. Unger (Malden, MA: Wiley, 2017), 7–20. Notable examples that stress continuity over rupture during this period include Rebecca Edwards, *New Spirits: Americans in the Gilded Age, 1865–1905* (New York: Oxford University Press, 2006); Leon Fink, *The Long Gilded Age: American Capitalism and the Lessons of a New World Order* (Philadelphia: University of Pennsylvania Press, 2015); T. Jackson Lears, *Rebirth of a Nation: The Making of Modern America, 1877–1920* (New York: HarperCollins, 2009); Daniel T. Rodgers, *Atlantic Crossings: Social Politics in a Progressive Age* (Cambridge, MA: Belknap Press of Harvard University Press, 1998); Robert Wiebe, *The Search for Order, 1877–1920* (New York: Hill and Wang, 1967).

7. In framing the study along these lines, I take inspiration from recent work on the material culture of science. See, for example, Samuel Alberti, *Nature and Culture: Objects, Disciplines and the Manchester Museum* (Manchester: Manchester University Press, 2009); D. Graham Burnett, *Trying Leviathan: The Nineteenth-Century New York Court Case That Put the Whale on Trial and Challenged the Order of Nature* (Princeton, NJ: Princeton University Press, 2007); Angela N. H. Creager, *Life Atomic: A History of Radioisotopes in Science and Medicine* (Chicago: University of Chicago Press, 2013); Sadiah Qureshi, *Peoples on Parade: Exhibitions, Empire, and Anthropology in Nineteenth-Century Britain* (Chicago: University of Chicago Press, 2011).

8. On circulation in the history of science, see, for example, Kapil Raj, *Relocating Modern Science: Circulation and the Construction of Knowledge in South Asia and Europe, 1650–1900* (Houndsmills: Palgrave Macmillan, 2007); James Secord, "Knowledge in Transit," *Isis* 95, no. 4 (2004): 654–672; Simon Schaffer et al., eds., *The Brokered World: Go-Betweens and Global Intelligence, 1770–1820* (Sagamore Beach, MA: Science History Publications, 2009). On popular science, see Peter J. Bowler, *Science for All: The Popularization of Science in Early Twentieth-Century Britain* (Chicago: University of Chicago Press, 2009); Carin Berkowitz and Bernard Lightman, eds., *Science Museums in Transition: Cultures of Display in Nineteenth-Century Britain and America* (Pittsburgh, PA: University of Pittsburgh Press, 2017); Aileen Fyfe and Bernard Lightman, *Science in the Marketplace: Nineteenth-Century Sites and Experiences* (Chicago: University of Chicago Press, 2007); Bernard V. Lightman, *Victorian Popularizers of Science: Designing Nature for New Audiences* (Chicago: University of Chicago Press, 2007); Lynn Nyhart, *Modern Nature: The Rise of the Biological Perspective in Germany* (Chicago: University of Chicago Press, 2009); James Secord, *Victorian Sensation: The Extraordinary Publication, Reception, and Secret Authorship of Vestiges of the Natural History of Creation* (Chicago: University of Chicago Press, 2000). On the popularization of knowledge about prehistory in particular, see Gowan Dawson, *Show Me the Bone* (Chicago: University of Chicago Press, 2016); Ralph O'Connor, *The Earth on Show: Fossils and the Poetics of Popular Science, 1802–1856* (Chicago: University of Chicago Press, 2007).

9. "Economic Growth and Structural Change," in Stanley L. Engerman and Robert E. Gallman, *The Cambridge Economic History of the United States,* vol. 2 (Cambridge: Cambridge University Press, 2000), 6.

10. Gavin Wright, "The Origins of American Industrial Success, 1879–1940," *American Economic Review* 80, no. 4 (1990): 651–668; Gavin Wright, "Natural Resources and the American Economy," in *Encyclopedia of the United States in the Twentieth Century,* ed. Stanley I. Kutler (New York: Scribner's, 1996); A. Paul David and Gavin Wright, "Increasing Returns and the Genesis of American Resource Abundance," *Industrial and Corporate Change* 6, no. 2 (1997): 203–245. For a longue-durée history of economic

extraction, see Edward Barbier, *Scarcity and Frontiers: How Economies Have Developed through Natural Resource Exploitation* (Cambridge: Cambridge University Press, 2011).

11. For more on the interconnectedness of urban centers and their rural hinterlands, see William Cronon, *Nature's Metropolis: Chicago and the Great West* (New York: Norton, 1991); Noam Maggor, *Brahmin Capitalism: Frontiers of Wealth and Populism in America's First Gilded Age* (Cambridge, MA: Harvard University Press, 2017). See also Richard White, *Railroaded: The Transcontinentals and the Making of Modern America* (New York: Norton, 2011).

12. Some readers will note that I do not follow the lead of paleontologists who synonymize *Brontosaurus* with *Apatosaurus*, judging that fossils assigned to the two genera actually hailed from the same type of organism. That is because almost all paleontologists during the late nineteenth and early twentieth centuries viewed *Brontosaurus* as a distinct genus, and I follow their practice throughout this book. The same also holds true for other biological taxa that have subsequently been consigned to the dustbin of history, such as *Laelaps* and *Trachodon*. My aim is not to use the most up-to-date terminology, but to remain faithful to the linguistic practices and epistemological paradigms of the historical actors I discuss.

13. Sven Beckert, *The Monied Metropolis: New York City and the Consolidation of the American Bourgeoisie, 1850–1896* (Cambridge: Cambridge University Press, 2001). See also Sven Beckert and Julia Rosenbaum, eds., *The American Bourgeoisie: Distinction and Identity in the Nineteenth Century* (New York: Palgrave Macmillan, 2010).

14. US Department of Labor, *Twenty-First Annual Report of the Commissioner of Labor* (Washington, DC: Government Printing Office, 1906), 15–16, quoted in Thomas C. Leonard, *Illiberal Reformers: Race, Eugenics, and American Economics in the Progressive Era* (Princeton, NJ: Princeton University Press, 2016), 4. See also Walter Licht, *Industrializing America: The Nineteenth Century* (Baltimore, MD: Johns Hopkins University Press, 1995), 173. For more on the experience of working people, see Herbert George Gutman, *Work, Culture, and Society in Industrializing America: Essays in American Working-Class and Social History* (New York: Knopf, 1976); David Montgomery, *Citizen Worker: The Experience of Workers in the United States with Democracy and the Free Market during the Nineteenth Century* (Cambridge: Cambridge University Press, 1993).

15. Stephen Skowronek, *Building a New American State: The Expansion of National Administrative Capacities, 1877–1920* (Cambridge: Cambridge University Press, 1982), 105; Richard Oestreicher, "Two Souls of American Democracy," in *The Social Construction of Democracy, 1870–1990*, ed. George Reid Andrews and Herrick Chapman (New York: New York University Press, 1995), 128.

16. Robert M. Fogelson, *America's Armories: Architecture, Society, and Public Order* (Cambridge, MA: Harvard University Press, 1989).

17. Andrew Carnegie, *The Gospel of Wealth, and Other Timely Essays* (New York: Century, 1900), 1, 12.

18. William Murphey, "Theodore Roosevelt and the Bureau of Corporation: Executive-Corporate Co-operation and the Advancement of the Regulatory State," *American Nineteenth Century History* 14, no. 1 (2013): 92, 101.

19. For more on the corporate reconstruction of North America's political economy, see Alfred D. Chandler, *The Visible Hand: The Managerial Revolution in American Business* (Cambridge, MA: Belknap Press of Harvard University Press, 1977); Naomi Lamoreaux, *The Great Merger Movement in American Business, 1895–1904* (Cambridge: Cambridge University Press, 1985); Martin J. Sklar, *The Corporate Reconstruction of American Capitalism, 1890–1916: The Market, the Law, and Politics* (Cambridge: Cambridge University Press, 1988). The trope that Gilded Age capitalists constituted a class of "robber barons" was largely due to Progressive journalists who fought to see them brought to account. See, for example, Ida M. Tarbell, *The History of the Standard Oil Company* (New York: McClure, Phillips, 1904); Matthew Josephson, *The Robber Barons: The Great American Capitalists, 1861–1901* (New York: Harcourt, Brace, 1934).

20. Robert K. Merton, *The Sociology of Science: Theoretical and Empirical Investigations* (Chicago: University of Chicago Press, 1973). See also Michael Polanyi, "The Republic of Science: Its Political and Economic Theory," *Minerva* 1, no. 1 (1962): 54–73; Max Weber, *The Vocation Lectures,* ed. David S. Owen and Tracy B. Strong, trans. Rodney Livingstone (Indianapolis: Hackett, 2004). For an influential account of how science became disinterested, see Steven Shapin, *A Social History of Truth: Civility and Science in Seventeenth-Century England* (Chicago: University of Chicago Press, 1994).

21. See, for example, Elizabeth Popp Berman, *Creating the Market University: How Academic Science Became an Economic Engine* (Princeton, NJ: Princeton University Press, 2012); Sally Smith Hughes, *Genentech: The Beginnings of Biotech* (Chicago: University of Chicago Press, 2011); Philip Mirowski, *Science-Mart: Privatizing American Science* (Cambridge, MA: Harvard University Press, 2011); David F. Noble, *America by Design: Science, Technology, and the Rise of Corporate Capitalism* (New York: Knopf, 1977); Steven Shapin, *The Scientific Life: A Moral History of a Late Modern Vocation* (Chicago: University of Chicago Press, 2008).

22. See, for example, Jack Ralph Kloppenburg Jr., *First the Seed: The Political Economy of Plant Biotechnology, 1492–2000* (Cambridge: Cambridge University Press, 1988); Michelle Murphy, *Sick Building Syndrome and the Problem of Uncertainty: Environmental Politics, Technoscience, and Women Workers* (Durham, NC: Duke University Press, 2006); Robert Proctor, *Golden Holocaust: Origins of the Cigarette Catastrophe and the Case for Abolition* (Berkeley: University of California Press, 2011); Londa Schiebinger, *Plants and Empire: Colonial Bioprospecting in the Atlantic World* (Cambridge, MA: Harvard University Press, 2004); Vandana Shiva, *Biopiracy: The Plunder of Nature and Knowledge* (Boston: South End Press, 1997).

23. See, for example, Daniel Lee Kleinman and Steven P. Vallas, "Science, Capitalism, and the Rise of the 'Knowledge Worker': The Changing Structure of Knowledge Production in the United States," *Theory and Society* 30, no. 4 (2001): 451–492; Hallam Stevens, "On the Means of Bio-Production: Bioinformatics and How to Make Knowledge in a High-Throughput Genomics Laboratory," *BioSocieties* 6, no. 2 (2011): 217–242; Kaushik Sunder Rajan, *Biocapital: The Constitution of Postgenomic Life* (Durham, NC: Duke University Press, 2006).

24. See, for example, Daniel Lee Kleinman, *Impure Cultures: University Biology and the World of Commerce* (Madison: University of Wisconsin Press, 2003); Naomi Oreskes and Erik M. Conway, *Merchants of Doubt: How a Handful of Scientists Obscured the Truth on Issues from Tobacco Smoke to Global Warming* (New York: Bloomsbury, 2010); Sergio Sismondo, "Medical Publishing and the Drug Industry: Is Medical Science for Sale?" *Learned Publishing* 25, no. 1 (2012): 7–15.

25. On capitalism as a mode of social organization wherein the marketplace is "disembedded" from other cultural institutions, see Karl Polanyi, *The Great Transformation* (New York: Rinehart, 1944). On the "market revolution" in early America, see Joyce Appleby, *The Relentless Revolution: A History of Capitalism* (New York: Norton, 2010); John Larson, *The Market Revolution in America: Liberty, Ambition, and the Eclipse of the Common Good* (New York: Cambridge University Press, 2010); Charles Sellers, *The Market Revolution: Jacksonian America, 1815–1846* (New York: Oxford University Press, 1991). On the Long Gilded Age as a time in which all value was reduced to money value, see William Leach, *Land of Desire: Merchants, Power, and the Rise of a New American Culture* (New York: Pantheon, 1993). Sociologists have pushed back against Polanyi's claim about the disembedding of markets. See Mark Granovetter, "Economic Action and Social Structure: The Problem of Embeddedness," *American Journal of Sociology* 91, no. 3 (1985): 481–510; Greta Krippner et al., "Polanyi Symposium: A Conversation on Embeddedness," *Socio-Economic Review* 2, no. 1 (2004): 109–135; Viviana Zelizer, *The Social Meaning of Money* (New York: Basic Books, 1994).

26. Jens Beckert, *Imagined Futures: Fictional Expectations and Capitalist Dynamics* (Cambridge, MA: Harvard University Press, 2016). Some scholars have even argued that capital is, essentially, a process of

forward-looking valuation, which turns everything into "an income-generating, money-making capital good or investment." See Eli Cook, *The Pricing of Progress: Economic Indicators and the Capitalization of American Life* (Cambridge, MA: Harvard University Press, 2017), 6. See also Jonathan Levy, "Capital as Process and the History of Capitalism," *Business History Review* 91, no. 3 (2017): 483–510; Fabian Muniesa et al., *Capitalization: A Cultural Guide* (Paris: Presses des Mines, 2017).

27. On quantitative techniques for predicting the future, see William Deringer, "Compound Interest Corrected: The Imaginative Mathematics of the Financial Future in Early Modern England," *Osiris* 33 (2018): 109–129. On legal and cultural institutions of capitalism, see Stephen H. Haber et al., eds., *Political Institutions and Financial Development* (Stanford, CA: Stanford University Press, 2008); Douglass C. North, "Institutions," *Journal of Economic Perspectives* 5, no. 1 (1991): 97–112; Douglass C. North, "Institutions, Ideology, and Economic Performance," *Cato Journal* 11, no. 3 (1992): 477–496. On the new temporal regimes more broadly, see Reinhart Koselleck, *Futures Past: On the Semantics of Historical Time* (Cambridge, MA: MIT Press, 1985).

28. See Martin J. Rudwick, *Bursting the Limits of Time: The Reconstruction of Geohistory in the Age of Revolution* (Chicago: University of Chicago Press, 2005); Martin J. Rudwick, *Worlds before Adam: The Reconstruction of Geohistory in the Age of Reform* (Chicago: University of Chicago Press, 2008).

1. Prospecting for Dinosaurs

1. Charles Schuchert and Clara Mae LeVene, *O. C. Marsh, Pioneer in Paleontology* (New Haven, CT: Yale University Press, 1940); John H. Ostrom and John S. McIntosh, *Marsh's Dinosaurs: The Collections from Como Bluff* (New Haven, CT: Yale University Press, 1966); Brent Breithaupt, "Biography of William Harlow Reed: The Story of a Frontier Fossil Collector," *Earth Sciences History* 9, no. 1 (1990): 6–13; Mark Jaffe, *The Gilded Dinosaur: The Fossil War between E. D. Cope and O. C. Marsh and the Rise of American Science* (New York: Crown, 2000). For Reed's version of the events, which differs from both the documentary record and the consensus of historians, see William H. Reed to Charles W. Gilmore, Apr. 1, 1914, folder 22, box 12, Record Unit 156, SIA.

2. Othniel Charles Marsh, "Siredon, a Larval Salamander," *American Naturalist* 2, no. 9 (1868): 493; William E. Carlin, "Observations on Siredon Lichenoides," *Science* 2, no. 58 (1881): 367–368.

3. Carlin and Reed to Marsh, July 19, 1877, Marsh Papers.

4. On the important role of the extractive industry in fueling the United States' economic expansion, see Gavin Wright, "The Origins of American Industrial Success, 1879–1940," *American Economic Review* 80, no. 4 (1990): 651–668.

5. On the importance of trust in the history of science, see Steven Shapin, *A Social History of Truth: Civility and Science in Seventeenth-Century England* (Chicago: University of Chicago Press, 1994); and Theodore Porter, *Trust in Numbers: The Pursuit of Objectivity in Science and Public Life* (Princeton, NJ: Princeton University Press, 1995). For a contrasting view, see Daniel Margocsy, *Commercial Visions: Science, Trade, and Visual Culture in the Dutch Golden Age* (Chicago: University of Chicago Press, 2014). On trust in economic transactions, see Diego Gambetta, ed., *Trust: Making and Breaking Cooperative Relations* (New York: B. Blackwell, 1988); Naomi Lamoreaux, Daniel Raff, and Peter Temin, "Beyond Markets and Hierarchies: Toward a New Synthesis of American Business History," *American Historical Review* 108, no. 2 (2003): 404–433; Georg Simmel, *Philosophie Des Geldes* (Leipzig: Duncker & Humbolt, 1900).

6. The mining frontier was far from unique in this regard, as deception and fraud were widespread in nineteenth century America. See, for example, James Cook, *The Arts of Deception: Playing with Fraud in the Age of Barnum* (Cambridge, MA: Harvard University Press, 2001); Ann Fabian, *Card Sharps, Dream Books, and Bucket Shops: Gambling in 19th-Century America* (Ithaca, NY: Cornell University Press, 1990);

Jane Kamensky, *The Exchange Artist: A Tale of High-Flying Speculation and America's First Banking Collapse* (New York: Viking, 2008); and Stephen Mihm, *A Nation of Counterfeiters: Capitalists, Con-Men, and the Making of the United States* (Cambridge, MA: Harvard University Press, 2007). See also Lukas Rieppel, "Hoaxes, Humbugs, and Fraud: Distinguishing Truth from Untruth in Early America," *Journal of the Early Republic* 38, no. 3 (2018): 501–529. For more on "charlatans" in natural history, see Irina Podgorny, *Charlatanes: Crónicas de remedios incurables* (Buenos Aires: Eterna Cadencia, 2012).

7. My account is informed by studies of market transactions under conditions of asymmetrically distributed information, especially George A. Akerlof, "The Market for 'Lemons': Quality Uncertainty and the Market Mechanism," *Quarterly Journal of Economics* 84, no. 3 (1970): 488–500. While these insights have not been applied to America's late nineteenth-century mineral industry, much less the market for vertebrate fossils, historians of science have noted the prevalence of deception and fraud in both contexts. See Paul Lucier, *Scientists and Swindlers: Consulting on Coal and Oil in America, 1820–1890* (Baltimore: Johns Hopkins University Press, 2008). For an examination of technical solutions to the core problem of valuing mineral claims, see Eric C. Nystrom, *Seeing Underground: Maps, Models, and Mining Engineering in America* (Reno: University of Nevada Press, 2014). My analysis is also inspired by Clifford Geertz, "The Bazaar Economy: Information and Search in Peasant Marketing," *American Economic Review* 68, no. 2 (1978): 28–32; Clifford Geertz, "Suq: The Bazaar Economy in Sefrou," in *Meaning and Order in Moroccan Society* (Cambridge: Cambridge University Press, 1979), 123–244.

8. Martin J. Rudwick, *Earth's Deep History* (Chicago: University of Chicago Press, 2014), 31. See also Martin J. Rudwick, *Bursting the Limits of Time: The Reconstruction of Geohistory in the Age of Revolution* (Chicago: University of Chicago Press, 2005); Martin J. Rudwick, *Worlds before Adam: The Reconstruction of Geohistory in the Age of Reform* (Chicago: University of Chicago Press, 2008).

9. See David Sepkoski, "The Earth as Archive: Contingency, Narrative, and the History of Life," in *Science in the Archives: Pasts, Presents, Futures,* ed. Lorraine Daston (Chicago: University of Chicago Press, 2017), 53–84.

10. On the American mastodon, see Paul Semonin, *American Monster: How the Nation's First Prehistoric Creature Became a Symbol of National Identity* (New York: New York University Press, 2000), 84. On *Megatherium,* see Juan Pimentel, *The Rhinoceros and the Megatherium: An Essay in Natural History,* trans. Peter Mason (Cambridge, MA: Harvard University Press, 2017); José M. López Piñero, "Juan Bautista Bru (1740–1799) and the Description of the Genus Megatherium," *Journal of the History of Biology* 21, no. 1 (1988): 147–163.

11. Georges Cuvier, "Memoir on the Species of Elephants, Both Living and Fossil," in *Georges Cuvier, Fossil Bones, and Geological Catastrophes: New Translations and Interpretations of the Primary Texts,* by Martin J. Rudwick (Chicago: University of Chicago Press, 1997), 18–24; Georges Cuvier and Alexandre Brongniart, "Essay on the Mineral Geography of the Environs of Paris," in Rudwick, 133–156.

12. On the new temporal regime in general, see François Hartog, *Regimes of Historicity: Presentism and Experiences of Time,* trans. Saskia Brown (New York: Columbia University Press, 2015); Reinhart Koselleck, *Futures Past: On the Semantics of Historical Time* (Cambridge, MA: MIT Press, 1985). On its importance for the history of capitalism, see Jens Beckert, *Imagined Futures: Fictional Expectations and Capitalist Dynamics* (Cambridge, MA: Harvard University Press, 2016). On capitalists as especially forward looking, see Jonathan Levy, "Capital as Process and the History of Capitalism," *Business History Review* 91, no. 3 (2017): 483–510. The quotations by Georges Cuvier are from his "Preliminary Discourse" to the *Recherches sur les ossemens fossiles,* in Rudwick, *Georges Cuvier,* 186, 193.

13. Rudwick, 183, 217. On Cuvier's "law of correlation," see Gowan Dawson, *Show Me the Bone* (Chicago: University of Chicago Press, 2016). On projection and planning, see Martin Giraudeau, "Proving

Future Profit: Business Plans as Demonstration Devices," *Osiris* 33, no. 1 (2018): 130–148; Martin Gireaudeau and Frédéric Graber, eds., *Les Projets: Une histoire politique* (Paris: Presses des mines, 2018).

14. Shelley Emling, *The Fossil Hunter: Dinosaurs, Evolution, and the Woman Whose Discoveries Changed the World* (New York: Palgrave Macmillan, 2009).

15. See Everard Home, "Some Account of the Fossil Remains of an Animal More Nearly Allied to Fishes Than Any of the Other Classes of Animals," *Philosophical Transactions of the Royal Society of London* 104 (Jan. 1, 1814): 571–577; Everard Home, "Additional Facts respecting the Fossil Remains of an Animal, . . . Showing That the Bones of the Sternum Resemble Those of the Ornithorhynchus Paradoxus," *Philosophical Transactions of the Royal Society of London* 108 (Jan. 1, 1818): 24–32; and Everard Home, "An Account of the Fossil Skeleton of the Proteo-Saurus," *Philosophical Transactions of the Royal Society of London* 109 (Jan. 1, 1819): 209–211. See also Charles König, *Synopsis of the Contents of the British Museum,* 11th ed. (London: British Museum, 1817), 54.

16. William Conybeare, "On the Discovery of an Almost Perfect Skeleton of the Plesiosaurus," *Transactions of the Geological Society of London,* 2nd ser., pt. 2, no. 1 (1824): 381–389. For earlier speculations about this creature, see William Conybeare and Henry De la Beche, "Notice of a Discovery of a New Fossil Animal, Forming a Link between the Ichthyosaurus and the Crocodile," *Transactions of the Geological Society of London,* no. 5 (1821): 558–594.

17. See Simon J. Knell, *The Culture of English Geology, 1815–1851: A Science Revealed through Its Collecting* (Aldershot, UK: Ashgate, 2000). On the price of specimens, see Gideon Mantell, "A Few Notes on the Price of Fossils," *London Geological Journal* 1 (1846): 13–17. See also W. D. Ian Rolfe, "Fossil Sales," in *Natural History Auctions, 1700–1972* (London: Sotheby Parke Bernet, 1976), 32–38; and W. D. Ian Rolfe, Angela C. Milner, and F. G. Hay, "The Price of Fossils," *Special Papers in Paleontology* 40 (1988): 139–171.

18. Thomas Hawkins, *The Book of the Great Sea-Dragons* (London: W. Pickering, 1840). See also Victoria Carroll, "'Beyond the Pale of Ordinary Criticism': Eccentricity and the Fossil Books of Thomas Hawkins," *Isis* 98 (2007): 225–265.

19. See Christopher McGowan, *The Dragon Seekers: How an Extraordinary Circle of Fossilists Discovered the Dinosaurs and Paved the Way for Darwin* (Cambridge, MA: Perseus, 2001), 140, 144. On the controversy over the Hawkins collection, see *Report from the Select Committee on the Condition, Management and Affairs of the British Museum,* Ordered by the House of Commons, Aug. 6, 1835. See also "Official Letters and Minutes Arising from Meetings of the Trustees, 1821–1849: Hawkins Collection," DF103/1, Minutes of the Trustees, Archives of the NHM London.

20. See Martin J. Rudwick, "Charles Darwin in London: The Integration of Public and Private Science," *Isis* 73, no. 2 (1982): 186–206; and Alistair Sponsel, *Darwin's Evolving Identity: Adventure, Ambition, and the Sin of Speculation* (Chicago: University of Chicago Press, 2018). On how this division of labor helped shape the market for specimens, see Lukas Rieppel, "Albert Koch's Hydrarchos Craze: Credibility, Identity, and Authenticity in 19th Century Natural History," in *Science Museums in Transition: Cultures of Display in Nineteenth-Century Britain and America,* ed. Carin Berkowitz and Bernard Lightman (Pittsburgh: University of Pittsburgh Press, 2017), 139–161.

21. Anne Secord, "Corresponding Interests: Artisans and Gentlemen in Nineteenth-Century Natural History," *British Journal for the History of Science* 27, no. 4 (1994): 383–408; Anne Secord, "Science in the Pub: Artisan Botanists in Early Nineteenth-Century Lancashire," *History of Science* 32 (Sept. 1994): 269–315.

22. Quoted in Hugh Torrens, "Mary Anning (1799–1847) of Lyme; 'the Greatest Fossilist the World Ever Knew,'" *British Journal for the History of Science* 28, no. 3 (1995): 268–269.

23. For Buckland's description of *Megalosaurus,* see William Buckland, "Notice on the Megalosaurus or Great Fossil Lizard of Stonesfield," *Transactions of the Geological Society of London* 2, no. 1 (1824): 390–396. For Mantell's early descriptions, see Gideon Mantell, "Notice on the Iguanodon, a Newly Discovered

Fossil Reptile, from the Sandstone of Tilgate Forest, in Sussex," *Philosophical Transactions of the Royal Society of London* 115 (1825): 179–186; and Gideon Mantell, *The Geology of the South-East of England* (London: Longman, Rees, Orme, Brown, Green and Longman, 1833). For Owen's invention of dinosaurs, see Richard Owen, *Report on British Fossil Reptiles,* pt. 2, Report of the British Association for the Advancement of Science (London: printed by R. and J. E. Taylor, 1842), 103.

24. Richard Owen, *Geology and Inhabitants of the Ancient World* (London: Crystal Palace Library, 1854), 17. See also James A. Secord, "Monsters at the Crystal Palace," in *Models: The Third Dimension of Science,* ed. Soraya de Chadarevian and Nick Hopwood (Stanford, CA: Stanford University Press, 2004), 157–158.

25. Joseph Leidy, "Remarks Concerning Hadrosaurus," *Proceedings of the Academy of Natural Sciences in Philadelphia* 10 (1858): 215–218. See also Joseph Leidy, "Cretaceous Reptiles of the United States," *Smithsonian Contributions to Knowledge* 14, no. 192 (1865): 76–97. For a historical overview, see Valerie Bramwell and Robert McCracken Peck, *All in the Bones: A Biography of Benjamin Waterhouse Hawkins* (Philadelphia: Academy of Natural Sciences of Philadelphia, 2008).

26. Chris Manias, "Reconstructing an Incomparable Organism: The Chalicothere in Nineteenth and Early-Twentieth Century Palaeontology," *History and Philosophy of the Life Sciences* 40, no. 1 (2018): 22; Adrian Desmond, "Designing the Dinosaur: Richard Owen's Response to Robert Edmond Grant," *Isis* 70, no. 2 (1979): 224–234. See also Adrian Desmond, *Archetypes and Ancestors: Palaeontology in Victorian London, 1850–1875* (Chicago: University of Chicago Press, 1982); and Adrian J. Desmond, *The Politics of Evolution: Morphology, Medicine, and Reform in Radical London* (Chicago: University of Chicago Press, 1989).

27. Leidy, "Cretaceous Reptiles," 97.

28. For a description of Cope's discoveries in the New Jersey marl, see Cope, "August 21st," *Proceedings of the Academy of Natural Sciences of Philadelphia* 18 (1866): 275–279. In some instances, *Laelaps* was explicitly described as a "kangaroo-dinosaur." See, for example, Wilhelm Bölsche, *Tiere der Urwelt* (Hamburg: Selbst-Verlag der Kakao-Compagnie Theodor Reichardt, 1900), plate 14.

29. Othniel Charles Marsh, "Notice of New Jurassic Reptiles," *American Journal of Science* 18, no. 3 (1879): 501–505; Othniel Charles Marsh, "Principal Characters of American Jurassic Dinosaurs Part VI: Restorations of Brontosaurus," *American Journal of Science* 21, no. 3 (1883): 81–85. As a comprehensive review by two distinguished paleontologists explains, "The Como discoveries were not the earliest even in North America, but they were the first of truly spectacular magnitude." See Ostrom and McIntosh, *Marsh's Dinosaurs,* 2.

30. In addition, there are also scattered reports of dinosaur bones being sold at local curio shops in Wyoming and Colorado during the early 1870s. See Edward Drinker Cope, "On the Existence of Dinosauria in the Transition Beds of Wyoming," *Proceedings of the American Philosophical Society* 12 (1872): 481–483; Beth Southwell, "J. D. Conley's Cabinet of Curiosities and Other Early Wyoming Museums," *Annals of Wyoming* 76, no. 3 (2004): 24–32; and Samuel Wendell Williston, "American Jurassic Dinosaurs," *Transactions of the Kansas Academy of Science* 6 (1877): 42–43.

31. James LaPointe, *Legends of the Lakota* (San Francisco: Indian Historian Press, 1976), 16–19.

32. Richard Erdoes and Alfonso Ortiz, eds., *American Indian Myths and Legends* (New York: Pantheon Books, 1984), 222. On Native American knowledge about dinosaurs more broadly, see Adrienne Mayor, *Fossil Legends of the First Americans* (Princeton, NJ: Princeton University Press, 2005).

33. Henry Fairfield Osborn, *Cope: Master Naturalist* (Princeton, NJ: Princeton University Press, 1931), 432.

34. The story is not one of dispossession alone, however, as Native American storytellers and oral historians such as James LaPointe often engage in acts of creative reappropriation by invoking the word "dinosaur" when describing creatures such as the *Unktehi.* For more on the temporal dispossession of

Native American tribes by white paleontologists, see Kyla Schuller, "The Fossil and the Photograph: Red Cloud, Prehistoric Media, and Dispossession in Perpetuity," *Configurations* 24, no. 2 (2016): 233. See also Allison Dussias, "Science, Sovereignty, and the Sacred Text: Paleontological Resources and Native American Rights," *Maryland Law Review* 55 (1996): 84–159; Adrienne Mayor, "Suppression of Indigenous Fossil Knowledge: From Claverack, New York, 1705 to Agate Springs, Nebraska, 2005," in *Agnotology: The Making and Unmaking of Ignorance,* ed. Robert N. Proctor and Londa Schiebinger (Stanford: Stanford University Press, 2008), 163–182. This history has also led to calls for the repatriation of fossils dug up on Native American reservation land. See Rex Dalton, "Laws under Review for Fossils on Native Land," *Nature News* 449, no. 7165 (2007): 952–953; and Lawrence Bradley, "Dinosaurs and Indians: Paleontology Resource Dispossession from Sioux Lands" (PhD diss., University of Nebraska–Lincoln, 2010).

35. See Augustine Brannigan, *The Social Basis of Scientific Discoveries* (Cambridge: Cambridge University Press, 1981); Robert K. Merton, "Priorities in Scientific Discovery: A Chapter in the Sociology of Science," *American Sociological Review* 22, no. 6 (1957): 635–659; and Simon Schaffer, "Scientific Discoveries and the End of Natural Philosophy," *Social Studies of Science* 16, no. 3 (1986): 387–420.

36. See, for example, Kapil Raj, *Relocating Modern Science: Circulation and the Construction of Knowledge in South Asia and Europe, 1650–1900* (Houndmills, UK: Palgrave Macmillan, 2007); James Secord, *Victorian Sensation: The Extraordinary Publication, Reception, and Secret Authorship of Vestiges of the Natural History of Creation* (Chicago: University of Chicago Press, 2000); James Secord, "Knowledge in Transit," *Isis* 95, no. 4 (2004): 654–672; Simon Schaffer et al., eds., *The Brokered World: Go-Betweens and Global Intelligence, 1770–1820* (Sagamore Beach, MA: Science History Publications, 2009).

37. For the "invention" of America's mining frontier, see Karen Clay and Gavin Wright, "Gold Rush Legacy: American Minerals and the Knowledge Economy," in *Property in Land and Other Resources,* ed. Daniel H. Cole and Elinor Ostrom (Cambridge, MA: Lincoln Institute of Land Policy, 2012), 67–95; and Kent A. Curtis, *Gambling on Ore: The Nature of Metal Mining in the United States, 1860–1910,* Mining the American West (Boulder: University Press of Colorado, 2013). On America's late nineteenth-century specimen trade, see Mark Barrow, "The Specimen Dealer: Entrepreneurial Natural History in America's Gilded Age," *Journal of the History of Biology* 33, no. 3 (2000): 493–534; and Steve Ruskin, "The Business of Natural History," *Historical Studies in the Natural Sciences* 45, no. 3 (2015): 357–396. For an examination of the fossil trade in early nineteenth-century Latin America, see Irina Podgorny, "Fossil Dealers, the Practices of Comparative Anatomy and British Diplomacy in Latin America, 1820–1840," *British Journal for the History of Science* 46, no. 4 (2013): 647–674.

38. See Michael R. Haines, "Population, by Region and Urban-Rural Residence: 1790–1990," table Aa36-92 in *Historical Statistics of the United States,* ed. Susan B. Carter et al. (Cambridge: Cambridge University Press, 2006).

39. This vision was actively reinforced by boosters like William Gilpin who sang the region's praises loud and wide, often because they had made speculative investments in the land there. See, for example, William Gilpin, *The Central Gold Region: The Grain, Pastoral and Gold Regions of North America: With Some New Views of Its Physical Geography; and Observations on the Pacific Railroad* (Philadelphia: Sower, Barnes, 1860). For more on western boosterism in the second half of the nineteenth century, see William Cronon, "Dreaming the Metropolis," in *Nature's Metropolis: Chicago and the Great West* (New York: Norton, 1991), 23–55.

40. Karen Clay and Gavin Wright, "Order without Law? Property Rights during the California Gold Rush," *Explorations in Economic History* 42, no. 2 (2005): 155–183.

41. On the legal infrastructure, see Benjamin Horace Hibbard, *A History of the Public Land Policies* (Madison: University of Wisconsin Press, 1965). On government surveys, see William H. Goetzmann, *Exploration and Empire: The Explorer and the Scientist in the Winning of the American West* (New York: Norton, 1978); Thomas G. Manning, *Government in Science: The U.S. Geological Survey, 1867–*

1894 (Lexington: University of Kentucky Press, 1967); Amy Lee Kohout, "From the Field: Nature and Work on American Frontiers, 1876–1909" (PhD diss., Cornell University, 2015); and Jeremy Vetter, *Field Life: Science in the American West during the Railroad Era* (Pittsburgh: University of Pittsburgh Press, 2016).

42. Richard White, *Railroaded: The Transcontinentals and the Making of Modern America* (New York: Norton, 2011). For travel times, see Charles Oscar Paullin, *Atlas of the Historical Geography of the United States* (Washington, DC: Carnegie Institution of Washington, 1932).

43. Jeremy Vetter, "Science along the Railroad: Expanding Field Work in the US Central West," *Annals of Science* 61, no. 2 (2004): 187.

44. See Jaffe, *Gilded Dinosaur;* Elizabeth Noble Shor, *The Fossil Feud between E. D. Cope and O. C. Marsh* (New York: Exposition, 1974); and David Rains Wallace, *The Bonehunters' Revenge: Dinosaurs, Greed, and the Greatest Scientific Feud of the Gilded Age* (Boston: Houghton Mifflin, 1999).

45. Cope to Jesup, Apr. 9, 1888, Early Administrative Records, Archives of the AMNH; Schuchert and LeVene, *O. C. Marsh,* 268–273.

46. On the importance of eponymy for the scientific community, see Merton, "Priorities in Scientific Discovery," 635; and Stephen M. Stigler, "Stigler's Law of Eponymy," *Transactions of the New York Academy of Sciences* 39, no. 1 (1980): 147–157.

47. Cope, "August 21st," 279.

48. Beth Simmons and Katherine Honda, "Arthur Lakes: Founder of the Famed Colorado School of Mines Geology Museum," *Rocks and Minerals* 84 (Oct. 2009): 426–430.

49. Lakes to Marsh, Apr. 2 and 20, 1877, Marsh Papers.

50. Lakes to Marsh, Apr. 26, 1877, Marsh Papers.

51. Lakes to Marsh, May 23, 1877, Marsh Papers.

52. Lakes to Marsh, June 16, 1877 [dated 1876 in error], Marsh Papers.

53. Lakes to Marsh, June 27, 1877, Marsh Papers.

54. Lakes to Marsh, May 15, 1878, Marsh Papers.

55. Howard W. Bell, "Fossil-Hunting in Wyoming," *The Cosmopolitan,* Jan. 1900, 267.

56. See, for example, Reed to Marsh, n.d. [probably late 1881], Marsh Papers; and Brown to Osborn, May 2 and June 14, 1897, folder 6, Field Correspondence, box 1 (1891–1900), Archives of the DVP. Another example is "float," which referred to loose bits of ore or fossil skeleton that accumulated at the base of a hillside. See, for example, Reed to Holland, Dec. 22, 1898, and May 26, 1899, Big Bone Room Archive; and Douglas Gilmore, "Vertebrate Collecting in the Jurassic," 1901, folder 11, box 10, S. H. Knight Papers.

57. On Brown's entry into vertebrate paleontology, see Brown to Marsh, July 2, 1882, Marsh Papers.

58. For a survey of academic consulting in the nineteenth-century mineral industry, see Paul Lucier, "Commercial Interests and Scientific Disinterestedness: Consulting Geologists in Antebellum America," *Isis* 86, no. 2 (1995): 245–267; and Lucier, *Scientists and Swindlers.*

59. Other prominent examples of learned naturalists who transitioned to mining include Cope and Alexander Agassiz, who fared much better than Ward. See Persifor Frazer, "The Life and Letters of Edward Drinker Cope," *American Geologist* 26, no. 2 (1900): 67–128; Henry Fairfield Osborn, *Cope: Master Naturalist* (Princeton, NJ: Princeton University Press, 1931); and Mary P. Winsor, *Reading the Shape of Nature: Comparative Zoology at the Agassiz Museum* (Chicago: University of Chicago Press, 1991).

60. Office of the Midas Mining Co. report to stockholders, May 29, 1867, file drawer 7, Henry Augustus Ward Papers, Department of Rare Books, Special Collections, and Preservation, River Campus Libraries, University of Rochester.

61. Ward to Mumford, June 10, 1867, file drawer 7, Henry Augustus Ward Papers.

62. Ward to Mumford, Dec. 2, 1867, file drawer 7, Henry Augustus Ward Papers. See also Roswell Howell Ward, *Henry A. Ward: Museum Builder to America* (Rochester, NY: Rochester Historical Society, 1948), 140–142, 151. On Ward and nineteenth-century natural history supply houses in general, see Sally Gregory Kohlstedt, "Henry A. Ward: The Merchant Naturalist and American Museum Development," *Journal of the Society for the Bibliography of Natural History* 9 (1980): 647–661.

63. On the mutual entanglement of industry and the academy in nineteenth-century geology, see Lucier, *Scientists and Swindlers.*

64. On how the market for mineral prospecting was structured, see Rodman Paul, *Mining Frontiers of the Far West, 1848–1880* (New York: Holt, Rinehart and Winston, 1963). See also Clay and Wright, "Order without Law?"; John Hittell, *Mining in the Pacific States of North America* (San Francisco: H. H. Bancroft, 1861); Rodman Paul, *California Gold: The Beginning of Mining in the Far West* (Cambridge, MA: Harvard University Press, 1947); Charles Howard Shinn, *Land Laws of Mining Districts,* Johns Hopkins University Studies in Historical and Political Science, 2nd ser., no. 12 (Baltimore: N. Murray, publication agent, Johns Hopkins University, 1884); and Frederick Jackson Turner, *The Frontier in American History* (Tucson: University of Arizona Press, 1986).

65. This made mineral prospects a species of what modern economists call "experience products." See A. Kirmani and A. R. Rao, "No Pain, No Gain: A Critical Review of the Literature on Signaling Unobservable Product Quality," *Journal of Marketing* 64, no. 2 (2000): 66–79; and Phillip Nelson, "Advertising as Information," *Journal of Political Economy* 82, no. 4 (1974): 729–754.

66. Richard Stretch, *Prospecting, Locating, and Valuing Mines* (New York: Scientific Publishing, 1899), 165.

67. Arthur Lakes, *Prospecting for Gold and Silver* (Scranton, PA: Colliery Engineer, 1895), 177, 250. See also Hittell, *Mining,* 208–209.

68. Breithaupt, "William Harlow Reed"; A. Dudley Gardner, *Forgotten Frontier: A History of Wyoming Coal Mining* (Boulder, CO: Westview, 1989); Lakes, *Prospecting for Gold and Silver,* 177; Carlin and Reed to Marsh, July 19, 1877, Marsh Papers.

69. Carlin and Reed to Marsh, July 19, 1877, Marsh Papers.

70. Carlin and Reed to Marsh, Aug. 16, 1877, Marsh Papers.

71. See Carlin and Reed to Marsh, Aug. 16, 1877, Marsh Papers.

72. Mash to Carlin and Reed, Oct. 20, 1877, Marsh Papers.

73. Carlin and Reed to Marsh, Oct. 28, 1877, Marsh Papers.

74. Carlin and Reed to Marsh, Oct. 31, 1877, Marsh Papers.

75. Elizabeth Noble Shor, *Fossils and Flies: The Life of a Compleat Scientist* (Norman: University of Oklahoma Press, 1971), 89.

76. Shor, 91.

77. Williston to Marsh, Nov. 16, 1877, Marsh Papers.

78. Carlin and Reed to Marsh, Sept. 19, 1877, Marsh Papers.

79. Williston to Marsh, Dec. 13, 1877, Marsh Papers.

80. Carlin to Marsh, Apr. 1, 1878, Marsh Papers.

81. Williston to Marsh, Dec. 18, 1877, Marsh Papers.

82. Reed to Marsh, Dec. 22, 1877, Marsh Papers.

83. Reed to Marsh, Dec. 22, 1877. In a last-ditch effort to make his dinosaurs pay, Reed claimed to have found another prospect, but according to Williston he refused to "give the slightest idea of their locality." Williston to Marsh, Dec. 23, 1877, Marsh Papers.

84. Reed to Marsh, Jan. 12, 1878, Marsh Papers.

85. Of the three 1877 discoveries, the one made by Oramel Lucas is the least well documented and therefore has not been discussed here. An informative glimpse of Lucas's pecuniary motivations in contacting Cope to announce his discovery is nonetheless available in the form of a short, autobiographical account that he dictated to his daughter near the end of his life. See Oramel Lucas and Ethel Lucas, "Discovering Dinosaur Bones in Colorado," Local History Center, Canon City Public Library, Canon City, CO.

86. See Naomi R. Lamoreaux and Daniel M. G. Raff, eds., *Coordination and Information: Historical Perspectives on the Organization of Enterprise* (Chicago: University of Chicago Press, 1995); Lamoreaux, Raff, and Temin, "Beyond Markets and Hierarchies"; Oliver E. Williamson, "Markets and Hierarchies: Some Elementary Considerations," *American Economic Review* 63, no. 2 (1973): 316–325; Oliver E. Williamson, "The Modern Corporation: Origins, Evolution, Attributes," *Journal of Economic Literature* 19, no. 4 (1981): 1537–1568; and Oliver Williamson, *Economic Organization: Firms, Markets, and Policy Control* (New York: New York University Press, 1986).

2. Tea with *Brontosaurus*

1. Walter L. Beasley, "The Giant Brontosaurus in Central Park," *Los Angeles Herald*, Sunday Supplement, Apr. 2, 1905.

2. "500 to Drink Tea under Big Dinosaur," *New York Times*, Feb. 15, 1905.

3. On sites for the circulation of knowledge among a popular audience, see, for example, Toby Appel, "Science, Popular Culture, and Profit: Peale's Philadelphia Museum," *Journal of the Society for the Bibliography of Natural History* 9 (1980): 619–634; Peter Bowler, *Science for All: The Popularization of Science in Early Twentieth-Century Britain* (Chicago: University of Chicago Press, 2009); Aileen Fyfe and Bernard Lightman, *Science in the Marketplace: Nineteenth-Century Sites and Experiences* (Chicago: University of Chicago Press, 2007); Bernard V. Lightman, *Victorian Popularizers of Science: Designing Nature for New Audiences* (Chicago: University of Chicago Press, 2007); and Joe Kember, John Plunkett, and Jill A. Sullivan, eds., *Popular Exhibitions, Science and Showmanship, 1840–1910* (London: Pickering and Chatto, 2012).

4. Lorraine Daston and Katharine Park, *Wonders and the Order of Nature, 1150–1750* (New York: Zone Books, 1998); Paula Findlen, *Possessing Nature: Museums, Collecting, and Scientific Culture in Early Modern Italy* (Berkeley: University of California Press, 1994); Mark Meadow, "Merchants and Marvels: Hans Jacob Fugger and the Origins of the Wunderkammer," in *Merchants and Marvels*, ed. Pamela Smith and Paula Findlen (London: Routledge, 2002), 182–200. See also Pierre Bourdieu, *Distinction: A Social Critique of the Judgment of Taste* (London: Routledge and Kegan Paul, 1986); Paul DiMaggio, "Cultural Entrepreneurship in Nineteenth-Century Boston," *Media, Culture, and Society* 4 (1982): 33–50. On philanthropy in particular, see Francie Ostrower, *Why the Wealthy Give: The Culture of Elite Philanthropy* (Princeton, NJ: Princeton University Press, 1995).

5. See Jeffrey Abt, "The Origins of the Public Museum," in *A Companion to Museum Studies*, ed. Sharon Macdonald (Malden, MA: Wiley Blackwell, 2006), 115–134; Andrew McClellan, *Inventing the Louvre: Art, Politics, and the Origins of the Modern Museum in Eighteenth-Century Paris* (Cambridge: Cambridge University Press, 1994); and Joel J. Orosz, *Curators and Culture: The Museum Movement in America, 1740–1870* (Tuscaloosa: University of Alabama Press, 1990). For a Foucauldian perspective, see Tony Bennett, *The Birth of the Museum: History, Theory, Politics* (London: Routledge, 1995).

6. See Duncan F. Cameron, "The Museum, a Temple or the Forum," *Curator: The Museum Journal* 14, no. 1 (1971): 11–24. See also Les Harrison, *The Temple and the Forum: The American Museum and Cultural Authority in Hawthorne, Melville, Stowe, and Whitman* (Tuscaloosa: University of Alabama Press, 2007).

7. Karen Rader and Victoria Cain, *Life on Display* (Chicago: University of Chicago Press, 2014); Steven Conn, *Museums and American Intellectual Life, 1876–1926* (Chicago: University of Chicago Press, 2000).

8. For a detailed account of this process, see Paul Brinkman, *The Second Jurassic Dinosaur Rush: Museums and Paleontology in America at the Turn of the Twentieth Century* (Chicago: University of Chicago Press, 2010).

9. Jeremy Atack and Fred Bateman, "Physical Output of Selected Manufactured Products: 1860–1997," table Dd366-436 in *Historical Statistics of the United States,* ed. Susan B. Carter et al. (New York: Cambridge University Press, 2006).

10. Richard Sutch, "Gross Domestic Product: 1790–2002," table Ca9-19 in Carter et al., *Historical Statistics.*

11. Peter H. Lindert, "Distribution of Household Wealth: 1774–1998," table Be39-46 in Carter et al., *Historical Statistics.* See also Thomas Piketty, *Capital in the Twenty-First Century,* trans. Arthur Goldhammer (Cambridge, MA: Belknap Press of Harvard University Press, 2014).

12. On the consolidation of an "American bourgeoisie," see Sven Beckert, *The Monied Metropolis: New York City and the Consolidation of the American Bourgeoisie, 1850–1896* (Cambridge: Cambridge University Press, 2001); and Michael Zakim and Gary John Kornblith, "Introduction: An American Revolutionary Tradition," in *Capitalism Takes Command: The Social Transformation of Nineteenth-Century America,* ed. Michael Zakim and Gary John Kornblith (Chicago: University of Chicago Press, 2012), 1–12. Histories of the American bourgeoisie draw inspiration from a literature on bourgeois class formation in Europe, especially Germany. See, for example, David Blackbourn and Richard J. Evans, eds., *The German Bourgeoisie: Essays on the Social History of the German Middle Class from the Late Eighteenth to the Early Twentieth Century* (London: Routledge, 1991); Jürgen Kocka, "Bürgertum und bürgerliche Gesellschaft im 19. Jahrhundert," in *Bürgertum im 19. Jahrhundert,* ed. Jürgen Kocka (Munich: Deutscher Taschenbuch Verlag, 1988); and Jürgen Kocka and Allan Mitchell, eds., *Bourgeois Society in Nineteenth-Century Europe* (Oxford: Berg, 1993).

13. Albert S. Bickmore, "An Autobiography with a Historical Sketch of the Founding and Early Development of the American Museum of Natural History," 1908, 1:22–23, Rare Books and Manuscript Division, Library of the AMNH.

14. Bickmore, 2:2.

15. Bickmore to Dodge, Nov. 13, 1868, Albert Bickmore Correspondence, folder 35, box 2, Archives of the AMNH.

16. See *The First Annual Report of the American Museum of Natural History* (New York: printed for the museum, by Major and Knapp Engraving, Manufacturing, and Lithographic, 1870), 17.

17. Sven Beckert and Julia Rosenbaum, introduction to *The American Bourgeoisie: Distinction and Identity in the Nineteenth Century,* ed. Sven Beckert and Julia Rosenbaum (New York: Palgrave Macmillan, 2010), 3.

18. James J. Ayers, *Gold and Sunshine, Reminiscences of Early California* (Boston: R. G. Badger, 1922), 281. See also Wayne Craven, *Gilded Mansions: Grand Architecture and High Society* (New York: W. W. Norton, 2009); and Mrs. Charles Morgan, ed., *Artistic Houses* (New York: D. Appleton, 1883). For a particularly extreme set of practices, see Ruth Brandon, *The Dollar Princesses: Sagas of Upward Nobility, 1870–1914* (New York: Knopf, 1980).

19. Harold John Cook, *Matters of Exchange,* 1–42; Katie Whitaker, "The Culture of Curiosity," in *Cultures of Natural History,* ed. Nicholas Jardine, James A. Secord, and E. C. Spary (Cambridge: Cambridge University Press, 1996), 75–90. On the long history of museums as tools for the performance of social distinction, see Lukas Rieppel, "Museums and Gardens," in *A Companion to the History of Science,* ed. Bernard V. Lightman (Hoboken, NJ: Wiley, 2015), 238–251.

20. On public museums of natural history in Europe, see, for example, Samuel Alberti, *Nature and Culture: Objects, Disciplines and the Manchester Museum* (Manchester: Manchester University Press, 2009); Nyhart, *Modern Nature;* Dorinda Outram, *Georges Cuvier: Vocation, Science, and Authority in Post-Revolutionary France* (Manchester: Manchester University Press, 1984); and Emma Spary, *Utopia's Garden: French Natural History from Old Regime to Revolution* (Chicago: University of Chicago Press, 2000).

21. Thorstein Veblen, *The Theory of the Leisure Class: An Economic Study of Institutions* (New York: Macmillan, 1899). See also DiMaggio, "Classification in Art"; Lawrence Levine, *Highbrow/Lowbrow: The Emergence of Cultural Hierarchy in America* (Cambridge, MA: Harvard University Press, 1988); and John Ott, "The Manufactured Patron," in Beckert and Rosenbaum, *American Bourgeoisie,* 257–276. For an analysis of the way philanthropy "defines the cultural identity and organizational boundaries of the elite," see Ostrower, *Why the Wealthy Give,* 28.

22. Quoted in Winifred Eva Howe, *A History of the Metropolitan Museum of Art* (New York: published for the Metropolitan Museum of Art by Columbia University Press, 1913), 200.

23. On objectivity as an epistemic virtue that became particularly widespread during the nineteenth century, see Lorraine Daston and Peter Galison, *Objectivity* (New York: Zone Books, 2007); and Theodore Porter, *Trust in Numbers: The Pursuit of Objectivity in Science and Public Life* (Princeton, NJ: Princeton University Press, 1995).

24. Jonathan Crary, *Suspensions of Perception: Attention, Spectacle, and Modern Culture* (Cambridge, MA: MIT Press, 1999), 13–14.

25. Sally Gregory Kohlstedt, *Teaching Children Science: Hands-on Nature Study in North America, 1890–1930* (Chicago: University of Chicago Press, 2010).

26. See Steven Shapin and Barry Barnes, "Science, Nature and Control: Interpreting Mechanics' Institutes," *Social Studies of Science* 7, no. 1 (1977): 31–74; and E. P. Thompson, "Time, Work-Discipline, and Industrial Capitalism," *Past and Present,* no. 38 (1967): 56–97.

27. *The First Annual Report of the American Museum of Natural History* (New York: printed for the museum, by Major and Knapp Engraving, Manufacturing, and Lithographic, 1870), 17.

28. See Roy Rosenzweig and Elizabeth Blackmar, *The Park and the People: A History of Central Park* (Ithaca: Cornell University Press, 1992).

29. *Sixth Annual Report of the Board of Commissioners of the Central Park* (New York: Wm. C. Bryant, 1863), 15.

30. On Green's negotiations with Hawkins, see the *Twelfth Annual Report of the Board of Commissioners of the Central Park for the Year Ending December 31, 1868* (New York: Evening Post Steam Presses, 1868), 132–134.

31. Thomas Jefferson, *Notes on the State of Virginia* (London: J. Stockdale, 1787), 71. For Buffon's critique, see Georges Louis Leclerc Buffon and Louis-Jean-Marie Dauberton, *Histoire naturelle, générale et particulière, avec la description du cabinet du roi* (Paris: L'Imprimerie royale, 1761), 9:87. For a general overview, see Paul Semonin, *American Monster: How the Nation's First Prehistoric Creature Became a Symbol of National Identity* (New York: New York University Press, 2000).

32. On the North American valorization of wilderness, see Roderick Nash, *Wilderness and the American Mind* (New Haven, CT: Yale University Press, 1967). On science and America's sense of inferiority to Europe, see Robert V. Bruce, *The Launching of Modern American Science, 1846–1876* (New York: Knopf, 1987); and George H. Daniels, *Science in American Society: A Social History* (New York: Knopf, 1971).

33. *Twelfth Annual Report,* 125, 129.

34. *First Annual Report of the Board of Commissioners of the Department of Public Parks for the Year Ending May 1, 1871* (New York: William C. Bryant, 1871), 19.

35. "Meeting of the Lyceum of Natural History—Prof. Waterhouse Hawkins' Report on the Paleozoic Museum at Central Park," *New York Times,* Mar. 7, 1871, 5. On the story, see Edwin H. Colbert and

Katharine Beneker, "The Paleozoic Museum in Central Park, or the Museum That Never Was," *Curator* 2 (1959): 137–150; and Adrian J. Desmond, "Central Park's Fragile Dinosaurs," *Natural History* 83 (1974): 65–71.

36. Dodge to William Haines, Mar. 2, 1869, quoted in Kennedy, "Philanthropy and Science," 43.

37. Beckert, *Monied Metropolis,* 172–204.

38. Bickmore, "Autobiography," 2:8–11.

39. "Reception of the American Museum of Natural History," *New York Times,* Mar. 6, 1872, 3.

40. For a survey of various amusements on offer in American urban centers such as New York at the time, see David Nasaw, *Going Out: The Rise and Fall of Public Amusements* (New York: Basic Books, 1993).

41. In her survey of nineteenth-century dime museums, Andrea Stulman Dennett lists over thirty-five such institutions in New York City alone. Andrea Stulman Dennett, *Weird and Wonderful: The Dime Museum in America* (New York: New York University Press, 1997), 152. See also Brooks McNamara, "A Congress of Wonders: The Rise and Fall of the Dime Museum," *Emerson Society Quarterly* 20, no. 3 (1974): 216–234. For more on Barnum's museum in particular, see Katherine Pandora, "The Permissive Precincts of Barnum's and Goodrich's Museums of Miscellaneity: Lessons in Knowing Nature for New Learners," in Lightman and Berkowitz, *Science Museums in Transition,* 36–66.

42. Phineas T. Barnum, *Life of P. T. Barnum, Including His Golden Rules for Money-Making* (Buffalo, NY: Courier, 1888), 63.

43. Barnum, 56–57. For another description of Barnum's museum at the time, see *Picture of New York in 1846* (New York: C. S. Francis, 1846).

44. Barnum, *Life of P. T. Barnum,* 57, 59, 62. See also *Barnum's American Museum Illustrated,* 1866, MWEZ nc 4025, Billy Rose Theater Division, NYPL.

45. *Illustrated and Descriptive History of Barnum's Museum,* 1866, v, MWEZ nc 4025, Billy Rose Theater Division.

46. Barnum, *Life of P. T. Barnum,* 265.

47. Letter reprinted in *New York World,* Mar. 19, 1867. See also P. T. Barnum, *Struggles and Triumphs: Or, Forty Years' Recollections of P. T. Barnum* (Hartford, CT: J. B. Burr, 1869), p. 566; Barnum to Agassiz, Feb. 13, 1882, Rare Books and Manuscripts Division, Ernst Mayr Library, MCZ. For comparison, see also Louis Agassiz, "A Living Whale," *The Boston Journal,* May 27, 1861, and the advertisement for the Boston Aquarial Gardens, *Boston Post,* Feb. 10, 1862, Rare Books and Manuscripts Division, Ernst Mayr Library, MCZ.

48. Andrew McClellan, "P. T. Barnum, Jumbo the Elephant, and the Barnum Museum of Natural History at Tufts University," *Journal of the History of Collections* 24, no. 1 (2012): 45. See also John Betts, "P. T. Barnum and the Popularization of Natural History," *Journal of the History of Ideas* 20, no. 3 (1959): 353–358.

49. "Dearth of Museum Freaks," *New York Times,* Oct. 15, 1899.

50. "Attacking Low Resorts," *New York Times,* Nov. 29, 1883.

51. *Phelps' Strangers and Citizens' Guide to New York City* (New York: Watson, 1865); "Their Experience in a Museum," *New York Times,* July 7, 1885.

52. See Jan Bondeson, *The Feejee Mermaid and Other Essays in Natural and Unnatural History* (Ithaca, NY: Cornell University Press, 1999).

53. Barnum, *Life of P. T. Barnum,* 65.

54. "A Grand National Museum in the Park," *New York Herald,* Apr. 30, 1866, 4; "A National Museum Wanted," *New York Herald,* July 12, 1866, 4; "A Museum without Humbug," *New York Times,* Mar. 18, 1868, 4.

55. See, for example, Johnson Jones Hooper, *Some Adventures of Captain Simon Suggs, Late of the Tallapoosa Volunteers; Together with "Taking the Census," and Other Alabama Sketches* (Philadelphia: Carey

and Hart, 1845); and Herman Melville, *The Confidence-Man: His Masquerade* (London: Longman, Brown, Green, Longmans and Roberts, 1857).

56. James Cook, *The Arts of Deception: Playing with Fraud in the Age of Barnum* (Cambridge, MA: Harvard University Press, 2001), 77–78. For a similar kind of interpretation, see Neil Harris, *Humbug: The Art of P. T. Barnum* (Boston: Little, Brown, 1973).

57. On the architecture of natural history museums, see Carla Yanni, *Nature's Museums: Victorian Science and the Architecture of Display* (Baltimore: Johns Hopkins University Press, 1999).

58. John David Wolfe to William A. Haines, Nov. 15, 1871, quoted in Kennedy, "Philanthropy and Science," 60.

59. Morris K. Jesup to Barnum, Apr. 20, 1889, Early Administrative Records, Archives of the AMNH.

60. R. P. Whitfield, *Visitors' Guide to the Collections of Shells, Minerals, and Fossils in the American Museum of Natural History* (New York: Steam Printing House of Wm. C. Martin, 1885).

61. John Edward Gray, "On Museums, Their Use and Improvement, and on the Acclimatization of Animals," *Annals and Magazine of Natural History* 14 (1864): 284.

62. On the history of the New Museum Idea, see Steven Conn, *Do Museums Still Need Objects?*, Arts and Intellectual Life in Modern America (Philadelphia: University of Pennsylvania Press, 2010); Nyhart, *Modern Nature;* and Rader and Cain, *Life on Display.*

63. "Museum of Natural History," *New York Times,* May 30, 1874, 2.

64. "A Great School of Science," *New York Times,* Apr. 15, 1877, 10.

65. "Natural History Museum," *New York Times,* June 3, 1874.

66. Othniel Charles Marsh, "Address of O. C. Marsh," *Popular Science Monthly* 12 (1878): 475–476.

67. See the exchange between Allen and Jesup in a series of letters dated Feb. 9, 11, 14, and 28, 1885, Early Administrative Records, Archives of the AMNH.

68. Newberry to Jesup, Nov. 12, 1881, Early Administrative Records, Archives of the AMNH.

69. Marsh, "Address," 476.

70. Quoted in W. E. Decrow, *Yale and "The City of Elms"* (Boston: W. E. Decrow, 1882).

71. Darwin to Marsh, Aug. 31, 1880, Marsh Papers.

72. Robert Parr Whitfield to Jesup, Feb. 6, 1889, Early Administrative Records, Archives of the AMNH.

73. Osborn to Jesup, Apr. 18, 1891, Correspondence Relative to the Formation and Organization of the DVP, Archives of the AMNH. See also Osborn to Seth Low, Mar. 1, 1890, box 117, Osborn Papers, AMNH; and Osborn to Jesup, Feb. 2, 1891, Correspondence Relative to the Formation and Organization of the DVP, Archives of the AMNH.

74. See letters between Jesup and Osborn, June 2 and June 9, 1891, Correspondence Relative to the Formation and Organization of the DVP, Archives of the AMNH.

75. Brown to Osborn, June 17, 1897, folder 6, box 1, Field Correspondence, Archives of the DVP. See also Brown to Osborn, May 2, 6, 8, and 21, 1897, folder 6, box 1, Field Correspondence, Archives of the DVP.

76. Osborn to Granger, July 13, 1899, folder 8, box 1, Field Correspondence, Archives of the DVP. For a more detailed account, see Brinkman, *Second Jurassic Dinosaur Rush,* 30–63.

77. *Annual Report of the President* (New York: American Museum of Natural History, 1904), 17. See also *Annual Report of the Department of Vertebrate Paleontology,* 1898, 1:1, box 1, Archives of the DVP.

78. W. D. Matthew, "Report of the Associate Curator," *Annual Report of the Department of Vertebrate Paleontology,* 1901, p. 41, box 1, Archives of the DVP.

79. *Annual Report of the President* (New York: American Museum of Natural History, 1898), 17. Writing about two dinosaur displays that went on exhibit at the same time, W. D. Matthew announced that they "attract much attention and comment from visitors." See W. D. Matthew, "Cataloguing and Exhibition," p. 2, in *Annual Report of the Department of Vertebrate Paleontology,* 1898.

80. See "Report of the Curators," *Proceedings of the Academy of Natural Sciences in Philadelphia* 21 (1869): 234–235.

81. W. S. W. Ruschenberger and George W. Tryon Jr., *Guide to the Museum of the Academy of Natural Sciences of Philadelphia* (Philadelphia: Academy of Natural Sciences of Philadelphia, 1876), 115.

82. "Report of the Curators," 235. See also Valerie Bramwell and Robert McCracken Peck, *All in the Bones: A Biography of Benjamin Waterhouse Hawkins* (Philadelphia: Academy of Natural Sciences of Philadelphia, 2008).

83. O. C. Marsh to S. F. Baird, Dec. 20, 1875, folder 23, box 29, RU 7002: Spencer Fullerton Baird Papers, SIA. See also Charles Schuchert and Clara Mae LeVene, *O. C. Marsh, Pioneer in Paleontology* (New Haven, CT: Yale University Press, 1940), 294–296.

84. Othniel Charles Marsh, "Restoration of Some European Dinosaurs, with Suggestions as to Their Place among the Reptilia," *American Journal of Science* 1, no. 299 (1895): 407–408.

85. *Annual Report of the President* (New York: American Museum of Natural History, 1905), 17, 29.

86. Attendance jumped from 458,451 in 1899 to 565,489 in 1905 before peaking at 1,043,582 in 1908. In subsequent years, it tended to hover around 800,000 or 900,000. See the annual reports of the American Museum of Natural History.

87. "Old and Young Call to See the Dinosaur," *New York Times,* Feb. 20, 1905, 12.

88. Franz Boas, "Some Principles of Museum Administration," *Science,* n.s., no. 25 (1907): 921, 922, 924.

89. *Annual Report of the President* (New York: American Museum of Natural History, 1906), 2.

90. "Fun in a Fossil," *Punch,* Jan. 21, 1854, 24. See also "The Crystal Palace at Sydenham," *Illustrated London News,* Jan. 7, 1854, 22.

91. James A. Secord, "Monsters at the Crystal Palace," in *Models: The Third Dimension of Science,* ed. Soraya de Chadarevian and Nick Hopwood (Stanford, CA: Stanford University Press, 2004), 145–146.

3. Andrew Carnegie's *Diplodocus*

1. See "Invitation" and "Ticket of Admission," May 12, 1905, DF1011, printed forms, nos. 550 and 551, Archives of the NHM London.

2. "The Presentation of a Reproduction of *Diplodocus carnegii,*" *Annals of the Carnegie Museum* 3 (1905), 452.

3. "Moral Reflections at the Natural History Museum," *Punch; or the London Charivari,* May 16, 1906, Press Cuttings, Vol. 2, Part I, DF5014/1/2, Archives of the NHM London. Andrew Carnegie interpreted his gesture differently, framing the donation as a celebration of the free trade in ideas because it showed that both sides of the Atlantic could "enrich each other without in the least depleting ourselves." See Presentation of *Diplodocus,* May 12, 1905, Press Cuttings, vol. 2, pt. 1, DF5014/1/2, Archives of the NHM London.

4. My account here and elsewhere is indebted to Ilja Nieuwland, *American Dinosaur Abroad: A Cultural History of Carnegie's Plaster Diplodocus* (Pittsburgh: University of Pittsburgh Press, 2019); Tom Rea, *Bone Wars: The Excavation and Celebrity of Andrew Carnegie's Dinosaur* (Pittsburgh: University of Pittsburgh Press, 2001).

5. For that reason, philanthropy deserves to be at the center of the history of capitalism. See Olivier Zunz, "Why Is the History of Philanthropy Not a Part of American History?," in *Philanthropy in Democratic Societies: History, Institutions, Values,* ed. Rob Reich, Lucy Bernholz, and Chiara Cordelli (Chicago: University of Chicago Press, 2016), 44–64. On the history of philanthropy in nineteenth-century

America more broadly, see Peter Dobkin Hall, *Inventing the Nonprofit Sector and Other Essays on Philanthropy, Voluntarism, and Nonprofit Organizations* (Baltimore: Johns Hopkins University Press, 1992); Helen Lefkowitz Horowitz, *Culture and the City: Cultural Philanthropy in Chicago from the 1880's to 1917* (Lexington: University Press of Kentucky, 1976); Kathleen D. McCarthy, *American Creed: Philanthropy and the Rise of Civil Society, 1700–1865* (Chicago: University of Chicago Press, 2003); Reich, Bernholz, and Cordelli, *Philanthropy in Democratic Societies;* and Olivier Zunz, *Philanthropy in America: A History* (Princeton, NJ: Princeton University Press, 2012).

6. Harold Joseph Laski, *The Dangers of Obedience and Other Essays* (New York, London: Harper and Brothers, 1930). See also Robert F. Arnove, ed., *Philanthropy and Cultural Imperialism: The Foundations at Home and Abroad* (Bloomington: Indiana University Press, 1982); and Barry Karl and Stanley Katz, "The American Private Philanthropic Foundation and the Public Sphere, 1890–1930," *Minerva* 19, no. 2 (1981): 236–270. For a similar critique of philanthropy in contemporary politics, see Jane Mayer, *Dark Money: The Hidden History of the Billionaires behind the Rise of the Radical Right* (New York: Doubleday, 2016); and Patrick Radden Keefe, "Empire of Pain," *New Yorker,* October 30, 2017.

7. See Charles Harvey et al., "Andrew Carnegie and the Foundations of Contemporary Entrepreneurial Philanthropy," *Business History* 53, no. 3 (2011): 425–450. The literature on the importance of social capital for allocating financial capital is vast, but see, for example, Naomi R. Lamoreaux, *Insider Lending: Banks, Personal Connections, and Economic Development in Industrial New England* (Cambridge: Cambridge University Press, 1994). For a more sociological perspective, see Howard E. Aldrich and Catherine Zimmer, "Entrepreneurship through Social Networks," in *The Art and Science of Entrepreneurship,* ed. Donald L. Sexton and Raymond W. Smilor (Cambridge, MA: Ballinger, 1986); Joel M. Podolny, "Market Uncertainty and the Social Character of Economic Exchange," *Administrative Science Quarterly* 39, no. 3 (1994): 458–483; and Scott Shane and Daniel Cable, "Network Ties, Reputation, and the Financing of New Ventures," *Management Science* 48, no. 3 (2002): 364–381.

8. Andrew Carnegie, "The Road to Business Success," in *The Empire of Business* (Toronto: William Briggs, 1902), 8.

9. On the way social ties mediated access to financial capital during the Gilded Age, see Susie Pak, *Gentlemen Bankers: The World of J. P. Morgan* (Cambridge, MA: Harvard University Press, 2013).

10. Reed to Marsh, Feb. 24, 1879, and Nov. 1881 [exact date unknown], Marsh Papers.

11. Reed to Samuel Wendell Williston, Feb. 18, 1879, and Reed to Marsh, Mar. 14, 1879, Marsh Papers.

12. Reed to Marsh, Oct. 23 and Nov. 22, 1882, Marsh Papers.

13. Reed to Marsh, July 18, 1884, Marsh Papers.

14. Reed to Marsh, Nov. 23, 1885, and Marsh to Reed, Feb. 24, 1886, Marsh Papers.

15. Reed to Marsh, Aug. 3 and Sept. 12, 1890, Marsh Papers.

16. For Knight's courses, see *Catalogue for 1897–98 and Announcements for 1898–99,* esp. 4, Grace Raymond Hebard Collection, University of Wyoming, Coe Library; and President's Annual Report, June 27, 1893, folder 26, box 12, President's Records, AHC. For his correspondence, see Samuel H. Knight Papers, esp. box 50, AHC. For his research output, see, for example, Wilbur Knight, "The Coal Mines of Wyoming," *Mineral Industry,* 1894; Wilbur Knight, "The Petroleum Industry of Wyoming," *American Manufacture and Iron World,* May 29, 1896; Wilbur Knight, "The Salt Creek Oil Field," *Engineering and Mining Journal,* January 1896; Wilbur Knight, *The Rich Oil Fields of Wyoming* (Evanston, WY, 1902).

17. President's Annual Report, June 26, 1894, folder 26, box 12, President's Records, AHC.

18. See Knight to Johnson, Apr. 13, 1896, folder 9, box 50, S. H. Knight Papers.

19. *Catalogue for 1897–98,* esp. 125.

20. Reed to Marsh, Sept. 17, 1895, Marsh Papers.

21. Marsh to Reed, Oct. 9, 1895, Marsh Papers.

22. Reed to Marsh, Nov. 22, 1895, Marsh Papers.

23. Report of the Executive Committee of the Board of Trustees, July 19, 1896, President's Records, AHC.

24. Reed to Marsh, Sept. 30, 1898, Marsh Papers.

25. Knight to Osborn, Mar. 23, 1897, folder 10, box 50, S. H. Knight Papers.

26. Brown to Osborn, Mar. 30, 1897, folder 6, box 1, Field Correspondence, Archives of the DVP.

27. Reed to Marsh, Sept. 30, 1898, Marsh Papers.

28. Reed to Marsh, Nov. 23, 1898, Marsh Papers.

29. See *Laramie Boomerang,* Dec. 3, 1898.

30. See, for example, *St. Louis Globe-Democrat,* Nov. 28, 1898; *Omaha World Herald,* Dec. 4 and 18, 1898; *Manawatu (New Zealand) Herald,* April 15, 1899; *Otago (New Zealand) Daily Times,* Apr. 17, 1899; and *Wanganui (New Zealand) Herald,* Apr. 22, 1899.

31. *New York Journal,* Dec. 11, 1898.

32. *New York Journal,* Dec. 11, 1898.

33. Clipping from *New York Post,* Dec. 1, 1898, Big Bone Room Archive.

34. For a detailed biography of Carnegie, see David Nasaw, *Andrew Carnegie* (New York: Penguin, 2006). For a much shorter introduction, see Harold C. Livesay, *Andrew Carnegie and the Rise of Big Business* (New York: Longman, 2000).

35. Lance Davis, "The Capital Markets and Industrial Concentration: The U.S. and the U.K., a Comparative Study," *Economic History Review* 19, no. 2 (1966): 264.

36. See Livesay, *Andrew Carnegie,* 93–94; Nasaw, *Andrew Carnegie,* 66–164; and Mary A. O'-Sullivan, "Finance Capital in Chandlerian Capitalism," *Industrial and Corporate Change* 19, no. 2 (2010): 577–578.

37. The classic accounts of the way symbolic capital functions in modern society remain Pierre Bourdieu, *Distinction: A Social Critique of the Judgment of Taste* (London: Routledge and Kegan Paul, 1986); and Pierre Bourdieu, "The Forms of Capital," in *Handbook of Theory and Research for the Sociology of Education,* ed. John G. Richardson (Westport, CT: Greenwood, 1986), 241–260.

38. Andrew Carnegie, "The Common Interest of Labor and Capital," in *The Empire of Business,* 71.

39. See Pierre Bourdieu, *The Field of Cultural Production: Essays on Art and Literature* (New York: Columbia University Press, 1993).

40. See, for example, Andrew Carnegie, *Our Coaching Trip: Brighton to Inverness* (New York: A. Carnegie, 1882).

41. Nasaw, *Andrew Carnegie,* 113.

42. Andrew Carnegie, *Triumphant Democracy, or, Fifty Years' March of the Republic* (New York: C. Scribner's Sons, 1886), v, 14, viii.

43. See Paul Krause, *The Battle for Homestead, 1880–1892: Politics, Culture, and Steel* (Pittsburgh: University of Pittsburgh Press, 1992). For a broader perspective, see Leon Fink, *In Search of the Working Class: Essays in American Labor History and Political Culture* (Urbana: University of Illinois Press, 1994); and David Montgomery, *The Fall of the House of Labor: The Workplace, the State, and American Labor Activism, 1865–1925* (Cambridge: Cambridge University Press, 1987).

44. Carnegie, *Autobiography of Andrew Carnegie* (Boston: Houghton Mifflin, 1920), 232.

45. Herbert Spencer, "The Development Hypothesis," [1852] in *Essays: Scientific, Political, and Speculative* (London: Williams and Norgate, 1868), 1:377–83; Spencer, "Recent Astronomy and the Nebular Hypothesis," *Westminster Review* 70 (July–Oct. 1858): 185–225.

46. Herbert Spencer, "Progress: Its Law and Cause," *Westminster Review* 67 (April 1857): 446–447.

47. For a classic example, see Richard Hofstadter, *Social Darwinism in American Thought, 1860–1915* (Philadelphia: University of Pennsylvania Press, 1945). For criticisms of Hofstadter's reading, see Robert C. Bannister, *Social Darwinism: Science and Myth in Anglo-American Social Thought* (Philadelphia: Temple University Press, 1979); and Carl Degler, *In Search of Human Nature: The Decline and Revival of Darwinism in American Social Thought* (New York: Oxford University Press, 1991).

48. See, for example, Herbert Spencer, *The Principles of Biology* (London: William and Norgate, 1864), 1:444. Spencer articulated his disagreement with Darwinian evolution by natural selection in especially clear terms in a debate with August Weismann. See Frederick Churchill, "The Weismann-Spencer Controversy over the Inheritance of Acquired Characters," in *Human Implications of Scientific Advance: Proceedings of the XVth International Congress of the History of Science,* ed. E. G. Forbes (Edinburgh: Edinburgh University Press, 1978), 451–468.

49. See also Camille Limoges, "Darwin, Milne-Edwards et le principe de divergence," in *Actes du XIIe Congrès International d'Histoire des Sciences* (Paris: Blanchard, 1968), 111–115; and Camille Limoges, "Milne-Edwards, Darwin, Durkheim and the Division of Labour: A Case Study in Reciprocal Conceptual Exchanges between the Social and Natural Sciences," in *The Natural Sciences and the Social Sciences: Some Critical and Historical Perspectives,* ed. I. Bernard Cohen (Dordrecht: Kluwer Academic, 1994), 317–343.

50. Herbert Spencer, "The Social Organism," in *Essays: Scientific, Political, and Speculative* (London: Williams and Norgate, 1868), 1:386.

51. On Spencer's social organism, see Naomi Beck, *La gauche evolutionniste: Spencer et ses lecteurs en France et en Italie* (Besançon, France: Presses universitaires de Franche Comté, 2014); James Elwick, "Herbert Spencer and the Disunity of the Social Organism," *History of Science* 41 (2003): 35–72; and James Elwick, "Containing Multitudes: Herbert Spencer, Organisms Social and Orders of Individuality," in *Herbert Spencer: Legacies,* ed. Mark Francis and Michael Taylor (London: Routledge, 2015), 89–110.

52. See Michael Ruse, *Monad to Man: The Concept of Progress in Evolutionary Biology* (Cambridge, MA: Harvard University Press, 1996).

53. Nasaw, *Andrew Carnegie,* 224–225.

54. Carnegie, *Autobiography,* 339.

55. Andrew Carnegie, "The Gospel of Wealth," in *The Gospel of Wealth, and Other Timely Essays* (New York: Century, 1900), 1–4.

56. Carnegie, 1, 12–15. For Carnegie's contrast between "evolutionary" and "revolutionary" socialism, see Andrew Carnegie, "The Long March Upward," in *Problems of To-day: Wealth, Labor, Socialism* (New York: Doubleday, 1908), 173–184.

57. See Carnegie Museum, *Annual Report of the Director for the Year Ending March 31, 1898* (Pittsburg: Murdoch-Kerr, 1898), 7–12; and *Dedication Souvenir of the Carnegie Library* (Pittsburg: J. Eichbaum Printers, 1895), 19.

58. *Dedication Souvenir,* 43.

59. Andrew Carnegie, "Presentation of the Carnegie Library to the People with a Description of the Dedicatory Exercises," reprinted as "The Best Use of Wealth," in *Miscellaneous Writings of Andrew Carnegie,* vol. 2, ed. Burton Hendrick (Garden City, NY: Doubleday, 1933), 203–218, on p. 217.

60. See Carnegie Museum, *Annual Report of the Director, 1898,* 11; and Carnegie Museum, *Annual Report of the Director for the Year Ending March 31, 1899* (Pittsburg: Murdoch-Kerr, 1899), 11.

61. Holland to Carnegie, Mar. 8, 1898, Carnegie Collection.

62. On Western resentment against Eastern capital, see Noam Maggor, "To Coddle and Caress These Great Capitalists: Eastern Money, Frontier Populism, and the Politics of Market-Making in the American West," *American Historical Review* 122, no. 1 (2017): 55–84.

63. Holland to Reed, Dec. 10, 1898, Big Bone Room Archive.

64. Reed to Holland, Dec. 22, 1898, Big Bone Room Archive.

65. Holland to Carnegie, Mar. 8, 1899, Carnegie Collection.

66. Holland to Reed, Mar. 27, 1899, box 4, Letterbooks, Holland Papers. See also Holland to Carnegie, Mar. 27, 1899, Big Bone Room Archive; and Reed Contract, March 27, 1899, Big Bone Room Archive.

67. See Salary List, 1899–1900, folder 24, box 12, President's Records, AHC.

68. See, for example, William Pratt Wade, *Manual of American Mining Law as Practiced in the Western States and Territories* (Saint Louis: F. H. Thomas, 1889), 214; and Joseph Wesley Thompson, *United States Mining Statutes Annotated* (Washington, DC: Government Printing Office, 1915), 55.

69. Holland to Carnegie, Mar. 27, 1899, Carnegie Collection.

70. See Stephen Downey to Holland, Apr. 20, 1899, Holland to Gramm, May 2, 1899, and Gramm to Holland, May 4, 1899, Big Bone Room Archive; and Holland to Carnegie, May 5, 1899, Carnegie Collection. See also Report of the Executive Committee of the Board of Trustees of the University of Wyoming, July 20, 1899, President's Records, AHC.

71. See "A Generous Offer," *Laramie Republican,* May 6, 1899.

72. Holland to Carnegie, May 5, 1899, Carnegie Collection.

73. Holland to Carnegie, Mar. 27 and May 5, 1899, Carnegie Collection.

74. Holland to Carnegie, Apr. 1, 1899, Carnegie Collection.

75. Holland to Reed, May 12, 1899, Big Bone Room Archive.

76. Wortman to Holland, June 28, 1899, Big Bone Room Archive.

77. Holland to Wortman, July 3, 1899, Big Bone Room Archive.

78. Quoted in Samuel Knight, "History of the Department of Geology," folder 16, box 5, Wilson Ober Clough Papers, AHC.

79. Union Pacific Railroad Company, *Some of Wyoming's Vertebrate Fossils* (Omaha: Union Pacific Railroad Company, 1899). For a detailed account of the competition for dinosaur specimens among different museums in the United States at the time, see Paul Brinkman, *The Second Jurassic Dinosaur Rush: Museums and Paleontology in America at the Turn of the Twentieth Century* (Chicago: University of Chicago Press, 2010).

80. See, for example, "Grave Robbing Expedition," *Omaha World Herald,* June 8, 1899, 8; "To Hunt for Dinosaurs," *Anaconda (MT) Standard,* June 17, 1899, 2; "The Fossil Wonderland of the World," *San Francisco Chronicle,* July 23, 1899, 6; "In Rich Fossil Fields," *New York Tribune,* Aug. 6, 1899, 10; and "Fossil-Hunting in the Rockies," *Springfield (MA) Republican,* Jan. 9, 1900, 5.

81. Holland to Hatcher, May 21, 1900, Letterbooks, box 6, Holland Papers.

82. Holland to Carnegie, Mar. 26, 1900, Letterbooks, box 5, Holland Papers.

83. Hatcher to Holland, May 9, 1900, J. B. Hatcher Papers, CMNH.

84. Hatcher to Holland, June 4, 1900, Big Bone Room Archive.

85. Holland to Hatcher, May 21, 1900.

86. Holland to Hatcher, June 12, 1900, Letterbooks, box 6, Holland Papers.

87. Reed to Nelson, Jan. 16 and Feb. 19, 1900, Aven Nelson Papers, AHC.

88. Holland to Hatcher, June 12, 1900.

89. "Editorial," *Annals of the Carnegie Museum* 1, no. 1 (1901): 2.

90. While my account highlights the changing social structure of the scientific community, other historians have stressed the "professionalization" of natural history at this time. See Elizabeth Keeney, *The Botanizers: Amateur Scientists in Nineteenth-Century America* (Chapel Hill: University of North Carolina Press, 1992); Robert Kohler, *All Creatures: Naturalists, Collectors, and Biodiversity, 1850–1950* (Princeton, NJ: Princeton University Press, 2006), 110; and Jeremy Vetter, "The Regional Development of Science: Knowledge, Environment, and Field Work in the US Central Plains and Rocky

Mountains, 1860–1920" (PhD diss., University of Pennsylvania, 2005), 106. On the complex history of professionalization in natural history, see Jim Endersby, *Imperial Nature: Joseph Hooker and the Practices of Victorian Science* (Chicago: University of Chicago Press, 2008); and Paul Lucier, "The Professional and the Scientist in Nineteenth-Century America," *Isis* 100, no. 4 (2009): 699–732.

91. John Bell Hatcher, "Diplodocus (Marsh): Its Osteology, Taxonomy, and Probable Habits, with a Restoration of the Skeleton," in *Memoirs of the Carnegie Museum of Natural History*, vol. 1, *1901–1904*, ed. W. J. Holland and J. B. Hatcher (Pittsburgh: Carnegie Institute, 1904), esp. 57.

92. Holland to Carnegie, May 15 and 28, 1901, Carnegie Collection.

93. Carnegie Museum, *Annual Report of the Director for the Year Ending March 31, 1903* (Pittsburg: Murdoch-Kerr, 1903), 27.

94. William J. Holland, "The Diplodocus Goes to Mexico," *Carnegie Magazine* 4, no. 3 (1930): 84.

95. See "Mr. Carnegie's Imitation Dinosaur 'Makes a Hit' in England," *New York Times*, June 4, 1905, 7.

96. Holland to Carnegie, Jan. 31, 1903, Carnegie Collection.

97. Lukas Rieppel, "Plaster Cast Publishing in 19th Century Paleontology," *History of Science* 53, no. 4 (2015): 456–491; Jeffrey Trask, *Things American: Art Museums and Civic Culture in the Progressive Era* (Philadelphia: University of Pennsylvania Press, 2012); Alan Wallach, "The American Cast Museum: An Episode in the History of the Institutional Definition of Art," in *Exhibiting Contradiction: Essays on the Art Museum in the United States* (Amherst: University of Massachusetts Press, 1998), 38–56. See also Rune Frederiksen and Eckart Marchand, *Plaster Casts: Making, Collecting, and Displaying from Classical Antiquity to the Present* (Berlin: De Gruyter, 2010).

98. *Dedication Souvenir*, 58.

99. "The Old Diplodocus and the New," *Daily News*, May 8, 1905, DF5014/1/2, Newspaper Clippings, Archives of the London Natural History Museum.

100. "A Visit to Diplodocus Carnegii," *Westminster Gazette* (London), May 3, 1905, DF5014/1/2, Newspaper Clippings, Archives of the London Natural History Museum.

101. "A Sixty-Foot Reptile," *Morning Leader* (London), Apr. 15, 1903, DF5014/1/2, Newspaper Clippings, Archives of the London Natural History Museum.

102. "Concerning Dragons," *Westminster Gazette* (London), May 15, 1905, DF5014/1/2, Newspaper Clippings, Archives of the London Natural History Museum.

103. "The Presentation of a Reproduction of *Diplodocus carnegii*," *Annals of the Carnegie Museum* 3 (1905): 445–446.

104. Sven Beckert, "American Danger: United States Empire, Eurafrica, and the Territorialization of Industrial Capitalism, 1870–1950," *The American Historical Review* 122, no. 4 (2017): 1137–70. A number of alarmist books warned of an American threat to European hegemony at the turn of the twentieth century. See, for example, Fred Arthur McKenzie, *The American Invaders: Their Plans, Tactics and Progress* (New York: Street and Smith, 1901); William Thomas Stead, *The Americanization of the World; or, The Trend of the Twentieth Century* (New York: H. Markley, 1902); and Benjamin Howarth Thwaite, *The American Invasion* (London: S. Sonnenschein, 1902). On the rise of Germany as an industrial empire, see David Blackbourn, *The Long Nineteenth Century: A History of Germany, 1780–1918* (New York: Oxford University Press, 1998), esp. 177–190, 313–336; Mansel G. Blackford, *The Rise of Modern Business: Great Britain, the United States, Germany, Japan, and China* (Chapel Hill: University of North Carolina Press, 2008); and Gerd Hohorst, Jürgen Kocka, and Gerhard A. Ritter, eds., *Sozialgeschichtliches Arbeitsbuch: Materialien zur Statistik des Kaiserreichs, 1870–1914* (Munich: C. H. Beck, 1975), 57–93.

105. See Volker Rolf Berghahn, *Germany and the Approach of War in 1914* (New York: St. Martin's, 1973); James Joll, *The Origins of the First World War* (London: Longman, 1984); and Annika Mombauer, *The Origins of the First World War: Controversies and Consensus* (London: Longman, 2002).

106. Nasaw, *Andrew Carnegie*, 689–693.

107. W. J. Holland, *To the River Plate and Back* (New York: G. P. Putnam's Sons, 1913), 12–13.

108. Hatcher, "Diplodocus (Marsh)," esp. 57–59. See also Lukas Rieppel, "Bringing Dinosaurs Back to Life: Exhibiting Prehistory at the American Museum of Natural History," *Isis* 103, no. 3 (2012): 472–475.

109. Oliver Perry Hay, "On the Habits and Pose of the Sauropodous Dinosaurs, Especially of Diplodocus," *American Naturalist* 42, no. 502 (1908): 679–680. See also Oliver P. Hay, "On the Restoration of Skeletons of Fossil Vertebrates," *Science*, n.s., 30, no. 759 (1909): 93–95; Oliver Perry Hay, "On the Manner of Locomotion of the Dinosaurs, Especially Diplodocus, with Remarks on the Origin of the Birds," *Proceedings of the Washington Academy of Sciences* 12, no. 1 (1910): 1–25; and Oliver Perry Hay, "Further Observations on the Pose of the Sauropodous Dinosaurs," *American Naturalist* 45, no. 535 (1911): 398–412.

110. Gustav Tornier, "Wie war der *Diplodocus carnegii* wirklich gebaut?," *Sitzungsberichte der Gesellschaft Naturforschender Freunde zu Berlin* 4 (Apr. 20, 1909): 193–209. See also Gustav Tornier, "Ernstes und Lustiges aus Kritiken über meine *Diplodocus*arbeit," *Sitzungsberichte der Gesellschaft Naturforschender Freunde zu Berlin* 9 (Nov. 9, 1909): 505–557; Gustav Tornier, "Über und gegen neue Diplodocus-Arbeiten," *Monatsberichten der Deutschen Geologischen Gesellschaft* 62, no. 8 (1910): 536–576.

111. Friedrich Drevermann, "*Diplodocus* und seine Stellung," *Sitzungsberichte der Gesellschaft Naturforschender Freunde zu Berlin* 10 (Apr. 1910): 400.

112. Othenio Abel, "Die Rekonstruktion des Diplodocus," *Abhandlungen des Kaiserlich- und Königlichen Zoologisch-Botanischen Gesellschaft in Wein* 5 (Oct. 1910): 1–60.

113. William J. Holland, "A Review of Some Recent Criticisms of the Restorations of Sauropod Dinosaurs," *American Naturalist* 44 (1910): 260–261. See also "The Attitude of Diplodocus," *Scientific American*, Nov. 6, 1909; and William Diller Matthew, "The Pose of Sauropodous Dinosaurs," *American Naturalist* 44, no. 525 (1910): 547–560. For a particularly personal and scathing dismissal of Tornier's scientific views, see Holland to Woodward, Sept. 2, 1910, DF100/49, Palaeontology Departmental Correspondence, NHM London.

114. Hermann Stremme, "Wie ist *Diplodocus* richtig aufzustellen?," *Naturwissenschaftliche Wochenschrift* 8, no. 6 (1909): 796–799; Gustav Tornier, "Reptilia," in *Handwörterbuch der Naturwissenschaften*, ed. E. Korschelt et al. (Jena, Germany: Gustav Fischer, 1913), 336–376.

115. Wilhelm Bölsche and Heinrich Harder, "Der Diplodokus," plate 17 in *Tiere der Urwelt*, ser. 1a (Wandsbek, Germany: Kakao Compagnie Reichardt, 1910).

116. Gustav Tornier, "Über den Erinnerungstag an das 150-jährige Bestehen der Gesellschaft: Rückblick auf die Palaeontologie," *Sitzungsberichte der Gesellschaft Naturforschender Freunde zu Berlin* 1925 (1927): 90–91.

117. Sven Sachs, "Diplodocus—Ein Sauropode aus dem Oberen Jura (Morrison-Formation) Nordamerikas," *Natur und Museum* 131, no. 5 (2001): 133–154; Christian Strunz, "Unsere Donner-Echse (Diplodocus) in Neuer Haltung," *Natur und Volk* 66, no. 8 (1936): 371–379; Archives of the Senckenberg Museum.

118. Ina Heumann et al., *Dinosaurierfragmente: Zur Geschichte Der Tendaguru-Expedition Und Ihrer Objekte, 1906–2017* (Göttingen: Wallstein, 2018); Gerhard Maier, *African Dinosaurs Unearthed: The Tendaguru Expeditions* (Bloomington: Indiana University Press, 2003).

119. *Führer durch die Schausammlungen des Museums für Naturkunde in Berlin*, vol. 2, *Geologisch-paläontologische Schausammlung* (Berlin: Hopfer, 1910), 46.

120. Schuchert to Branca, Jan. 8, 1915, Akte Ausland A–Z, Historische Bild- und Schriftgutsammlungen des Museums für Naturkunde Berlin. See also Lukas Rieppel, "Ein Amerikaner in Deutschland," in *Wissensdinge: Geschichten Aus Dem Naturkundemuseum*, ed. Anita Hermannstädter, Ina Heumann, and Kerstin Pannhorst (Berlin: Nicolai, 2015), 128–31.

121. On the relationship between *Brachiosaurus* and Prussian nationalism, see Marco Tamborini, "'If the Americans Can Do It, So Can We': How Dinosaur Bones Shaped German Paleontology," *History of Science* 54, no. 3 (2016): 225–256.

122. Holland to Carnegie, Feb. 3, 1912, Carnegie Collection.

4. Accounting for Dinosaurs

1. John Joseph Flinn, *Official Guide to the World's Columbian Exposition* (Chicago: Columbian Guide, 1893), 9, 51. See also Robert W. Rydell, *World of Fairs: The Century-of-Progress Expositions* (Chicago: University of Chicago Press, 1993); and Alan Trachtenberg, *The Incorporation of America: Culture and Society in the Gilded Age* (New York: Hill and Wang, 1982), 208–234.

2. On the early history of the Field Museum, and its connection to the 1893 World's Fair, see Paul Brinkman, "Frederic Ward Putnam, Chicago's Cultural Philanthropists, and the Founding of the Field Museum," *Museum History Journal* 2, no. 1 (2009): 73–100; Robert H. Kargon, "The Counterrevolution of Progress," in *Urban Modernity: Cultural Innovation in the Second Industrial Revolution,* ed. Miriam R. Levin et al. (Cambridge, MA: MIT Press, 2010), 133–166. On the history of vertebrate paleontology in Chicago, see Paul D. Brinkman, "Establishing Vertebrate Paleontology at Chicago's Field Columbian Museum, 1893–1898," *Archives of Natural History* 27, no. 1 (2000): 81–114; Paul Brinkman, *The Second Jurassic Dinosaur Rush: Museums and Paleontology in America at the Turn of the Twentieth Century* (Chicago: University of Chicago Press, 2010).

3. See Oliver C. Farrington, "A Brief History of the Field Museum from 1893 to 1930," *Field Museum News* 1, no. 1 (1930): 1, 3 and continuing in each succeeding issue concluding in v. 2, no. 2 (February 1931): 4.

4. For example, the vertebrate paleontologist Oliver Perry Hay privately accused Skiff of being "wholly unfitted in every way for such a position." See Hay to George Baur, 1 July 1896, Archives of the MCZ.

5. Quoted in Kevin J. Fernlund, *William Henry Holmes and the Rediscovery of the American West* (Albuquerque: University of New Mexico Press, 2000), 161. As the historian Paul Brinkman summarizes the affair, curators "resented any interference by non-scientists into what they felt was their rightful domain." Paul D. Brinkman, "The 'Chicago Idea': Patronage, Authority, and Scientific Autonomy at the Field Columbian Museum, 1893–97," *Museum History Journal* 8, no. 2 (2015): 182.

6. F. J. V. Skiff, *Annual Report of the Director to the Board of Trustees for the Year 1900–1901,* Field Columbian Museum Publication 14, Report Series, vol. 1, no. 2 (Chicago: Field Columbian Museum, 1896), 111; F. J. V. Skiff, *Annual Report of the Director to the Board of Trustees for the Year 1900–1901,* Field Columbian Museum Publication 62, Report Series, vol. 2, no. 1 (Chicago: Field Columbian Museum, 1901), 7.

7. As Walter Licht has written, "large, corporately owned, bureaucratically managed, multifunctional," and "capital-intensive" firms began to market "mass-produced items nationally and even internationally." See Walter Licht, *Industrializing America: The Nineteenth Century* (Baltimore: Johns Hopkins University Press, 1995), 3. See also Louis Galambos, "The Emerging Organizational Synthesis in Modern American History," *Business History Review* 44, no. 3 (1970): 279–290; Glenn Porter, *The Rise of Big Business, 1860–1920* (Wheeling, IL: Harlan Davidson, 2006); Daniel T. Rodgers, "In Search of Progressivism," *Reviews in American History* 10, no. 4 (1982): 113–132; and Robert Wiebe, *The Search for Order, 1877–1920* (New York: Hill and Wang, 1967).

8. Naomi Lamoreaux, *The Great Merger Movement in American Business, 1895–1904* (Cambridge: Cambridge University Press, 1985). See also Martin J. Sklar, *The Corporate Reconstruction of American Capitalism, 1890–1916: The Market, the Law, and Politics* (Cambridge: Cambridge University Press, 1988).

On the merger movement's cultural implications, see Alan Trachtenberg, *The Incorporation of America: Culture and Society in the Gilded Age* (New York: Hill and Wang, 1982).

9. On corporate management strategies, see Alfred D. Chandler, *The Visible Hand: The Managerial Revolution in American Business* (Cambridge, MA: Belknap Press of Harvard University Press, 1977). However, the functionalist underpinnings of Chandler's narrative have rightly been criticized as too rosy by half. See Naomi R. Lamoreaux and William J. Novak, eds., *Corporations and American Democracy* (Cambridge, MA: Harvard University Press, 2017); William Roy, *Socializing Capital: The Rise of the Large Industrial Corporation in America* (Princeton, NJ: Princeton University Press, 1997); and Richard White, *Railroaded: The Transcontinentals and the Making of Modern America* (New York: Norton, 2011). On America as a nation of investors, see Julia C. Ott, *When Wall Street Met Main Street* (Cambridge, MA: Harvard University Press, 2011). On the invention of the economy through statistical indicators, see Daniel Hirschman, "Inventing the Economy (or, How We Learned to Stop Worrying and Love the GDP)" (PhD diss., University of Michigan, 2016); Timothy Mitchell, "Fixing the Economy," *Cultural Studies* 12, no. 1 (1998): 82–101; and Timothy Shenk, "Inventing the American Economy" (PhD diss., Columbia University, 2016).

10. Quoted in John Kennedy, "Philanthropy and Science in New York City: The American Museum of Natural History, 1868–1968" (PhD diss., Yale University, 1968), 65. Nearly a quarter century later, the president of the American Association of Museums made a similar argument, stating that "business men . . . have devoted themselves to the upbuilding of museums . . . because of a deep-seated conviction . . . that money provided for their support is safely invested." Hermon C. Bumpus, "The Museum as a Factor in Education," *Independent* 61 (1906): 269.

11. Sociologists describe such convergences using the terms "institutional isomorphism" and "organizational imprinting." See, for example, Paul J. DiMaggio and Walter W. Powell, "The Iron Cage Revisited: Institutional Isomorphism and Collective Rationality in Organizational Fields," *American Sociological Review* 48, no. 2 (1983): 147–160; and Victoria Johnson, "What Is Organizational Imprinting? Cultural Entrepreneurship in the Founding of the Paris Opera," *American Journal of Sociology* 113, no. 1 (2007): 97–127.

12. On value neutrality, see Max Weber, *The Vocation Lectures,* ed. David S. Owen and Tracy B. Strong, trans. Rodney Livingstone (Indianapolis: Hackett, 2004). See also Robert Proctor, *Value-Free Science: Purity and Power in Modern Knowledge* (Cambridge, MA: Harvard University Press, 1991). On objectivity, see Lorraine Daston and Peter Galison, *Objectivity* (New York: Zone Books, 2007). On the long history of disinterestedness in science, see Steven Shapin, *A Social History of Truth: Civility and Science in Seventeenth-Century England* (Chicago: University of Chicago Press, 1994).

13. Henry to Bache, Aug. 9, 1838, in *The Papers of Joseph Henry,* ed. Nathan Reingold (Washington, DC: Smithsonian Institution Press, 1981), 4:95–100. For a similar argument, see Joseph Henry, "Address to the American Association for the Advancement of Science, 1850," in *The Papers of Joseph Henry,* ed. Nathan Reingold (Washington, DC: Smithsonian Institution Press, 1998), 8:89–94.

14. See, for example, Alexander Dallas Bache, "Anniversary Address before the American Institute of the City of New York," in *Transactions of the American Institute in the City of New York* (Albany, NY: C. van Benthuysen, 1857), 103. On the policing of community composition in early American science, see Lukas Rieppel, "Hoaxes, Humbugs, and Fraud: Distinguishing Truth from Untruth in Early America," *Journal of the Early Republic* 38, no. 3 (Fall 2018): 501–529. See also Sally Gregory Kohlstedt, *The Formation of the American Scientific Community: The American Association for the Advancement of Science, 1848–60* (Urbana: University of Illinois Press, 1976).

15. H. A. Rowland, "A Plea for Pure Science," *Science* 2, no. 29 (1883): 242–250. For an earlier example, see Benjamin Apthorp Gould, "Address of Ex-President Gould," *Proceedings of the American Association for the Advancement of Science* 18 (Aug. 1869): 1–37. See also "Incentives to the Pursuit of Science," *Popular Science Monthly* 22 (Apr. 1883): 844–848.

16. For more on the history of "pure science" in the United States, see Andrew Jewett, *Science, Democracy, and the American University: From the Civil War to the Cold War* (Cambridge: Cambridge University Press, 2012), esp. 39–46; and Daniel J. Kevles, *The Physicists: The History of a Scientific Community in Modern America* (Cambridge, MA: Harvard University Press, 1987), 1–59.

17. Paul Lucier, "The Origins of Pure and Applied Science in Gilded Age America," *Isis* 103, no. 3 (2012): 536.

18. Rowland, "Plea for Pure Science," 247, 244.

19. Jonathan Levy, "Altruism and the Origins of Nonprofit Philanthropy," in *Philanthropy and American Higher Education,* ed. John R. Thelin and Richard W. Trollinger (New York: Palgrave Macmillan, 2014), 37. See also Peter Dobkin Hall, "A Historical Overview of Philanthropy, Voluntary Association, and Nonprofit Organizations in the United States, 1600–2000," in *The Nonprofit Sector: A Research Handbook,* ed. Walter W. Powell and Richard Steinberg (New Haven, CT: Yale University Press, 2006); and Peter Dobkin Hall, *Inventing the Nonprofit Sector and Other Essays on Philanthropy, Voluntarism, and Nonprofit Organizations* (Baltimore: Johns Hopkins University Press, 1992).

20. Charles H. Sternberg, *The Life of a Fossil Hunter* (New York: H. Holt, 1909), 248, 31, 45. See also Katherine Rogers, *The Sternberg Fossil Hunters: A Dinosaur Dynasty* (Missoula, MT: Mountain, 1999). Finally, see the rich correspondence between Sternberg and Henry Fairfield Osborn, folder 25, box 92, General Correspondence, Archives of the DVP; as well as that between Sternberg and Alexander Agassiz, sMu 2674.10.5, Rare Books and Manuscripts Division, Ernst Mayr Library, MCZ.

21. While the "big three" philanthropic museums of natural history—in Chicago, Pittsburgh, and New York—were slightly too old to be chartered as nonprofit corporations explicitly, they were clearly designed to project an altruistic intent. Eventually, they were reclassified as nonprofit entities, and today they exist as tax-exempt 501(c)(3) corporations. See, for example, "Articles of Incorporation," in *Annual Report of the Director to the Board of Trustees for the Year 1894–95,* Field Columbian Museum Publication 6, Report Series, vol. 1, no. 1 (Chicago: Field Columbian Museum, 1895), 52–53; "An Act to Incorporate the American Museum of Natural History," in *The First Annual Report of the American Museum of Natural History* (New York: printed for the museum, by Major and Knapp Engraving, Manufacturing, and Lithographic, 1870), 10–11; and *Dedication Souvenir of the Carnegie Library* (Pittsburg: J. Eichbaum, 1895).

22. On the notion of "action at a distance" in science, see Bruno Latour, *Science in Action* (Cambridge, MA: Harvard University Press, 1987), 215–257.

23. The original list of instructions is reproduced in Charles Schuchert and Clara Mae LeVene, *O. C. Marsh, Pioneer in Paleontology* (New Haven, CT: Yale University Press, 1940), 172–173.

24. For a more recent example, see Harold Riggs's instructions and printed labels for the Field Museum's 1922–1923 expedition to Argentina, Collections Management Office, FMNH Department of Geology. For a compelling example from early twentieth-century Germany, see Marco Tamborini and Mareike Vennen, "Disruptions and Changing Habits: The Case of the Tendaguru Expedition," *Museum History Journal* 10, no. 2 (2017): 183–199. For a discussion of similar techniques in botany, see Daniela Bleichmar, *Visible Empire: Botanical Expeditions and Visual Culture in the Hispanic Enlightenment* (Chicago: University of Chicago Press, 2012); Müller-Wille, "Lists as Research Technologies"; and Müller-Wille and Charmantier, "Natural History."

25. See David Sepkoski, "The Earth as Archive: Contingency, Narrative, and the History of Life," in Daston, *Science in the Archives,* 53–84. See also Gowan Dawson, "Paleontology in Parts: Richard Owen, William John Broderip, and the Serialization of Science in Early Victorian Britain," *Isis* 103, no. 4 (2012): 637–667; Martin J. S. Rudwick, "Georges Cuvier's Paper Museum of Fossil Bones," *Archives of Natural History* 27, no. 1 (2000): 51–68; and David Sepkoski, "Towards 'a Natural History of Data': Evolving Practices and Epistemologies of Data in Paleontology, 1800–2000," *Journal of the History of Biology* 46, no. 3 (2013): 401–444.

26. Adam Hermann, "Modern Laboratory Methods in Vertebrate Paleontology," *Bulletin of the American Museum of Natural History* 26 (1909): 284.

27. Remnants of these paper technologies abound at the Yale Peabody Museum of Natural History (Yale University, New Haven, CT). For example, see VP Doc. 1136 in the Vertebrate Paleontology Archives for the original shipping label with associated notes for portions of a pterodactyl collected by Reed at Como on June 17, 1879. See also the map of quarry 10 at Como Bluff that Reed drew up in December 1879, VP Doc. 0140, YPM 1980, Vertebrate Paleontology Archives. Finally, see Specimen No. 1020, box 4, Yale Peabody Museum, Vertebrate Paleontology Collections, for the "map" of an individual fragmented sauropod limb bone and accompanying documentation collected by Benjamin Mudge in the Morrison Formation at Garden Park, Colorado, in 1877.

28. Receipt books and accession ledgers are in the archives of the Yale Peabody Museum, whereas the specimens, printed labels, and other ephemera generated out in the field are housed in the museum's Vertebrate Paleontology Collections.

29. See Hatcher to Marsh, Aug. 17, 1884, Aug. 31, 1884, and so on, Marsh Papers.

30. Hatcher to Marsh, Aug. 17, 1884. It is not clear whether Hatcher developed this grid method independently, but similar techniques do not appear to have been used in archeological field work until Mortimer Wheeler, drawing upon the innovative field practices of Pitt Rivers, began using a "box method" of "horizontal" excavation. See Mortimer Wheeler, *Archaeology from the Earth* (London: Oxford University Press, 1954); Gavin Lucas, *Critical Approaches to Fieldwork: Contemporary and Historical Archaeological Practice* (London: Routledge, 2001), 36–43.

31. See Brinkman, *Second Jurassic Dinosaur Rush,* 46–47; and Hermann, "Modern Laboratory Methods," 284.

32. These techniques thus resemble the "mechanical objectivity" described by Daston and Galison, *Objectivity;* Theodore Porter, *Trust in Numbers: The Pursuit of Objectivity in Science and Public Life* (Princeton, NJ: Princeton University Press, 1995).

33. On the changing role of the museum curator, including a sample questionnaire designed to establish whether a potential employee was up to the task, see A. R. Crook, "The Training of Museum Curators," *Proceedings of the American Association of Museums* 4 (1910): 59–64. For a biography of one such assistant, see Elizabeth Noble Shor, *Fossils and Flies: The Life of a Compleat Scientist* (Norman: University of Oklahoma Press, 1971).

34. For Hatcher's frustrations in particular, see Hatcher to Marsh, Dec. 20, 1890, and Jan. 8, 1891, Marsh Papers; and Tom Rea, *Bone Wars: The Excavation and Celebrity of Andrew Carnegie's Dinosaur* (Pittsburgh: University of Pittsburgh Press, 2001), 125–130.

35. Schuchert and LeVene, *O. C. Marsh,* 308, 274, 310.

36. See Chandler, *Visible Hand,* for the classic account of this history. My own thinking especially draws inspiration from recent accounting history, particularly Jonathan Levy, "Accounting for Profit and the History of Capital," *Critical Historical Studies* 1, no. 2 (2014): 171–214. See also Anthony G. Hopwood, *Accounting and Human Behaviour,* Accountancy Age Books (London: Haymarket, 1974); and Theodore Porter, "Quantification and the Accounting Ideal in Science," *Social Studies of Science* 22, no. 4 (1992): 633–651.

37. James Howard Bridge, *The "Carnegie Millions and the Men Who Made Them": Being the Inside History of the Carnegie Steel Company* (London: Limpus, Baker, 1903), 169.

38. Alfred Marshall, *Principles of Economics* (London: Macmillan, 1890), 322.

39. J. A. Hobson, *The Evolution of Modern Capitalism: A Study of Machine Production* (London: Walter Scott, 1895).

40. Marshall, *Principles of Economics,* 345.

41. Later economists stripped Marshall's insight of its organicist language and reformulated it in terms of what they described as "diminishing returns to management." See Nicholas Kaldor, "The Equilibrium

of the Firm," *Economic Journal* 44, no. 173 (1934): 60–76; and Austin Robinson, "The Problem of Management and the Size of Firms," *Economic Journal* 44, no. 174 (1934): 242–257. For an especially influential account that framed the issue in terms of "transaction costs," see Ronald Coase, "The Nature of the Firm," *Economica* 16, no. 4 (1937): 386–405.

42. JoAnne Yates, *Control through Communication: The Rise of System in American Management* (Baltimore: Johns Hopkins University Press, 1989). See also James Beniger, *The Control Revolution: Technological and Economic Origins of the Information Society* (Cambridge, MA: Harvard University Press, 1989); Naomi R. Lamoreaux and Daniel M. G Raff, eds., *Coordination and Information: Historical Perspectives on the Organization of Enterprise* (Chicago: University of Chicago Press, 1995); and Peter Temin, ed., *Inside the Business Enterprise: Historical Perspectives on the Use of Information* (Chicago: University of Chicago Press, 1991).

43. On the history of management accounting, see H. Thomas Johnson and Robert Kaplan, *Relevance Lost: The Rise and Fall of Management Accounting* (Boston: Harvard Business School Press, 1991); and Margaret Levenstein, *Accounting for Growth: Information Systems and the Creation of the Large Corporation* (Stanford, CA: Stanford University Press, 1998). An indication of the importance of accounting at the turn of the twentieth century is that its precise role in the historical development of capitalism was a heated and productive debate among scholars at that time already. See Werner Sombart, *Der moderne Kapitalismus* (Leipzig: Duncker and Humblot, 1902); and Max Weber, *Die protestantische Ethik, und der Geist des Kapitalismus* (Tübingen, Germany: J. C. B. Mohr, 1905).

44. Quoted in Bridge, *"Carnegie Millions,"* 85. On the history of cost accounting, see H. Thomas Johnson, ed., *System and Profits: Early Management Accounting at Du Pont and General Motors* (New York: Arno, 1980); Levy, "Accounting for Profit"; Michael Power, "Counting, Control and Calculation: Reflections on Measuring and Management," *Human Relations* 57, no. 6 (2004): 765–783; Gary John Previts, *A History of Accountancy in the United States: The Cultural Significance of Accounting* (Columbus: Ohio State University Press, 1998); and Hendrik Vollmer, "Bookkeeping, Accounting, Calculative Practice: The Sociological Suspense of Calculation," *Critical Perspectives on Accounting* 14, no. 3 (2003): 353–381.

45. David Nasaw, *Andrew Carnegie* (New York: Penguin, 2006), 608, 715.

46. Winifred Eva Howe, *A History of the Metropolitan Museum of Art* (New York: published for the Metropolitan Museum of Art by Columbia University Press, 1913), 53–54.

47. Henry W. Kent, "Some Business Methods in the Metropolitan Museum of Art," *Proceedings of the American Association of Museums* 5 (1911): 32. On the transformation of the Metropolitan Museum into a center for art education that was run with bureaucratic efficiency, see Jeffrey Trask, *Things American: Art Museums and Civic Culture in the Progressive Era* (Philadelphia: University of Pennsylvania Press, 2012).

48. William T. Blodgett to anonymous [probably Bickmore], May 31, 1869, Early Administrative Records, Archives of the AMNH.

49. See letters from the American Auditing Co. to George H. Sherwood, folder 212, Central Archives of the AMNH; and Lybrand, Ross Bros., and Montgomery to John C. Christensen, Dec. 2, 1941, folder 1139, Central Archives of the AMNH. See also Wayne M. Faunce to F. Trubee Davison, Apr. 24, 1934, Davison to Edward Stettinius, May 12, 1938, and Ford, Bacon, and Davis to Davison, June 1, 1938, folder 1121, Central Archives of the AMNH.

50. See Osborn's obituary in the *New York Times,* Mar. 5, 1894. On Osborn's management practices at the Illinois Central, see Yates, *Control through Communication,* 101–133. On Osborn as a scion of the New York wealthy elite, see Ronald Rainger, *An Agenda for Antiquity: Henry Fairfield Osborn and Vertebrate Paleontology at the American Museum of Natural History, 1890–1935* (Tuscaloosa: University of Alabama Press, 1991).

51. Conference between Osborn and Bumpus, Feb. 13, 1908, box 4, Administrative Records, Archives of the AMNH.

52. Conference between Osborn and Bumpus, Mar. 9, 1908, box 4, Administrative Records, Archives of the AMNH. On Bumpus's career as a bureaucratic administrator, see Kathrinne Duffy, "The Dead Curator: Education and the Rise of Bureaucratic Authority in Natural History Museums, 1870–1915," *Museum History Journal* 10, no. 1 (2017): 29–49.

53. See, for example, Paul M. Rea, "Museum Records," *Proceedings of the American Association of Museums* 1 (1907): 77–82; E. L. Morris, "Museum Catalogs," *Proceedings of the American Association of Museums* 5 (1911): 35–38; and Charles Louis Pollard "Double Card Entry Museum Catalog," in *Proceedings of the American Association of Museums* 5 (1911): 38–42. See also Bernadette Callery, "Evolutionary Change in the Accession Record in Three American Natural History Museums" (PhD diss., University of Pittsburgh, 2002).

54. For a comprehensive overview of the collection, see the contract of sale between Ward's Natural Science Establishment and the Columbian Museum, Nov. 18, 1893, accession no. 9, Archives of the FMNH. See also accession nos. G1–G106, Archives of the FMNH.

55. See *Annual Report of the Director to the Board of Trustees for the Year 1906,* Field Museum of Natural History Publication 119, Report Series, vol. 3, no. 1 (Chicago: Field Museum of Natural History, 1907).

56. Charles Schuchert, "Museum Methods: On the Arrangements of Great Paleontological Collections," *Science* 3, no. 68 (1896): 578–579.

57. *Annual Report of the Director to the Board of Trustees for the Year 1897–98,* Field Columbian Museum Publication 29, Report Series, vol. 1, no. 4 (Chicago: Field Columbian Museum, 1898), 270.

58. *Annual Report of the Director to the Board of Trustees for the Year 1898–99,* Field Columbian Museum Publication 42, Report Series, vol. 1, no. 5 (Chicago: Field Columbian Museum, 1899), 272, 359.

59. [Illegible] to Robert Parr Whitfield, Apr. 1, 1887, Letterpress Book 10, Archives of the AMNH.

60. Allen to Morris Ketchum Jesup, Apr. 8, 1887, Early Administrative Records, Archives of the AMNH.

61. Farrington to Skiff, May 10, 1894, Accession Records, G-1 to G-106, Archives of the FMNH.

62. Carnegie Museum, *Annual Report of the Director for the Year Ending March 31, 1900* (Pittsburg: Murdoch-Kerr, 1900), 8. For a similar accounting exercise that appraised the "aggregate cost" of the American Museum of Natural History's entire vertebrate paleontology collection at $233,787 in 1911, see Henry F. Osborn, *The American Museum of Natural History* (New York: Irving, 1911), 80.

63. Accession Records, 1894–1926, Archives of the DVP.

64. See, for example, the entry for the British Museum, London, in folder 2, Loans, Exchanges, Gifts—Foreign, A–H, box 1, Casts, Reproductions, Sales, and Exchanges, Archives of the DVP. A similar practice was used nearly a century earlier among US farmers who did not have access to large volumes of cash and thus bartered for goods directly. See Emily Pawley, "Accounting with the Fields: Chemistry and Value in Nutriment in American Agricultural Improvement, 1835–1860," *Science as Culture* 19, no. 4 (2010): 461–482.

65. See *Annual Report of the Department of Vertebrate Paleontology,* 1910, Archives of the DVP.

66. W. D. Matthew, "Cost-Sheets of Illustration, Supervision, and Research Work," *Annual Report for the Year 1917, Department of Vertebrate Paleontology,* Archives of the DVP.

67. *Annual Report of the Department of Vertebrate Paleontology,* 1910.

68. Matthew explained that the Sensiba specimen's value could be arrived at by "dividing up the total cost of expeditions (including all field salaries) among the principle specimens of exhibition value." See *Annual Report of the Department of Vertebrate Paleontology,* 1910.

69. Matthew, "Cost-Sheets," Archives of the DVP.

70. See Max Weber, "Bureaucracy," in *From Max Weber: Essays in Sociology,* ed. Hans Heinrich Gerth and C. Wright Mills (New York: Routledge, 2009), 196–266; and Max Weber, *Wirtschaft und Gesellschaft*

(Tübingen, Germany: J. C. B. Mohr, 1922). See also Werner Sombart, *Der moderne Kapitalismus* (Leipzig: Duncker and Humblot, 1902). For a more recent, and less triumphalist, account, see Ben Kafka, *The Demon of Writing: Powers and Failures of Paperwork* (New York: Zone Books, 2012).

71. Christopher A. Bayly, *Empire and Information: Intelligence Gathering and Social Communication in India, 1780–1870* (Cambridge: Cambridge University Press, 1996). On bureaucracy as rule via paperwork, see "The Character of Calculability," in Timothy Mitchell, *Rule of Experts: Egypt, Techno-Politics, Modernity* (Berkeley: University of California Press, 2002), 80–119. On the proliferation of ambitious "archival" projects to hoard knowledge in eighteenth-century Europe, see, for example, Lorraine Daston, ed., *Science in the Archives: Pasts, Presents, Futures* (Chicago: University of Chicago Press, 2017); Ian Hacking, *The Taming of Chance* (Cambridge: Cambridge University Press, 1990); Lisbet Koerner, *Linnaeus: Nature and Nation* (Cambridge, MA: Harvard University Press, 1999). On museums as elements of colonial bureaucracy, see Jessica Ratcliff, "The East India Company, the Company's Museum, and the Political Economy of Natural History in the Early Nineteenth Century," *Isis* 107, no. 3 (2016): 495–517.

72. On vertical integration as a strategy to reduce transaction costs, see Naomi Lamoreaux, Daniel Raff, and Peter Temin, "Beyond Markets and Hierarchies: Toward a New Synthesis of American Business History," *The American Historical Review* 108, no. 2 (April 2003): 404–433. On the use of information technologies inside industrial enterprises, see Naomi R. Lamoreaux and Daniel M. G. Raff, eds., *Coordination and Information: Historical Perspectives on the Organization of Enterprise* (Chicago: University of Chicago Press, 1995); JoAnne Yates, *Control Through Communication: The Rise of System in American Management* (Baltimore: Johns Hopkins University Press, 1989). On archival practices in a scholarly and a commercial context, see Markus Friedrich, *Die Geburt des Archivs: Eine Wissensgeschichte* (Munich: Oldenbourg Verlag, 2013); Markus Krajewski, *Paper Machines: About Cards and Catalogs, 1548–1929* (Cambridge, MA: MIT Press, 2011).

73. The vertical integration of American natural history has been previously suggested by Lynn Nyhart, *Modern Nature: The Rise of the Biological Perspective in Germany* (Chicago: University of Chicago Press, 2009), 358–368. For more on the importance of paperwork at the museum, see Steven D. Lubar, *Inside the Lost Museum: Curating, Past and Present* (Cambridge, MA: Harvard University Press, 2017), 110–128. On the importance of paper technologies in the history of science more broadly, see Elena Aronova, Christine von Oertzen, and David Sepkoski, eds., "Data Histories," special issue, *Osiris* 32 (2017); Ann Blair, *Too Much to Know: Managing Scholarly Information before the Modern Age* (New Haven, CT: Yale University Press, 2010). On the importance of paperwork in the history of capitalism, see Seth Rockman, "Introduction: The Paper Technologies of Capitalism," *Technology and Culture* 58, no. 2 (2017): 487–505; Michael Zakim, *Accounting for Capitalism* (Chicago: University of Chicago Press, 2018). On natural history as an accounting practice, see Anke te Heesen, "Accounting for the Natural World: Double-Entry Bookkeeping in the Field," in *Colonial Botany: Science, Commerce, and Politics in the Early Modern World*, ed. Londa Schiebinger and Claudia Swan (Philadelphia: University of Pennsylvania Press, 2005), 237–251. See also Janet Browne, "The Natural Economy of Households: Charles Darwin's Account Books," in *Aurora Torealis*, ed. Marco Beretta, Karl Grandin, and Svante Lindqvist (Sagamore Beach, MA: Science History, 2008), 87–110.

74. William Leach, *Land of Desire: Merchants, Power, and the Rise of a New American Culture* (New York: Pantheon Books, 1993), 3. See also Simon J. Bronner, ed., *Consuming Visions: Accumulation and Display of Goods in America, 1880–1920* (New York: Norton, 1989); Jackson Lears, "The Mysterious Power of Money," in *Rebirth of a Nation* (New York: Harper Perennial, 2009), 51–92; T. J. Jackson Lears, *Fables of Abundance: A Cultural History of Advertising in America* (New York: Basic Books, 1994); and James Livingston, *Pragmatism and the Political Economy of Cultural Revolution, 1850–1940* (Chapel Hill: University of North Carolina Press, 1994). My argument in this chapter is more closely aligned with the

work of historians who trace the so-called market revolution back to the beginning of the nineteenth century, in that I have characterized the beginning of the twentieth century as a period that witnessed a partial backlash against mass commodification. On the "market revolution" hypothesis, see John Larson, *The Market Revolution in America: Liberty, Ambition, and the Eclipse of the Common Good* (New York: Cambridge University Press, 2010); and Charles Sellers, *The Market Revolution: Jacksonian America, 1815–1846* (New York: Oxford University Press, 1991).

75. The friction that often results when different regimes of value—commercial and otherwise—are brought into contact has given rise to an entire subfield in economic sociology often described as "valuation studies." See, for example, Patrik Aspers and Jens Beckert, "Value in Markets," in *The Worth of Goods: Valuation and Pricing in the Economy,* ed. Jens Beckert and Patrik Aspers (Oxford: Oxford University Press, 2011), 3–38; and Claes-Fredrik Helgesson and Fabian Muniesa, "For What It's Worth: An Introduction to Valuation Studies," *Valuation Studies* 1, no. 1 (2013): 1–10. For a fascinating study of the valuation of nature in particular, see Marion Fourcade, "Cents and Sensibility: Economic Valuation and the Nature of 'Nature,'" *American Journal of Sociology* 116, no. 6 (2011): 1721–1777. And for an instructive account of the friction between different regimes of value that focuses on the history of life insurance in nineteenth-century America, see Viviana Zelizer, *Morals and Markets: The Development of Life Insurance in the United States* (New York: Columbia University Press, 1979).

76. On managing "centers of accumulation" in the history of science, see Lissa Roberts, "Accumulation and Management in Global Historical Perspective," *History of Science* 52, no. 3 (2014): 227–246.

5. Exhibiting Extinction

1. "Mining for Mammoths in the Bad Lands," *New York Times,* Dec. 3, 1905, SM1. See also "Real King of Beasts," *Washington Post,* Nov. 26, 1905, A7; "The Prize Fighter of Antiquity Discovered and Restored," *New York Times,* Dec. 30, 1906, 21; and Henry Fairfield Osborn, "Tyrannosaurus and Other Cretaceous Carnivorous Dinosaurs," *Bulletin of the American Museum of Natural History* 21 (1905): 259–265. See also the correspondence between Hornaday, Brown, and Osborn, May 29, 1902, to Sept. 22, 1905, Field Correspondence, box 2, folder 16, Archives of the DVP; and Lowell Dingus and Mark Norell, *Barnum Brown: The Man Who Discovered Tyrannosaurus Rex* (Berkeley: University of California Press, 2010), 90–110.

2. Barnum Brown, "Tyrannosaurus, a Cretaceous Carnivorous Dinosaur," *Scientific American* 113, no. 15 (1915): 322. See also *1912 Annual Report of the Curator,* Annual Reports, DVP; Matthew to Brown, Nov. 18, 1912, Field Correspondence, box 3, folder 10, Archives of the DVP; and Henry Fairfield Osborn, "Tyrannosaurus: Restoration and Model of the Skeleton," *Bulletin of the American Museum of Natural History* 32 (1913): 91–92. On the second specimen, see the correspondence between Brown and Osborn from July 15 to Aug. 10, 1908, Field Correspondence, box 3, folder 4, Archives of the DVP.

3. *The American Museum of Natural History Seventy-Third Annual Report for the Year 1941* (New York: American Museum of Natural History, 1942); Brown to Avinoff, Oct. 24, 1941, Big Bone Room Archive.

4. For an explicit statement of the museum's aims to produce a paleontological response to the habitat diorama, see the 1907 Department of Vertebrate Paleontology's Annual Report of the Associate Curator, Archives of the DVP. On the museum's exhibition strategy at this time, see "Opening of the Dinosaur Hall, 1905," AMNH Vertical Files, Hall: Dinosaurs, Archives of the AMNH.

5. William Diller Matthew, *Dinosaurs, with Special Reference to the American Museum Collections* (New York: American Museum of Natural History, 1915), 42, 83. See also Frederic A. Lucas, *The Story of*

Museum Groups, Guide Leaflet Series 53 (New York: American Museum of Natural History, 1921); W. D. Matthew, "Report of the Associate Curator," in *1907 Annual Report of the Department of Vertebrate Paleontology,* Archives of the DVP; and "A Carnivorous Dinosaur and Its Prey," July 19, 1938, box 1, folder 10, Rachel Nichols File, Archives of the DVP.

6. For this and many other examples of dinosaur paintings by Charles Knight, see Sylvia Czerkas and Donald F. Glut, *Dinosaurs, Mammoths, and Cavemen: The Art of Charles R. Knight* (New York: E. P. Dutton, 1982).

7. Ronald Rainger, *An Agenda for Antiquity: Henry Fairfield Osborn and Vertebrate Paleontology at the American Museum of Natural History, 1890–1935* (Tuscaloosa: University of Alabama Press, 1991), 163; Hilde Hein, "The Art of Displaying Science: Museum Exhibitions," in *The Elusive Synthesis: Aesthetics and Science,* ed. Alfred I. Tauber, Boston Studies in the Philosophy of Science 182 (Dordrecht: Kluwer, 1996), 277.

8. W. J. Thomas Mitchell, *The Last Dinosaur Book: The Life and Times of a Cultural Icon* (Chicago: University of Chicago Press, 1998), 143. Similarly, Brian Noble argues, "Just as modern white man was reckoned as ultimate in the evolutionary order of the contemporary moment of Euro-American privileged social worlds, *Tyrannosaurus* became the ultimate in the evolutionary order of the Mesozoic." Brian Noble, *Articulating Dinosaurs: A Political Anthropology* (Toronto: University of Toronto Press, 2016), 69.

9. On the "corporate reconstruction" of the United States' political economy, see Louis Galambos and Joseph A. Pratt, *The Rise of the Corporate Commonwealth: U.S. Business and Public Policy in the Twentieth Century* (New York: Basic Books, 1988); James Livingston, *Origins of the Federal Reserve System: Money, Class, and Corporate Capitalism, 1890–1913* (Ithaca, NY: Cornell University Press, 1986); Martin J. Sklar, *The Corporate Reconstruction of American Capitalism, 1890–1916: The Market, the Law, and Politics* (Cambridge: Cambridge University Press, 1988); Alan Trachtenberg, *The Incorporation of America: Culture and Society in the Gilded Age* (New York: Hill and Wang, 1982); and Olivier Zunz, *Making America Corporate, 1870–1920* (Chicago: University of Chicago Press, 1990).

10. See Henry Fairfield Osborn, "Aristogenesis, the Observed Order of Biomechanical Evolution," *PNAS* 19, no. 7 (1933): 699–703; and Henry Fairfield Osborn, "Aristogenesis, the Creative Principle in the Origin of Species," *American Naturalist* 68, no. 716 (1934): 193–235.

11. For specific uses of the term, see Madison Grant, *The Passing of the Great Race; or, The Racial Basis of European History* (New York: C. Scribner, 1916), 78, 125, 187.

12. On the history of Progressivism, I draw especially on Jonathan Levy, *Freaks of Fortune: The Emerging World of Capitalism and Risk in America* (Cambridge, MA: Harvard University Press, 2012), 264-307; Daniel T. Rodgers, *Atlantic Crossings: Social Politics in a Progressive Age* (Cambridge, MA: Belknap Press of Harvard University Press, 1998); and Robert Wiebe, *The Search for Order, 1877–1920* (New York: Hill and Wang, 1967). See also Howard Brick, *Transcending Capitalism: Visions of a New Society in Modern American Thought* (Ithaca, NY: Cornell University Press, 2006).

13. Matthew, *Dinosaurs,* 42.

14. In advancing this argument, I draw inspiration from Donna Haraway, "Teddy Bear Patriarchy: Taxidermy in the Garden of Eden, New York City, 1908–1936," *Social Text,* no. 11 (1985): 20–64. See also Gail Bederman, *Manliness and Civilization* (Chicago: University of Chicago Press, 1995); Gregg Mitman, *The State of Nature: Ecology, Community, and American Social Thought, 1900–1950* (Chicago: University of Chicago Press, 1992); and Gregg Mitman, "Defining the Organism in the Welfare State: The Politics of Individuality in American Culture, 1890–1950," in *Biology as Society, Society as Biology: Metaphors,* ed. Sabine Maasen, Everett Mendelson, and Peter Weingart, Sociology of the Sciences (Dordrecht: Springer, 1995), 249–278. Finally, this chapter is indebted to the work of historians who stress the "non-Darwinian" theories of American paleontologists, especially Peter Bowler, *The Eclipse of Darwinism: Anti-Darwinian Evolution Theories in the Decades around 1900* (Baltimore: Johns Hopkins University Press, 1983); and

Peter Bowler, *The Non-Darwinian Revolution: Reinterpreting a Historical Myth* (Baltimore: Johns Hopkins University Press, 1988). In contrast to Bowler's stress on degeneration and decline, this chapter follows Michael Ruse's characterization of orthogenesis as a progressivist theory: Michael Ruse, *Monad to Man: The Concept of Progress in Evolutionary Biology* (Cambridge, MA: Harvard University Press, 1996).

15. See Richard W. Burkhardt, *The Spirit of System: Lamarck and Evolutionary Biology* (Cambridge, MA: Harvard University Press, 1977); Pietro Corsi, *The Age of Lamarck: Evolutionary Theories in France, 1790–1830* (Berkeley: University of California Press, 1988); and Snait Gissis and Eva Jablonka, eds., *Transformations of Lamarckism: From Subtle Fluids to Molecular Biology* (Cambridge, MA: MIT Press, 2011).

16. See Jean-Baptiste de Lamarck, *Philosophie Zoologique* (Paris: Libraire Dentu, 1809); and Jean Baptiste Lamarck, *Zoological Philosophy*, trans. Hugh Elliott (London: Macmillan, 1914).

17. See Tobias Cheung, *Die Organisation des Lebendigen: Die Entstehung des biologischen Organismusbegriffs bei Cuvier, Leibniz und Kant* (Frankfurt: Campus, 2000).

18. See Stephen Jay Gould, *Ontogeny and Phylogeny* (Cambridge, MA: Belknap Press of Harvard University Press, 1977), 1–209. On the "Meckel-Serres Law," see also Johann Friedrich Meckel, *Beyträge zur vergleichenden Anatomie* (Leipzig: Carl Heinrich Reclam, 1808); Johann Friedrich Meckel, *Handbuch der pathologischen Anatomie* (Leipzig: Carl Heinrich Reclam, 1812); Antoine Étienne Serres, "Principes d'embryogenie, de zoogenie, et de tetratogenie," *Memoires de l'Académie des Sciences* 25 (1860): 834; and Antoine Étienne Serres, "Anatomie transcendente—quatrième mémoire: Loi de symétrie et de conjugaison du système sanguin," *Annales Sciences Naturales* 21 (1830): 48. A similar view was earlier expressed in Lorenz Oken, *Lehrbuch Der Naturphilosophie* (Jena: Friedrich Frommann, 1809). But for a dissenting view, see Karl Ernst von Baer, *Über Entwickelungsgeschichte der Thiere: Beobachtung und Reflexion* (Königsberg, Germany: Bei den Gebrüdern Bornträger, 1828), 224.

19. Louis Agassiz, *An Essay on Classification* (London: Longman, Brown, Green, Longmans and Roberts, 1859), 168, 175.

20. [Robert Chambers], *Vestiges of the Natural History of Creation* (New York: Wiley and Putnam, 1845), 174, 161, 162, 155.

21. Ernst Haeckel, *Generelle Morphologie der Organismen*, vol. 1, *Allgemeine Anatomie der Organismen* (Berlin: G. Reimer, 1866), 300; Ernst Haeckel, *Natürliche Schöpfungsgeschichte* (Berlin: Reimer, 1868), 276. See also Nick Hopwood, *Haeckel's Embryos: Images, Evolution, and Fraud* (Chicago: University of Chicago Press, 2015).

22. See Charles Darwin, *On the Origin of Species by Means of Natural Selection, or, The Preservation of Favoured Races in the Struggle for Life* (London: John Murray, 1859), 279–345. For early and influential articulations of the orthogenetic theory in Germany, see Theodor Eimer, *Die Entstehung der Arten auf Grund von Vererben erwobener Eigenschaften nach den Gesetzen organischen Wachsens: Ein Beitrag zur einheitlichen Auffassung der Lebewelt* (Jena, Germany: G. Fischer, 1888); and Theodor Eimer, *Über bestimmt gerichtete Entwicklung (Orthogenesis) und über Ohnmacht der Darwin'schen Zuchtwahl bei der Artbildung C.R.3. Congr. Internat. Zool* (Leiden, 1896). For an introduction to orthogenetic theories of evolution more broadly, see Georgy Levit and Lennart Olsson, "'Evolution on Rails': Mechanisms and Levels of Orthogenesis," *Annals for the History and Philosophy of Biology* 11 (2006): 97–136.

23. On the contrast to Darwinian Evolution, see Edward Drinker Cope, *The Origin of the Fittest: Essays on Evolution* (New York: D. Appleton, 1887), esp. 16. On "growth force," see Edward Drinker Cope, *The Primary Factors of Organic Evolution* (Chicago: Open Court, 1896), esp. 473–475.

24. Ernst Haeckel, *Generelle Morphologie der Organismen*, vol. 2, *Allgemeine Entwickelungsgeschichte der Organismen* (Berlin: G. Reimer, 1866), 299, 322–323.

25. Alpheus Hyatt, "Cycle in the Life of the Individual (Ontogeny) and in the Evolution of Its Own Group (Phylogeny)," *Science* 5 (1897): 166. See also Alpheus Hyatt, *On the Parallelism between the Different Stages of Life in the Individual and Those in the Entire Group of Molluscous Order Tetrabranchiata,*

Memoirs of the Boston Society of Natural History 1 (Boston: Boston Society of Natural History, 1866); and Alpheus Hyatt, "Phylogeny of an Acquired Characteristic," *Proceedings of the American Philosophical Society* 32, no. 143 (1894): 349–647.

26. Cope, *Primary Factors of Organic Evolution,* 201, 473–493.

27. Cope, 173–174.

28. Henry Fairfield Osborn, *The Origin and Evolution of Life* (New York: C. Scribner's Sons, 1917), 242, 219–221. Another favorite example were large and extremely strange-looking ungulates—called titanotheres—from the Eocene that evolved increasingly outlandish protuberances growing out of their heads. See Henry Fairfield Osborn, *The Titanotheres of Ancient Wyoming, Dakota, and Nebraska,* U.S. Geographical Survey Monograph 55 (Washington, DC: Government Printing Office, 1929).

29. On Osborn's ideas about extinction, see Henry Fairfield Osborn, "The Causes of Extinction of Mammalia," *American Naturalist* 40, no. 479 (1906): 769–795. For an especially fascinating account of "evolutionary inertia," see Othenio Abel, "Das biologische Trägheitsgesetz," *Palaeontologische Zeitschrift* 11, no. 1 (1929): 7–17.

30. F. B. Loomis, "Momentum in Variation," *American Naturalist* 39, no. 467 (1905): 839–843.

31. Arthur Smith Woodward, "Address of the President to the Geological Section," *Science* 30, no. 767 (1909): 325–327. For a review article that also stressed the tendency of dinosaur anatomy to become "so much exaggerated as to become handicaps," see Innokenty P. Tolmachoff, "Extinction and Extermination," *Geological Society of America Bulletin* 39, no. 4 (1928): 1134.

32. See, for example, Richard Swann Lull, "Recapitulation, Racial Old Age, Extinctions," in *Organic Evolution: A Text Book* (New York: Macmillan, 1917), 213–228.

33. Cope, *Primary Factors of Organic Evolution,* 173–174.

34. Bowler, *Eclipse of Darwinism,* 127.

35. See Marc Kirschner and John Gerhart, "Evolvability," *Proceedings of the National Academy of Sciences* 95, no. 15 (1998): 8420–8427.

36. Cope, *Primary Factors of Organic Evolution,* 174.

37. Henry Fairfield Osborn, *The Age of Mammals in Europe, Asia and North America* (New York: Macmillan, 1910), 97.

38. On the history of progressivist theories of evolution, see Ruse, *Monad to Man.*

39. Othniel Charles Marsh, *Introduction and Succession of Vertebrate Life in America: An Address Delivered before the American Association for the Advancement of Science, at Nashville, Tenn., Aug. 30, 1877* (New Haven, CT: Tuttle, Morehouse and Taylor, 1877), 54; and "Proceedings of the Spencer Banquet," in *Herbert Spencer on the Americans and the Americans on Herbert Spencer,* ed. Edward Livingston Youmans (New York: D. Appleton, 1883), 48.

40. Marsh articulated this law in a number of places, expanding and refining it over time. For a clear articulation, see Othniel Charles Marsh, *Dinocerata: A Monograph of an Extinct Order of Gigantic Mammals* (Washington, DC: Government Printing Office, 1886), 58–60.

41. Marsh, *Introduction and Succession,* 55. See also Peter J. Bowler, "American Palaeontology and the Reception of Darwinism," *Studies in History and Philosophy of Science Part C: Studies in History and Philosophy of Biological and Biomedical Sciences* 66 (December 1, 2017): 3–7.

42. E. Ray Lankester, *The Kingdom of Man* (New York: H. Holt, 1907). See also E. Ray Lankester, *Degeneration: A Chapter in Darwinism* (London: Macmillan, 1880); and Chris Manias, "Progress in Life's History: Linking Darwinism and Palaeontology in Britain, 1860–1914," *Studies in History and Philosophy of Science Part C: Studies in History and Philosophy of Biological and Biomedical Sciences* 66 (December 1, 2017): 18–26.

43. Lankester, *Kingdom of Man,* 15.

44. Osborn, *Origin and Evolution of Life,* 214–215.

45. Richard Swann Lull, *The Ways of Life* (New York: Harper and Brothers, 1925), 139, 143–144, 152, 158–159, 176.

46. Cope, *Primary Factors of Organic Evolution,* 475.

47. Lester F. Ward, "Mind as a Social Factor," *Mind* 9, no. 36 (1884): 569, 573. See also Lester Frank Ward, *The Psychic Factors of Civilization* (Boston: Ginn, 1893).

48. George Perkins Marsh, *Man and Nature; or, Physical Geography as Modified by Human Action* (New York: C. Scribner, 1864), 36, 39–41, 44, 47. See also David Lowenthal, *George Perkins Marsh: Prophet of Conservation* (Seattle: University of Washington Press, 2000).

49. See, for example, Bernhard Gissibl, Sabine Höhler, and Patrick Kupper, eds., *Civilizing Nature: National Parks in Global Historical Perspective* (New York: Berghahn Books, 2012); Richard A. Grusin, *Culture, Technology, and the Creation of America's National Parks* (Cambridge: Cambridge University Press, 2004); Roderick Nash, *Wilderness and the American Mind* (New Haven, CT: Yale University Press, 1967), 108–140; Randall K. Wilson, *America's Public Lands: From Yellowstone to Smokey Bear and Beyond* (Lanham, MD: Rowman and Littlefield, 2014); and Dorceta E. Taylor, *The Rise of the American Conservation Movement: Power, Privilege, and Environmental Protection* (Durham, NC: Duke University Press, 2016).

50. George Miller Beard, *American Nervousness, Its Causes and Consequences* (New York: Putnam, 1881), vi, 66, 118, 92. On gender anxieties among office workers during this period more broadly, see Julie Berebitsky, *Sex and the Office: A History of Gender, Power, and Desire* (New Haven, CT: Yale University Press, 2012).

51. See S. Weir Mitchell, *Wear and Tear; or, Hints for the Overworked* (Philadelphia: J. B. Lippincott, 1871), 58. A practitioner's handbook from the early twentieth century recommended that particularly recalcitrant women should be confined to a single room that was "darkened, and kept quiet as possible," adding that "visits from relations should be short and few in number." John Michell Clarke, *Hysteria and Neurasthenia* (London: J. Lane, 1905), 267–268. For a well-known literary description of "rest cures," see Charlotte Perkins Stetson, "The Yellow Wall-Paper," *New England Magazine,* 1892.

52. S. Weir Mitchell, *Wear and Tear,* 5, 18. On conventional ideas about biological sex differences at the time, see Carroll Smith-Rosenberg and Charles Rosenberg, "The Female Animal: Medical and Biological Views of Woman and Her Role in Nineteenth-Century America," *Journal of American History* 60, no. 2 (1973): 332–356. See also Katherine Williams, ed., *Women on the Verge: The Culture of Neurasthenia in Nineteenth-Century America* (Stanford, CA: Iris and B. Gerald Cantor Center for Visual Arts at Stanford University, 2004).

53. On nervous exhaustion more broadly, see T. Jackson Lears, *No Place of Grace: Antimodernism and the Transformation of American Culture, 1880–1920* (New York: Pantheon Books, 1981), 47–58; Anson Rabinbach, *The Human Motor: Energy, Fatigue, and the Origins of Modernity* (New York: Basic Books, 1990), 146–178; and David G. Schuster, *Neurasthenic Nation: America's Search for Health, Happiness, and Comfort, 1869–1920* (New Brunswick, NJ: Rutgers University Press, 2011).

54. Theodore Roosevelt and George Bird Grinnell, "The Boone and Crockett Club," in *American Big-Game Hunting: The Book of the Boone and Crockett Club,* ed. Theodore Roosevelt and George Bird Grinnell (New York: Forest and Stream, 1893), 9.

55. George Bird Grinnell, "Zoological Report," in *Report of a Reconnaissance from Carroll, Montana Territory, on the Upper Missouri, to the Yellowstone National Park, and Return, Made in the Summer of 1875,* by William Ludlow (Washington, DC: Government Printing Office, 1876), 66.

56. "Constitution and By-Laws of the Club," in Roosevelt and Grinnell, *American Big-Game Hunting,* 338; "Constitution of the Boone and Crockett Club," in *American Big Game in Its Haunts: The Book of the Boone and Crockett Club,* ed. George Bird Grinnell (New York: Forest and Stream, 1904), 486. See also

Christine Bold, "Boone and Crockett Writers," in *The Frontier Club: Popular Westerns and Cultural Power, 1880–1924* (New York: Oxford University Press, 2013), 14–54.

57. Lacey also sponsored the famous Lacey Act of 1900, which made it illegal to possess, transport, trade, or sell illegally obtained plants and animals. On Grinnell's conservation work, see John F. Reiger, *American Sportsmen and the Origins of Conservation* (Corvallis: Oregon State University Press, 2001). See also Richard Conniff, *House of Lost Worlds: Dinosaurs, Dynasties, and the Story of Life on Earth* (New Haven, CT: Yale University Press, 2016), 95–104.

58. William Temple Hornaday, "The Passing of the Buffalo," *The Cosmopolitan* 4 (1887): 85. See also Hannah Rose Shell, "Last of the Wild Buffalo," *Smithsonian Magazine* 30, no. 11 (2000): 26–30.

59. Quoted in Mark V. Barrow, *Nature's Ghosts: Confronting Extinction from the Age of Jefferson to the Age of Ecology* (Chicago: University of Chicago Press, 2009), 123.

60. William T. Hornaday, *Our Vanishing Wild Life: Its Extermination and Preservation* (New York: New York Zoological Society, 1913), 7.

61. Gifford Pinchot, *The Fight for Conservation* (New York: Doubleday, Page, 1910), 42, 44, 50.

62. George Bird Grinnell and Charles Sheldon, eds., *Hunting and Conservation: The Book of the Boone and Crockett Club* (New Haven, CT: Yale University Press, 1925), 202. These sentiments led the historian Samuel P. Hays to describe conservation as a "scientific movement" whose "essence was rational planning to promote efficient development and use of all natural resources." Samuel P. Hays, *Conservation and the Gospel of Efficiency: The Progressive Conservation Movement, 1890–1920* (Pittsburgh: University of Pittsburgh Press, 1999), 2.

63. Henry Fairfield Osborn, "Preservation of the World's Animal Life," *American Museum Journal* 12, no. 4 (1912): 123.

64. Lull, *Ways of Life,* 295; Henry Fairfield Osborn, *Man Rises to Parnassus: Critical Epochs in the Prehistory of Man* (Princeton, NJ: Princeton University Press, 1927), 219.

65. See, for example, Daniel J. Kevles, *In the Name of Eugenics: Genetics and the Uses of Human Heredity* (New York: Knopf, 1985); Kenneth M. Ludmerer, *Genetics and American Society: A Historical Appraisal* (Baltimore: Johns Hopkins University Press, 1972); and Diane B. Paul, *Controlling Human Heredity: 1865 to the Present* (Atlantic Highlands, NJ: Humanities, 1995). On the connection between wilderness conservation and the eugenics movement, see Garland E. Allen, "'Culling the Herd': Eugenics and the Conservation Movement in the United States, 1900–1940," *Journal of the History of Biology* 46, no. 1 (2012): 31–72; and Jonathan Peter Spiro, *Defending the Master Race: Conservation, Eugenics, and the Legacy of Madison Grant* (Burlington: University of Vermont Press, 2009).

66. As Davenport's right-hand man, Harry Laughlin, explained, it was hoped that before long, "social customs will make . . . hereditary potentialities marriage assets, valued along with—if not above—money, position, and charming personal qualities." See Harry H. Laughlin, "The Eugenics Record Office at the End of Twenty-Seven Months Work," *Report of the Eugenics Record Office* 1 (1913): 10–11. See also Garland E. Allen, "The Eugenics Record Office at Cold Spring Harbor, 1910–1940: An Essay in Institutional History," *Osiris* 2 (1986): 225–264.

67. On sterilization programs, see, for example, Harry Hamilton Laughlin, *Eugenical Sterilization in the United States* (Chicago: Psychopathic Laboratory of the Municipal Court of Chicago, 1922). See also Randall Hansen and Desmond S. King, *Sterilized by the State: Eugenics, Race, and the Population Scare in Twentieth-Century North America* (New York: Cambridge University Press, 2013); Wendy Kline, *Building a Better Race: Gender, Sexuality, and Eugenics from the Turn of the Century to the Baby Boom* (Berkeley: University of California Press, 2001); and Johanna Schoen, *Choice and Coercion: Birth Control, Sterilization, and Abortion in Public Health and Welfare* (Chapel Hill: University of North Carolina Press, 2005).

68. Ross warned, "The fewer the brains they have to contribute, and the lower the place they take among us, the faster they multiply." Edward A. Ross, *The Old World in the New: The Significance of Past*

and Present Immigration to the American People (New York: Century, 1914), 299. Similarly, Madison Grant from the Bronx Zoo claimed that admixture among geographically distinct "races" produced dysgenic and morphologically "disharmonic" offspring, and Davenport counseled, "Before any one person is admitted to citizenship let something be learned concerning his family history and his personal history on the other side of the ocean." Grant, *The Passing of the Great Race,* 11–32; Charles Benedict Davenport, *Heredity in Relation to Eugenics* (New York: H. Holt, 1911), 222.

69. Harry H. Laughlin, *Biological Aspects of Immigration: Statement of Harry H. Laughlin at the Hearings before the Committee on Immigration, U.S. House of Representatives, 16–17 April 1920* (Washington, DC: Government Printing Office, 1921), 17.

70. See Paul, *Controlling Human Heredity,* 97–105. On Laughlin's role in particular, see Frances Hassencahl, "Harry H. Laughlin, 'Expert Eugenics Agent' for the House Committee on Immigration and Naturalization, 1921 to 1931" (PhD diss., Case Western Reserve University, 1970).

71. As Garland Allen has put it, eugenic reform efforts were predicated on a "belief that modern society could no longer afford to function on a completely *laissez-faire* principle, but had to be planned and rationalized." Garland E. Allen, "Eugenics and American Social History, 1880–1950," *Genome* 31, no. 2 (1989): 886. See also Allen, "Eugenics Record Office," 255–60; and Michael Freeden, "Eugenics and Progressive Thought: A Study in Ideological Affinity," *Historical Journal* 22, no. 3 (1979): 645–671.

72. William T. Hornaday, *Our Vanishing Wild Life: Its Extermination and Preservation* (New York: New York Zoological Society, 1913), 63, 94–113.

73. Henry Fairfield Osborn, *Creative Education in School, College, University, and Museum: Personal Observation and Experience of the Half-Century 1877–1927* (New York: C. Scribner, 1927), 255.

74. Mary Anne Andrei, "Nature's Mirror: How the Taxidermists of Ward's Natural Science Establishment Transformed Wildlife Display in American Natural History Museums and Fought to Save Endangered Species" (PhD diss., University of Minnesota, 2006), 185. For a later example of ornithologists using the museum as an institutional base for conservation efforts, see Barrow, *Nature's Ghosts,* 234–260.

75. A. R. Crook, "The Museum and the Conservation Movement," in *Proceedings of the American Association of Museums* (Charleston, SC: Waverly, 1915), 94.

76. Oliver C. Farrington, "The Rise of Natural History Museums," *Science* 42, no. 1076 (1915): 207; Osborn, *Creative Education,* 270. On the habitat diorama as an "ecological" display, see Lynn Nyhart, *Modern Nature: The Rise of the Biological Perspective in Germany* (Chicago: University of Chicago Press, 2009), 251–292; and Karen Rader and Victoria Cain, *Life on Display* (Chicago: University of Chicago Press, 2014), 64–84.

77. "Wise matings and high fecundity mean racial progress," while "unfit matings mean racial degeneracy," Laughlin's exhibits explained. The noted anthropologist Aleš Hrdlička also curated an exhibition on race and migration that included a series of Native American skulls to demonstrate "the persistence to this day of Neanderthaloid forms and other primitive features." Harry Hamilton Laughlin, *The Second International Exhibition of Eugenics* (Baltimore: Williams and Wilkins, 1923), 22, 42.

78. Robert W. Rydell, *World of Fairs: The Century-of-Progress Expositions* (Chicago: University of Chicago Press, 1993), 38–58. See also Robert Rydell, Christina Cogdell, and Mark Largent, "The Nazi Eugenics Exhibit in the United States, 1934–43," in *Popular Eugenics: National Efficiency and American Mass Culture in the 1930s,* ed. Susan Currell and Christina Cogdell (Athens: Ohio University Press, 2006), 359–384; and Devon Stillwell, "Eugenics Visualized: The Exhibit of the Third International Congress of Eugenics, 1932," *Bulletin of the History of Medicine* 86, no. 2 (2012): 206–236.

79. Henry Fairfield Osborn, "The Hall of the Age of Man in the American Museum," *Natural History* 20, no. 1 (1920): 235–236. See also Henry F. Osborn, "The Hall of the Age of Man in the American Museum," *Nature* 107, no. 2686 (1921): 239.

80. Haraway, "Teddy Bear Patriarchy," 53.

81. Leonard Darwin, "Aims and Methods of Eugenical Societies," and Charles B. Davenport, "Research in Eugenics," in Charles B. Davenport et al., eds., *Eugenics, Genetics and the Family* (Baltimore: Williams and Wilkins, 1923), 15, 19, 28.

82. Osborn, *Man Rises to Parnassus,* 221.

83. Grant, *Passing of the Great Race,* xvi, 5–6.

84. Henry Fairfield Osborn, "Address of Welcome," in Davenport et al., eds., *Eugenics, Genetics and the Family,* 2–3.

85. Osborn, *Man Rises to Parnassus,* 220–221.

86. Osborn to Hanfstaengel, Nov. 18, 1934, folder 6, box 5, Osborn Papers.

87. Naomi Lamoreaux, *The Great Merger Movement in American Business, 1895–1904* (Cambridge: Cambridge University Press, 1985), 1–2.

88. According to Thomas C. Leonard, for example, progressive reformers went on a "crusade to dismantle laissez-faire and remake American economic life through the agency of the administrative state." Thomas C. Leonard, *Illiberal Reformers: Race, Eugenics, and American Economics in the Progressive Era* (Princeton, NJ: Princeton University Press, 2016), 10. On the development of an administrative state, see, for example, Daniel P. Carpenter, *The Forging of Bureaucratic Autonomy: Reputations, Networks, and Policy Innovation in Executive Agencies, 1862–1928* (Princeton, NJ: Princeton University Press, 2001); Brian J. Cook, *Bureaucracy and Self-Government: Reconsidering the Role of Public Administration in American Politics* (Baltimore: Johns Hopkins University Press, 2014); John A. Garraty, *The New Commonwealth, 1877–1890* (New York: Harper and Row, 1968); and Richard J. Stillman, *Creating the American State: The Moral Reformers and the Modern Administrative World They Made* (Tuscaloosa: University of Alabama Press, 1998).

89. On "ruinous competition" see Oswald Knauth, "Capital and Monopoly," *Political Science Quarterly* 31, no. 2 (1916): 246–250. Similarly, John Bates Clark and John Maurice Clark agreed with the "general belief . . . that price warfare of a ruinous sort was almost unavoidable" and "combinations offered a way of deliverance which, at first, was not altogether unwelcome." John Bates Clark and John Maurice Clark, *The Control of Trusts* (New York: Macmillan, 1912), 4. See also Spurgeon Bell, "Fixed Costs and Market Price," *Quarterly Journal of Economics* 32, no. 3 (1918): 507–524; Arthur Twining Hadley, *Economics: An Account of the Relations between Private Property and Public Welfare* (New York: G. P. Putnam's Sons, 1896); Alfred Marshall, "Some Aspects of Competition," *Journal of the Royal Statistical Society* 53, no. 4 (1890): 612–643; Jack High, "Economic Theory and the Rise of Big Business in America, 1870–1910," *Business History Review* 85, no. 1 (2011): 85–112; and Herbert Hovenkamp, *Enterprise and American Law, 1836–1937* (Cambridge, MA: Harvard University Press, 1991), 308–323. On "natural monopolies," see Manuela Mosca, "On the Origins of the Concept of Natural Monopoly: Economies of Scale and Competition," *European Journal of the History of Economic Thought* 15, no. 2 (2008): 317–353. For a broader consideration of antimonopoly sentiments among progressive reformers, see Richard R. John, "Robber Barons Redux: Antimonopoly Reconsidered," *Enterprise and Society* 13, no. 1 (2012): 1–38.

90. Andrew Carnegie, "Popular Illusions about Trusts," *Century* 60 (May 1900): 143–145. For an earlier version of similar, though somewhat less stridently articulated, views, see Andrew Carnegie, "The Bugaboo of Trusts," *North American Review* 148, no. 387 (1889): 141–150. Insofar as Carnegie's ideas reflected the developmental progressivism of orthogenetic theory, he did not simply fail to understand the nuances of Darwinian evolution, as previous historians have sometimes suggested. See, for example, Robert C. Bannister, *Social Darwinism: Science and Myth in Anglo-American Social Thought* (Philadelphia: Temple University Press, 1979), 57–79. See also Carl Degler, *In Search of Human Nature: The Decline and Revival of Darwinism in American Social Thought* (New York: Oxford University Press, 1991).

91. James Howard Bridge, "Introduction: America's Commercial Primacy and the Trust," in *The Trust: Its Book,* by Charles R. Flint et al., ed. James Howard Bridge (New York: Doubleday, Page, 1902), xiii, xxx, xxix, xxxv, xxxiii.

92. Indeed, given that biological evolution was often associated with socialism during the nineteenth century, they may have even been on the vanguard in this respect. See Adrian J. Desmond, *The Politics of Evolution: Morphology, Medicine, and Reform in Radical London* (Chicago: University of Chicago Press, 1989); Piers J. Hale, *Political Descent: Malthus, Mutualism, and the Politics of Evolution in Victorian England* (Chicago: University of Chicago Press, 2014); and Daniel Philip Todes, *Darwin without Malthus: The Struggle for Existence in Russian Evolutionary Thought* (New York: Oxford University Press, 1989).

93. Henry George, *Progress and Poverty: An Inquiry into the Cause of Industrial Depressions and of Increase of Want with Increase of Wealth* (n.p.: National Single Tax League, 1879), 485, 511, 515, 535, 540. Compare this with Edward Bellamy's best-selling utopian novel, in which the social and industrial strife of the Gilded Age is finally overcome once society comes to "recognize and cooperate" with a "process of industrial revolution which could not have terminated otherwise." Once it has been "completed by the final consolidation of the entire capital of the nation," this process of "logical evolution" is able "to open a golden future to humanity." Edward Bellamy, *Looking Backward, 2000–1887* (Boston: Ticknor, 1888), 67, 77. A similar, though far less utopian, reasoning also informed Thorstein Veblen's claim that "the emergence of a leisure class" was part of "the sequence of cultural evolution." Thorstein Veblen, *The Theory of the Leisure Class: An Economic Study of Institutions* (New York: Macmillan, 1899), 22.

94. Charles Whiting Baker, *Monopolies and the People* (New York: G. P. Putnam's Sons, 1889), 187.

95. Baker also believed that monopolistic corporations had "grown out of the chaos with competition, or, as the student of the natural sciences expresses it, the survival of the fittest, as its mainspring." See Charles Whiting Baker, "Public Control," in *Monopolies and the People* (New York: G. P. Putnam's Sons, 1889), 150–154.

96. Parsons therefore suggested the "*public ownership* of monopolies," which he believed would make it possible to "retain the economies of concentration and remove the evils of overgrown private power"—that is, to "keep all that is good" and "kill only what is evil." Frank Parsons, *The Public Ownership of Monopolies* (Philadelphia: Bureau of Nationalist Literature, 1894), quoted in William Dwight Porter Bliss, "Monopolies," in *The Encyclopedia of Social Reform* (New York: Funk and Wagnells, 1897), 891.

97. Arthur Jerome Eddy, *The New Competition* (New York: D. Appleton, 1912), 13.

98. Lee Benson, *The Concept of Jacksonian Democracy* (Princeton, NJ: Princeton University Press, 1961); Walter Edward Hugins, *Jacksonian Democracy and the Working Class: A Study of the New York Workingmen's Movement, 1829–1837,* Stanford Studies in History, Economics, and Political Science 19 (Stanford, CA: Stanford University Press, 1960); Charles Sellers, *The Market Revolution: Jacksonian America, 1815–1846* (New York: Oxford University Press, 1991); Sean Wilentz, *Chants Democratic: New York City and the Rise of the American Working Class, 1788–1850* (New York: Oxford University Press, 1984).

99. For a particularly dire warning about the "curse of bigness," see Louis D. Brandeis, *Other People's Money: And How the Bankers Use It* (New York: F. A. Stokes, 1914), 162.

100. Levy, *Freaks of Fortune,* 279; and Martin J. Sklar, *The United States as a Developing Country: Studies in U.S. History in the Progressive Era and the 1920s* (Cambridge: Cambridge University Press, 1992), 110.

101. Frederick Jackson Turner, *The Frontier in American History* (New York: H. Holt, 1920), 32.

102. Indeed, the People's Party itself largely grew out of late nineteenth-century farmers' alliances and other cooperative ventures that pooled individual resources to stand up to the power of corporate conglomerates. As the historian Charles Postel argues, Populists thus sought to advance "an alternative

capitalism in which private enterprise coalesced with both cooperative and state-based economies." Charles Postel, *The Populist Vision* (Oxford: Oxford University Press, 2007), 5.

103. Theodore Roosevelt, *The Works of Theodore Roosevelt,* ed. Hermann Hagedorn, vol. 15, *State Papers as Governor and President, 1899–1909* (New York: C. Scribner's Sons, 1926), 171–172.

104. Theodore Roosevelt, *Theodore Roosevelt: An Autobiography* (New York: Macmillan, 1913), 462–464, 470–473, 480–481.

105. The historian William Murphey describes Gary's attitude as follows: "If businesses could be legitimized by governmental supervision, they would be vaccinated from legal or public onslaughts." William Murphey, "Theodore Roosevelt and the Bureau of Corporation: Executive-Corporate Cooperation and the Advancement of the Regulatory State," *American Nineteenth Century History* 14, no. 1 (2013): 101. See also Sklar, *Corporate Reconstruction,* 228–285.

106. "Cooperation, Not Competition: The Life of Trade," *Trade: An Independent Weekly Journal for Merchants* 18, no. 5 (1911): 24.

107. See John A. Garraty, *Right-Hand Man: The Life of George W. Perkins* (New York: Harper, 1960).

108. Henry Holt, "Competition," in *Morals in Modern Business: Addresses Delivered in the Page Lecture Series, 1908, before the Senior Class of the Sheffield Scientific School, Yale University,* ed. Edward D. Page et al. (New Haven, CT: Yale University Press, 1909), 52–53.

109. Arthur M. Lewis, *Evolution: Social and Organic* (Chicago: Charles H. Kerr, 1909), 149.

110. John Bates Clark, *The Philosophy of Wealth: Economic Principles Newly Formulated* (Boston: Ginn, 1886), 151.

111. Joseph A. Schumpeter, *Capitalism, Socialism, and Democracy* (New York: Harper and Brothers, 1942), 81, 83.

112. Osborn, "Hall of the Age of Man" (1921).

113. William Diller Matthew, "The Ground Sloth Group," *American Museum Journal* 11, no. 4 (1911): 118–119, 115.

6. Bringing Dinosaurs Back to Life

1. "Dinosaurs Cavort in Film for Doyle," *New York Times,* June 3, 1922, 1, 4; "His Dinosaur Film a Hoax, Says Doyle," *New York Times,* June 4, 1922, 18. On this episode, see Russell Miller, *The Adventures of Arthur Conan Doyle* (London: Harvill Secker, 2008), 418–419.

2. That said, from time to time explorers did claim to have seen live dinosaurs in remote parts of the world. See, for example, Carl Hagenbeck, *Beasts and Men,* trans. Hugh Samuel Roger Elliot and Arthur Gordon Thacker (London: Longmans, Green, 1909), 95–97; "36 Feet of Neck on Shoulders 20 Feet HIGH," *Washington Herald,* Dec. 27, 1914; "Hunter Says He Saw Prehistoric Monster," *New York Times*, Dec. 13, 1919, 3; and "The African Brontosaurus," *New York Times,* Jan. 7, 1920, 18.

3. Arthur Conan Doyle, *The Coming of the Fairies* (New York: George H. Doran, 1922).

4. "His Dinosaur Film a Hoax, Says Doyle," *New York Times,* June 4, 1922, 18. Houdini would go on to characterize the obsession of spiritualists such as Doyle as being either "the result of deluded brains or those which were too intensely willing to believe." Harry Houdini, *A Magician among the Spirits* (New York: Harper, 1924), xvii–xviii, xix. On Doyle's enthusiasm for spirit photography, see Kelvin I. Jones, *Conan Doyle and the Spirits: The Spiritualist Career of Sir Arthur Conan Doyle* (Wellingborough, UK: Aquarian, 1989); and Alex Owen, "'Borderland Forms': Arthur Conan Doyle, Albion's Daughters, and the Politics of the Cottingley Fairies," *History Workshop,* no. 38 (1994): 48–85.

5. For an alternative account that argues for continuity between filmmakers and scientists, see David A. Kirby, *Lab Coats in Hollywood: Science, Scientists, and Cinema* (Cambridge, MA: MIT Press, 2011). On

the fraught but productive relationship between science and motion picture film during this period, see Scott Curtis, *The Shape of Spectatorship: Art, Science, and Early Cinema in Germany* (New York: Columbia University Press, 2015); and Oliver Gaycken, *Devices of Curiosity: Early Cinema and Popular Science* (New York: Oxford University Press, 2015).

6. See Chris Manias, "The Lost Worlds of Messmore & Damon: Science, Spectacle and Prehistoric Monsters in Early-Twentieth Century America," *Endeavour* 40, no. 3 (2016): 163–177. The torture chamber and "Crusader's Bride" are described in a 1947 *Mechanix Illustrated* article, "Man of the Monsters," Scrapbook A, Messmore & Damon Company Records, National Museum of American History Archives, Washington, DC. For numerous complaints about Messmore & Damon's presence at the World's Fair, see box 321, series 1-10210, Century of Progress records.

7. See E. D. Langworthy to Andrews, Dec. 10, 1932, and Andrews to Langworthy, Dec. 14, 1932, as well as the attendant promotional material, in box 11, folder 38, R. C. Andrews Administrative Papers, AMHN. Within the museum, Andrews was criticized as more showman than scientist, more interested in furthering his own celebrity than contributing to the stock of reliable knowledge. See, for example, "Coxcomb Roy," a letter to the editor of a staff publication at the museum, the *Red Fossil,* vol. 1, no. 7 (May 1, 1937), Rare Books and Special Collections Library, AMNH.

8. William Diller Matthew, *Dinosaurs, with Special Reference to the American Museum Collections* (New York: American Museum of Natural History, 1915), 116.

9. In making this argument, this chapter draws on the work of previous historians who have shown how representations of nature hid the human labor that was involved in their production. On wildlife documentaries, see Gregg Mitman, *Reel Nature: America's Romance with Wildlife on Films* (Cambridge, MA: Harvard University Press, 1999). On zoos and aquaria, see Elizabeth Hanson, *Animal Attractions: Nature on Display in American Zoos* (Princeton, NJ: Princeton University Press, 2002); Christina Wessely, *Künstliche Tiere: Zoologische Gärten und urbane Moderne* (Berlin: Kulturverlag Kadmos, 2008); and Lynn Nyhart, *Modern Nature: The Rise of the Biological Perspective in Germany* (Chicago: University of Chicago Press, 2009). On the construction of authenticity in particular, see Lynn K. Nyhart, "Science, Art, and Authenticity in Natural History Displays," in *Models: The Third Dimension of Science,* ed. Soraya de Chadarevian and Nick Hopwood (Stanford, CA: Stanford University Press, 2004), 307–335; and Michael Rossi, "Fabricating Authenticity: Modeling a Whale at the American Museum of Natural History, 1906–1974," *Isis* 101, no. 2 (2010): 338–361.

10. A number of historians have argued that although photographs may be taken, their reputation as a truth-telling medium had to be made. See, for example, Jordan Bear, *Disillusioned: Victorian Photography and the Discerning Subject* (University Park: Pennsylvania State University Press, 2015); John Tagg, *The Burden of Representation: Essays on Photographies and Histories* (Amherst: University of Massachusetts Press, 1988); Jennifer Tucker, *Nature Exposed: Photography as Eyewitness in Victorian Science* (Baltimore: Johns Hopkins University Press, 2005); Sandra S. Phillips et al., *Police Pictures: The Photograph as Evidence* (San Francisco: San Francisco Museum of Modern Art/Chronicle Books, 1997); and Phillip Prodger, *Time Stands Still: Muybridge and the Instantaneous Photography Movement* (New York: Oxford University Press with the Iris and B. Gerald Cantor Center for Visual Arts at Stanford University, 2001).

11. The philosopher Kendall Walton has even argued that photographs are transparent. "We see the world through them," Walton insists, much like we do with a telescope or a pair of eyeglasses. Kendall L. Walton, "Transparent Pictures: On the Nature of Photographic Realism," *Critical Inquiry* 11, no. 2 (1984): 251. But Walton's claims have elicited intense criticism. See, for example, Gregory Currie, *Image and Mind: Film, Philosophy and Cognitive Science* (Cambridge: Cambridge University Press, 1995); Edwin Martin, "On Seeing Walton's Great-Grandfather," *Critical Inquiry* 12, no. 4 (1986): 796–800; and Nigel Warburton, "Seeing through 'Seeing through Photographs,'" *Ratio* 1, no. 1 (1988): 64–74. See also Catharine Abell, "The Epistemic Value of Photographs," in *Philosophical Perspectives on Depiction,* ed. Catharine Abell and Katerina Bantinaki (Oxford: Oxford University Press, 2010), 81–103; and Scott Walden, "Objectivity

in Photography," *British Journal of Aesthetics* 45, no. 3 (2005): 258–272. For a good overview of these debates, see Katherine Thomson-Jones, *Aesthetics and Film* (New York: Continuum, 2008), 16–39.

12. In making this claim, I draw on the work of film theorists who emphasize the way animations appear to endow what are unquestionably fictional characters with life. See Scott Bukatman, *The Poetics of Slumberland: Animated Spirits and the Animating Spirit* (Berkeley: University of California Press, 2012), esp. 106–134; Sergei Eisenstein, *Eisenstein on Disney,* ed. Jay Leyda, trans. Alan Upchurch (London: Methuen, 1988); and Dan North, *Performing Illusions: Cinema, Special Effects and the Virtual Actor* (London: Wallflower, 2008).

13. Pierce's distinction informs a large literature on the history and philosophy of photography and cinematography. See, for example, André Bazin, *What Is Cinema?* (Berkeley: University of California Press, 1967); Stanley Cavell, *The World Viewed: Reflections on the Ontology of Film* (New York: Viking, 1971); Tom Gunning, "What's the Point of an Index? or, Faking Photographs," *Nordicom Review* 25, no. 1 (2004): 39–49; Peter Wollen, "The Semiology of the Cinema," in *Signs and Meaning in the Cinema* (Bloomington: University of Indiana Press, 1969), 116–154. For an argument that fossils resemble photographs because both are indexical, see Lukas Rieppel, "Plaster Cast Publishing in Nineteenth Century Paleontology," *History of Science* 53, no. 4 (2015): 456–491; and Kyla Schuller, "The Fossil and the Photograph: Red Cloud, Prehistoric Media, and Dispossession in Perpetuity," *Configurations* 24, no. 2 (2016): 229–261. On how fossils and other traces from the deep past might help us to formulate an alternative media history that does not privilege the human, see Jussi Parikka, *A Geology of Media* (Minneapolis: University of Minnesota Press, 2015).

14. Charles S. Peirce, *Writings of Charles S. Peirce: A Chronological Edition,* vol. 2, *1867–1871* (Bloomington: Indiana University Press, 1982), 56.

15. Charles S. Peirce, *Writings of Charles S. Peirce: A Chronological Edition,* vol. 3, *1872–1878* (Bloomington: Indiana University Press, 1986), 67.

16. Lorraine Daston and Peter Galison, *Objectivity* (New York: Zone Books, 2007), 120–121.

17. My argument here is informed by Tom Gunning's skepticism about the unique power of indexicality to produce cinema's uncanny realism. However, whereas Gunning draws on the history of animation to develop an alternative genealogy of the cinema, this chapter uses the history of stop-motion trick films to argue for an alternative phenomenology of the index. See Tom Gunning, "The Transforming Image: The Roots of Animation in Metamorphosis and Motion," in *Pervasive Animation,* ed. Suzanne Buchan (New York: Routledge, 2013), 52–69; and Tom Gunning, "Moving Away from the Index: Cinema and the Impression of Reality," *Differences* 18, no. 1 (2007): 29–52.

18. *Strand,* Mar. 1912, 360.

19. Arthur Conan Doyle, "The Lost World," *Strand,* Apr. 1912, 1. Compare to the frontispiece in H. Rider Haggard, *King Solomon's Mines* (London: Cassell, 1885).

20. On the novel, its critical reception, and its relationship to traditional adventure fiction, see Patrick Scott Belk, *Empires of Print: Adventure Fiction in the Magazines, 1899–1919* (New York: Routledge, 2017), 129–162; Rosamund Dalziell, "The Curious Case of Sir Everard Im Thurn and Sir Arthur Conan Doyle: Exploration and the Imperial Adventure Novel, *The Lost World,*" *English Literature in Transition (1880–1920)* 45, no. 2 (2002): 131–157; Ross G. Forman, "Room for Romance: Playing with Adventure in Arthur Conan Doyle's *The Lost World,*" *Genre* 43, no. 1–2 (2010): 27–59; and Amy R. Wong, "Arthur Conan Doyle's 'Great New Adventure Story': Journalism in *The Lost World,*" *Studies in the Novel* 47, no. 1 (2015): 60–79.

21. Arthur Conan Doyle, *The Lost World* (New York: Hodder and Stoughton, 1912), 7, 13, 51–53, 58.

22. Doyle, 70, 84, 163–164, 222, 238, 289, 291–306.

23. Arthur Conan Doyle, in *Arthur Conan Doyle: A Life in Letters,* ed. Jon L. Lellenberg, Daniel Stashower, and Charles Foley (New York: Penguin, 2007), 582–583.

24. William Henry Fox Talbot, *The Pencil of Nature* (London: Longman, Brown, Green and Longmans, 1844), 1. See also Mirjam Brusius, Katrina Dean, and Chitra Ramalingam, eds., *William Henry Fox*

Talbot: Beyond Photography (New Haven, CT: Yale Center for British Art, 2013); Douglas R. Nickel, "Nature's Supernaturalism: William Henry Fox Talbot and Botanical Illustration," in *Intersections: Lithography, Photography, and the Traditions of Printmaking,* ed. Kathleen Stewart Howe (Albuquerque: University of New Mexico Press, 1998), 15–23; Douglas R. Nickel, "Talbot's Natural Magic," *History of Photography* 26, no. 2 (2002): 132–140; and Stephen C. Pinson, *Speculating Daguerre: Art and Enterprise in the Work of L. J. M. Daguerre* (Chicago: University of Chicago Press, 2012).

25. As the historian Jennifer Tucker has argued, "Although nineteenth-century faith in photography was powerful, the idea that people over a hundred years ago accepted photographs at face value is exaggerated and misleading." Tucker, *Nature Exposed,* 4. See also Mia Fineman, *Faking It: Manipulated Photography before Photoshop* (New York: Metropolitan Museum of Art, 2012).

26. See Tom Gunning, "Phantom Images and Modern Manifestations: Spirit Photography, Magic Theater, Trick Films, and Photography's Uncanny," in *Fugitive Images: From Photography to Video,* ed. Patrice Petro, Theories of Contemporary Culture 16 (Bloomington: Indiana University Press, 1995), 42–71; and Tucker, *Nature Exposed,* 65–125. See also the special issue of *Art Journal* titled "Close Encounters," especially Louis Kaplan, "Where the Paranoid Meets the Paranormal: Speculations on Spirit Photography," *Art Journal* 62, no. 3 (2003): 18–27; and Karl Schoonover, "Ectoplasms, Evanescence, and Photography," *Art Journal* 62, no. 3 (2003): 30–41.

27. As Tom Gunning has put it, "Early audiences went to exhibitions to see machines demonstrated, . . . rather than to view films." Tom Gunning, "The Cinema of Attraction: Early Film, Its Spectator and the Avant-Garde," *Wide Angle* 8, no. 3 (1986): 65–66.

28. Ray Harryhausen and Tony Dalton, *A Century of Stop Motion Animation: From Méliès to Aardman* (New York: Watson-Guptill, 2008), 40. On the use of stop-motion and other techniques of the trick film in popular science, see Gaycken, *Devices of Curiosity,* 54–90.

29. North, *Performing Illusions,* 66, 192–193. Early films that depicted dinosaurs also included W. D. Griffith's *Of Primal Tribes/In Prehistoric Days* (1912–1913), Charlie Chaplin's *His Prehistoric Past* (1914), L. M. Glackens's *Stone Age Adventure* (1915), Buster Keaton's *Three Ages* (1923), Howard Hawks's *Fig Leaves* (1926), and Laurel and Hardy's *Flying Elephants* (1928), as well as the cartoon animations *Felix the Cat Trifles in Time* (1925), *Betty Boop's Museum* (1932), and *Daffy Duck and the Dinosaur* (1939).

30. Harryhausen and Dalton, *Century of Stop Motion Animation;* Steve Archer, *Willis O'Brien: Special Effects Genius* (Jefferson, NC: McFarland, 1993). On the dispute between O'Brien and Dawley, see Stephen Czerkas, "O'Brien vs. Dawley: The First Great Rivalry in Visual Effects," *Cinefex* 138 (July 2014): 14–26. See also Cathrine Curtis Corporation Records 1920–1927, AHC, esp. Dawley to Curtis, June 13, 1922, box 2, folder 2, and "Memorandum of Plaintiff's Cause of Action," box 1, folder E1. Finally, see Herbert M. Dawley, US Patent 1347993 A, Articulated Effigy, granted July 27, 1920.

31. As far away as Shanghai, for example, a newspaper journalist reported, "We sat there and fairly clutched our chairs as wild, weird, monstrous animals of prehistoric times lashed their tails in fury, knocked down towering trees and engaged in terrific combat to the death." See Geraldin Sartain, "Lost World Film Sends Shivers down Spine," *China Press,* Apr. 23, 1926, 3.

32. Harryhausen and Dalton, *Century of Stop Motion Animation,* 57.

33. Promotional material for *The Ghost of Slumber Mountain,* box 1, folder D, Cathrine Curtis Corporation Records, AHC.

34. Promotional flyer for *The Lost World,* box 1, folder D, Cathrine Curtis Corporation Records, AHC.

35. "The 'Lost World' at Tremont Temple," *Boston Daily Globe,* Feb. 3, 1925, 15; Regina Cannon, "See 'The Lost World' and Get the Thrill of a Lifetime," *New York Evening Graphic,* Feb. 9, 1925; "The Misty Heons: A Real Wonder Film 'THE LOST WORLD,'" *Los Angeles Times,* Feb. 25, 1925, C7; Mordaunt Hall, "How Miniature Replicas of Monsters Were Filmed," *New York Times,* Feb. 15, 1925, X5.

36. Hall, X5.

37. Colin Williamson, *Hidden in Plain Sight: An Archaeology of Magic and the Cinema* (New Brunswick, NJ: Rutgers University Press, 2015). See also Warren Buckland, "Between Science Fact and Science Fiction: Spielberg's Digital Dinosaurs, Possible Worlds, and the New Aesthetic Realism," *Screen* 40, no. 2 (1999): 177–192; and Michele Pierson, *Special Effects: Still in Search of Wonder* (New York: Columbia University Press, 2002).

38. See John Canemaker, *Winsor McCay: His Life and Art* (New York: Harry N. Abrams, 2005). On the role of lightning sketches in the early history of animation, see Malcolm Cook, "The Lightning Cartoon: Animation from Music Hall to Cinema," *Early Popular Visual Culture* 11, no. 3 (2013): 237–254.

39. See Donald Crafton, *Before Mickey: The Animated Film, 1898–1928* (Cambridge, MA: MIT Press, 1982).

40. Quoted in Crafton, 110.

41. Frederic A. Lucas, *The Story of Museum Groups,* Guide Leaflet Series 53 (New York: American Museum of Natural History, 1921), 26–27.

42. Akeley to Osborn, May 29, 1912, Osborn Papers, AMNH. On the history of the habitat diorama, see Karen Rader and Victoria Cain, "The Drama of the Diorama," in *Life on Display: Revolutionizing U.S. Museums of Science and Natural History in the Twentieth Century* (Chicago: University of Chicago Press, 2014), 51–90; and Karen Wonders, *Habitat Dioramas: Illusions of Wilderness in Museums of Natural History,* Acta Universitatis Upsaliensis 25 (Uppsala: Almqvist and Wiksell, 1993). See also Helmut Gernsheim and Alison Gernsheim, *L. J. M. Daguerre: The History of the Diorama and the Daguerreotype* (New York: Dover, 1968).

43. "Vital Museums of the New Era," *New York Times Magazine,* Mar. 20, 1932, 12–13; "Natural History That Rivals Nature," *New York Times,* Feb. 1, 1925, 15. See also Karen Wonders, "The Illusionary Art of Background Painting in Habitat Dioramas," *Curator: The Museum Journal* 33, no. 2 (1990): 91–93.

44. See Mary Anne Andrei, "Nature's Mirror: How the Taxidermists of Ward's Natural Science Establishment Transformed Wildlife Display in American Natural History Museums and Fought to Save Endangered Species" (PhD diss., University of Minnesota, 2006).

45. Andrew McClellan, *The Art Museum from Boullée to Bilbao* (Berkeley: University of California Press, 2008), 204.

46. Charlotte Klonk, "Patterns of Attention: From Shop Windows to Gallery Rooms in Early Twentieth-Century Berlin," *Art History* 28, no. 4 (2005): 468–496.

47. Osborn, November 1919, Box 130, Central Archives, AMNH; John Cotton Dana, *The Gloom of the Museum,* New Museum Series 2 (Woodstock, VT: Elm Tree, 1917), 23–24, 29. See also Victoria Cain, "'Attraction, Attention, and Desire': Consumer Culture as Pedagogical Paradigm in Museums in the United States, 1900–1930," *Paedagogica Historica* 48, no. 5 (2012): 745–769; Neil Harris, "Museums, Merchandising, and Popular Taste: The Struggle for Influence," in *Cultural Excursions: Marketing Appetites and Cultural Tastes in Modern America* (Chicago: University of Chicago Press, 1990), 56–110; Michelle Henning, *Museums, Media and Cultural Theory* (Maidenhead, UK: Open University Press, 2006), 37–69; and William Leach, *Land of Desire: Merchants, Power, and the Rise of a New American Culture* (New York: Pantheon Books, 1993), 153–190.

48. See, for example, the advertising circulars in Scrapbook 19, folder 1, box 2, Scrapbooks, Archives of the AMNH. Some philanthropically funded museums, such as the Field Museum in Chicago did charge a modest price for admission, but they waived it on weekends and other designated free days. See *Annual Report of the Director to the Board of Trustees for the Year 1898–99,* Field Columbian Museum Publication 42, Report Series, vol. 1, no. 5 (Chicago: Field Columbian Museum, 1899), 293.

49. Debra Singer Kovach, "Developing the Museum Experience: Retailing in American Museums 1945–91," *Museum History Journal* 7, no. 1 (2014): 103–121.

50. Quoted in Lucas, *Story of Museum Groups,* 5.

51. George Dorsey, "The Anthropological Exhibits at the American Museum of Natural History," *Science* 25 (June 1907): 584; Dresden zoologist quoted in Nyhart, "Science, Art, and Authenticity," 309.

52. My claim that philanthropic museums engaged in a practice of institutional demarcation is informed by Thomas Gieryn, "Boundary-Work and the Demarcation of Science from Non-science," *American Sociological Review* 48, no. 6 (1983): 781–795; and Thomas Gieryn, *Cultural Boundaries of Science: Credibility on the Line* (Chicago: University of Chicago Press, 1999). In contrast, historians who emphasize continuities across different exhibition spaces often draw on Tony Bennett's notion of an "exhibitionary complex." See Tony Bennett, "The Exhibitionary Complex," *New Formations* 4 (Spring 1988): 73–102; and Tony Bennett, *The Birth of the Museum: History, Theory, Politics* (London: Routledge, 1995).

53. Harlan I. Smith to Osborn, n.d., and Franz Boas to Bashford Dean, Oct. 4, 1910, box 20, Osborn Papers, NYHS.

54. "The Recent Administration of Dr. Bumpus," Osborn's report to the AMNH Board of Trustees, n.d., Osborn Papers, NYHS.

55. "Bumpus Out of Museum," *New York Tribune,* Jan. 21, 1911.

56. "Bumpus and Museum Part," *New York Post,* Jan. 20, 1911.

57. "Dr. Bumpus Is Out of History Museum," *New York Times,* Jan. 21, 1911.

58. Oliver C. Farrington, "The Rise of Natural History Museums," *Science* 42, no. 1076 (1915): 207.

59. Henry Fairfield Osborn, *Creative Education in School, College, University, and Museum: Personal Observation and Experience of the Half-Century 1877–1927* (New York: C. Scribner, 1927), 237.

60. Lucas, *Story of Museum Groups,* 34.

61. Farrington, "Rise of Natural History Museums," 207.

62. Osborn, *Creative Education,* 252–253.

63. John Edward Gray to Samuel Leigh Sotheby, Dec. 28, 1854, Sotheby Scrapbook, vol. 1, Central Library, Bromley, Kent. See also James A. Secord, "Monsters at the Crystal Palace," in Chadarevian and Hopwood, *Models,* 158.

64. "WB Scott Address at the Joseph Leidy Centenary (1923)," Collection 12 (Leidy Memorials), Philadelphia Academy of Natural Sciences, Philadelphia, PA.

65. William Diller Matthew, *The Mounted Skeleton of Brontosaurus in the American Museum of Natural History,* Guide Leaflet Series 18 (New York: American Natural History Museum, 1905): 6–8, reprinted in the *American Museum Journal* 5, no. 2 (1905): 63–70. For other takes on the same problem, see Othenio Abel, *Geschichte und Methode der Rekonstruktion vorzeitlicher Wirbeltiere* (Jena, Germany: G. Fischer, 1925); and Frederic A. Lucas, "The Restoration of Extinct Animals," in *Annual Report of the Board of Regents of the Smithsonian Institution* (Washington, DC: Government Printing Office, 1901), 479–492.

66. Henry Fairfield Osborn, "Skull and Skeleton of the Sauropodous Dinosaurs, Morosaurus and Brontosaurus," *Science,* n.s., 22, no. 560 (1905): 374–376; Henry Fairfield Osborn, "The Skeleton of Brontosaurus and Skull of Morosaurus," *Nature* 73, no. 1890 (1906): 282–284.

67. William J. Holland, "Heads and Tails: A Few Notes Relating to the Structure of the Sauropod Dinosaurs," *Annals of the Carnegie Museum* 9 (1915): 277.

68. David Berman and John McIntosh, *Skull and Relationship of the Upper Jurassic Sauropod Apatosaurus (Reptilia, Saurischia),* Bulletin of the Carnegie Museum of Natural History 8 (Pittsburgh: Carnegie Museum of Natural History, 1978). See also Keith Parsons, *Drawing Out Leviathan: Dinosaurs and the Science Wars* (Bloomington: Indiana University Press, 2001), 1–22.

69. Henry Fairfield Osborn and Charles Hazelius Sternberg, "A Mounted Skeleton of Naosaurus, a Pelycosaur from the Permian of Texas," *Bulletin of the American Museum of Natural History* 23, no. 14 (1907): 265.

70. Matthew to Charles Hezelius Sternberg, Oct. 9, 1918, box 93, folder 29, General Correspondence, Archives of the DVP. For the public description, see "The Naosaurus, or 'Ship-Lizard,'" *American Museum Journal* 7, no. 3 (1907): 36–41.

71. Rachel Nichols, "Skeletal Restorations in the AMNH," n.d., box 2, folder 5, Rachel Nichols Files, Archives of the DVP.

72. William Diller Matthew, "The Collection of Fossil Vertebrates: A Leaflet Guide to the Exhibition Halls of Vertebrate Paleontology in the American Museum of Natural History," *Supplement to the American Museum Journal* 3, no. 5 (1903): 9.

73. Nichols, "Skeletal Restorations."

74. Oliver P. Hay, "On the Restoration of Skeletons of Fossil Vertebrates," *Science*, n.s., 30, no. 759 (1909): 93–95.

75. Erwin Barbour, "Notes on the Paleontological Laboratory of the United States Geological Survey under Professor Marsh," *American Naturalist* 24, no. 280 (1890): 388–397.

76. *Annual Report of the Department of Vertebrate Paleontology,* 1907, box 1, Archives of the DVP; Lucas, *Story of Museum Groups,* 19.

77. *Annual Report of the Department of Vertebrate Paleontology,* 1908, box 1, Archives of the DVP.

78. Matthew, *Dinosaurs,* 83–85.

79. "Annual Report of the President," in *Fortieth Annual Report of the American Museum of Natural History for the Year 1908* (New York: American Museum of Natural History, 1909), 31.

80. Charles Knight, "American Museum's Murals of Prehistoric Animals," 1924, Charles R. Knight Papers, box 8, folder 1, Rare Books and Manuscripts Division, NYPL.

81. Knight to Stephen Chapman Simms, Charles R. Knight Contract for Paintings for Ernest R. Graham Hall, 1926–1981, file folder 2, folder 6, Archives of the FMNH.

82. "The Career of Charles Knight," *Curator,* 1961, 364–365.

83. "Transcript of Knight Interview with Miss Eslajean Geyer for *The Hour of Living Art,* a Radio Broadcast on W.F.A.S.," n.d., box 8, folder 5, Charles R. Knight Papers.

84. Osborn to Annie Knight, Sept. 22, 1932, Record Group 1262, Central Archives of the AMNH.

85. The formal arrangement between the two parties evolved to become increasingly rigid over the years, but for a good example, see Lucas to Knight, Apr. 27, 1922, Record Group 249, Central Archives of the AMNH. See also Marianne Sommer, "Seriality in the Making: The Osborn-Knight Restorations of Evolutionary History," *History of Science* 48, no. 3–4 (2010): 462–471.

86. Adam Hermann, "Modern Laboratory Methods in Vertebrate Paleontology," *Bulletin of the American Museum of Natural History* 26 (1909): 324.

87. William J. Holland, *Tenth Annual Report of the Director for the Year Ending March 31, 1907* (Pittsburgh: Carnegie Museum, 1907), 25.

88. Hermann, "Modern Laboratory Methods," 324.

89. Matthew, *Dinosaurs,* 85.

90. "A Carnivorous Dinosaur and Its Prey," box 1, folder 10, Rachel Nichols Files, Archives of the DVP. See also William Diller Matthew, "Allosaurus, a Carnivorous Dinosaur and Its Prey," *American Museum Journal* 8, no. 1 (1908): 4–5.

91. Doyle, *Lost World,* 51–53, 58.

92. See Rachel Nichols, "Skeletons and Restorations of Fossil Vertebrates in the American Museum of Natural History," box 2, folder 5, Rachel Nichols Files, Archives of the DVP.

93. In this respect, dinosaur exhibits illustrate what Gregg Mitman and Kelly Wilder describe as the "documentary impulse" of both science and photography, arguing that a photograph's "evidentiary status" is a function of more than "the indexical nature of the single image." Rather, it has been produced through an act of institutional scaffolding, which often expresses itself through the "incredible surfeit" of evidentiary

material contained across a collection of documents stored in an archive. Gregg Mitman and Kelley E. Wilder, introduction to *Documenting the World: Film, Photography, and the Scientific Record,* ed. Gregg Mitman and Kelley E. Wilder (Chicago: University of Chicago Press, 2016), 13.

Conclusion

1. "The Sinclair Dinosaur Book," 1934, Library of the DVP. On the world's fair exhibit, see *Picture News: Chicago World's Fair Edition,* folder 16-267, box 18, Century of Progress Publications #16, Century of Progress records. See also box 321, series 1-10210, and box 422, series 1-13608, Century of Progress records. For Sinclair's relationship to the museum, see file nos. 1145, 1147, and 1147.5 in Central Archives of the AMNH.

2. A French Catholic priest named Jean-Jacques Pouech had discovered dinosaur eggshells as early as 1859, but he did not describe them as such, and his find slipped into obscurity. See Eric Buffetaut and Jean Le Loeuf, "The Discovery of Dinosaur Eggshells in Nineteenth-Century France," in *Dinosaur Eggs and Babies,* ed. Kenneth Carpenter, Karl F. Hirsch, and John R. Horner (Cambridge: Cambridge University Press, 1994), 31–35. On the expedition, see the highly partisan but nonetheless useful Charles Gallenkamp, *Dragon Hunter: Roy Chapman Andrews and the Central Asiatic Expeditions* (New York: Viking, 2001). See also Roy Chapman Andrews, *The New Conquest of Central Asia* (New York: American Museum of Natural History, 1932).

3. Alexander G. Ruthven, "A Study of the American Museum of Natural History," January 13, 1941, file 1121, Central Archives of the AMNH.

4. "Statement of Operations, *The World a Million Years Ago,*" Sept. 5, 1934, and Gebhardt to Owings, Sept. 15, 1934, box 321, series 1-10210, Century of Progress records.

5. See Paul Farber, *Finding Order in Nature: The Naturalist Tradition from Linnaeus to E. O. Wilson* (Baltimore: Johns Hopkins University Press, 2000); Joel Hagen, *An Entangled Bank: The Origins of Ecosystem Ecology* (New Brunswick, NJ: Rutgers University Press, 1992); and Robert Kohler, *All Creatures: Naturalists, Collectors, and Biodiversity, 1850–1950* (Princeton, NJ: Princeton University Press, 2006). On the history of fieldwork in American science more broadly, see Robert E. Kohler, *Landscapes and Labscapes: Exploring the Lab-Field Border in Biology* (Chicago: University of Chicago Press, 2002); and Jeremy Vetter, *Field Life: Science in the American West during the Railroad Era* (Pittsburgh: University of Pittsburgh Press, 2016).

6. For the transition of nineteenth-century natural history to modern biology, see Keith R. Benson, "From Museum Research to Laboratory Research: The Transformation of Natural History into Academic Biology," in *The American Development of Biology,* ed. Ronald Rainger, Keith R. Benson, and Jane Maienschein (New Brunswick, NJ: Rutgers University Press, 1991), 49–84; and Lynn K. Nyhart, "Natural History and the 'New' Biology," in *Cultures of Natural History,* ed. Nicholas Jardine, James A. Secord, and E. C. Spary (Cambridge: Cambridge University Press, 1996), 426–446. For the more recent resurgence of natural history practices in silico, see Bruno Strasser, "The Experimenter's Museum: GenBank, Natural History, and the Moral Economies of Biomedicine," *Isis* 102, no. 1 (2011): 60–96; and Bruno Strasser, "Data-Driven Sciences: From Wonder Cabinets to Electronic Databases," *Studies in History and Philosophy of Biological and Biomedical Sciences* 43, no. 1 (2012): 85–87. On the uses of databases in recent biology more broadly, see Hallam Stevens, *Life Out of Sequence* (Chicago: University of Chicago Press, 2013).

7. George Gaylord Simpson, *Tempo and Mode in Evolution* (New York: Columbia University Press, 1944).

8. John Maynard Smith, "Paleontology at the High Table," *Nature* 309 (1984): 401. On the modern synthesis, see Ernst Mayr and William B. Provine, eds., *The Evolutionary Synthesis: Perspectives on*

the Unification of Biology (Cambridge, MA: Harvard University Press, 1998); and Vassiliki Betty Smocovitis, *Unifying Biology: The Evolutionary Synthesis and Evolutionary Biology* (Princeton, NJ: Princeton University Press, 1996). On the importance of theoretical population genetics in particular, see William B. Provine, *The Origins of Theoretical Population Genetics* (Chicago: University of Chicago Press, 1971); and William B. Provine, *Sewall Wright and Evolutionary Biology* (Chicago: University of Chicago Press, 1986). The synthesis and its aftermath continue to be an area of impassioned debate among working biologists. See Kevin Laland et al., "Does Evolutionary Theory Need a Rethink?," *Nature* 514, no. 7521 (2014): 161–164; Kevin N. Laland et al., "The Extended Evolutionary Synthesis: Its Structure, Assumptions and Predictions," *Proceedings of the Royal Society B* 282, no. 1813 (2015): 10–19; and Massimo Pigliucci and Gerd B. Müller, *Evolution: The Extended Synthesis* (Cambridge, MA: MIT Press, 2010).

9. Smith, "Paleontology at the High Table," 402.

10. See David Sepkoski and Michael Ruse, eds., *The Paleobiological Revolution: Essays on the Growth of Modern Paleontology* (Chicago: University of Chicago Press, 2009); and David Sepkoski, *Rereading the Fossil Record: The Growth of Paleobiology as an Evolutionary Discipline* (Chicago: University of Chicago Press, 2012).

11. On the importance of theoretical models in paleobiology, see Stephen Jay Gould, "The Promise of Paleobiology as a Nomothetic, Evolutionary Discipline," *Paleobiology* 6, no. 1 (1980): 96–118; Thomas J. M. Schopf, *Models in Paleobiology* (San Francisco: Freeman, Cooper, 1972); and Thomas J. M. Schopf, "Theory in Paleoecology," *Paleobiology* 1, no. 1 (1975): 129–131.

12. Thomas Henry Huxley, "On the Animals Which Are Most Nearly Intermediate between Birds and Reptiles," *Annals and Magazine of Natural History*, no. 2 (1868): 66–75; Thomas Henry Huxley, "Further Evidence of the Affinities Between the Dinosaurian Reptiles and Birds," *Quarterly Journal of the Geological Society of London*, no. 26 (1870): 12–31. See also Adrian Desmond, *The Hot-Blooded Dinosaurs: A Revolution in Palaeontology* (London: Blond and Briggs, 1975).

13. J. H. Ostrom, "The Ancestry of Birds," *Nature* 242, no. 5393 (1973): 136. See also John H. Ostrom, *Osteology of* Deinonychus antirrhopus, *an Unusual Theropod from the Lower Cretaceous of Montana,* Bulletin of the Peabody Museum of Natural History 30 (New Haven, CT: Peabody Museum of Natural History, 1969); John H. Ostrom, "*Archaeopteryx* and the Origin of Flight," *Quarterly Review of Biology* 49, no. 1 (1974): 27–47; John H. Ostrom, "The Origin of Birds," *Annual Review of Earth and Planetary Sciences* 3, no. 1 (1975): 55–77; and John H. Ostrom, "*Archaeopteryx* and the Origin of Birds," *Biological Journal of the Linnean Society* 8, no. 2 (1976): 91–182.

14. See Robert T. Bakker, *The Dinosaur Heresies: New Theories Unlocking the Mystery of the Dinosaurs and Their Extinction* (New York: William Morrow, 1986); Desmond, *Hot-Blooded Dinosaurs;* and John R. Horner, "Dinosaurs at the Table," in *The Paleobiological Revolution: Essays on the Growth of Modern Paleontology,* ed. David Sepkoski and Michael Ruse (Chicago: University of Chicago Press, 2009), 111–121.

15. Robert T. Bakker, "The Dinosaur Renaissance," *Scientific American* 232, no. 4 (1975): 58, 77. See also Robert T. Bakker, "The Superiority of Dinosaurs," *Discovery: The Magazine of the Peabody Museum of Natural History* 3, no. 2 (1968): 11–22.

16. Joel Cracraft, "The Origin and Early Diversification of Birds," *Paleobiology* 12, no. 4 (1986): 383–399; Jacques Gauthier, "Saurischian Monophyly and the Origin of Birds," *Memoirs of the California Academy of Sciences* 8 (1986): 1–55; Jacques Gauthier and Kevin Padian, "Phylogenetic, Functional, and Aerodynamic Analyses of the Origin of Birds and Their Flight," in *The Beginnings of Birds: Proceedings of the International Archaeopteryx Conference* (Eichstätt, Germany: Freunde des Jura-Museums Eichstatt, 1984), 185–197; Kevin Padian, "Macroevolution and the Origin of Major Adaptations: Vertebrate Flight as a Paradigm for the Analysis of Patterns," in *Third North American Paleontological Convention: Proceedings,* ed. Bernard Mamet and M. J. Copeland (Toronto: Business and Economic Service, 1982), 2:387–392. For an overview of these developments, see Luis M. Chiappe, "The First 85 Million Years of Avian Evolution,"

Nature 378, no. 6555 (1995): 349–355; Philip J. Currie, ed., *Feathered Dragons: Studies on the Transition from Dinosaurs to Birds* (Bloomington: Indiana University Press, 2004); Jacques Gauthier and Lawrence F. Gall, eds., *New Perspectives on the Origin and Early Evolution of Birds* (New Haven, CT: Yale University Press, 2001); David E. Fastovsky and David B. Weishampel, *The Evolution and Extinction of the Dinosaurs* (Cambridge: Cambridge University Press, 1996); and Lawrence M. Witmer, "Perspectives on Avian Origins," in *Origins of the Higher Groups of Tetrapods: Controversy and Consensus,* ed. Linda Trueb and Hans-Peter Schultze (Ithaca, NY: Comstock, 1991), 427–466.

17. See, for example, Luis M. Chiappe, *Glorified Dinosaurs: The Origin and Early Evolution of Birds* (Hoboken, NJ: Wiley, 2007); Lowell Dingus and Timothy Rowe, *The Mistaken Extinction: Dinosaur Evolution and the Origin of Birds* (New York: Freeman, 1998); and Gareth Dyke and Gary W. Kaiser, eds., *Living Dinosaurs: The Evolutionary History of Modern Birds* (Hoboken, NJ: Wiley-Blackwell, 2011).

18. Lowell Dingus, *Next of Kin: Great Fossils at the American Museum of Natural History* (New York: Rizzoli, 1996), 13, 66–96.

19. "Museum's T. rex Roars Back in New Stance," *New York Times,* June 16, 2008, A13.

20. See Christopher Brochu, *Osteology of* Tyrannosaurus rex: *Insights from a Nearly Complete Skeleton,* Memoir of the Society of Vertebrate Paleontology 7 (Northbrook, IL: Society of Vertebrate Paleontology, 2003).

21. Dan Chinoy, "Walking among Dinosaurs," *China Daily,* Nov. 24, 2009.

22. Strictly speaking, birds such as *Archaeopteryx* are now widely regarded as dinosaurs, but that theory remained highly contested during the 1990s. See Qiang Ji and Shu'an Ji, "On the Discovery of the Earliest Fossil Bird in China (*Sinosauropteryx* gen. nov.) and the Origin of Birds," *Chinese Geology* 233 (1996): 30–33. See also Qiang Ji and Shu'an Ji, "*Protoarchaeopteryx,* a New Genus of Archaeopterygidae in China," *Chinese Geology* 238 (1997): 38–41; and Pei-ji Chen, Zhi-ming Dong, and Shuo-nan Zhen, "An Exceptionally Well-Preserved Theropod Dinosaur from the Yixian Formation of China," *Nature* 391, no. 6663 (1998): 147–152. Subsequent discoveries in China have even led some paleontologists to conclude that *Archaeopteryx* was a nonavian dinosaur, meaning that it did not even constitute a true bird. See Lawrence M. Witmer, "Palaeontology: An Icon Knocked from Its Perch," *Nature* 475, no. 7357 (2011): 458–459.

23. Quoted in John Pickrell, *Flying Dinosaurs: How Fearsome Reptiles Became Birds* (New York: Columbia University Press, 2014), 19.

24. Malcolm W. Browne, "Feathery Fossil Hints Dinosaur-Bird Link," *New York Times,* Oct. 19, 1996.

25. Mee-Mann Chang, introduction to *The Jehol Fossils: The Emergence of Feathered Dinosaurs, Beaked Birds and Flowering Plants,* ed. Mee-Mann Chang et al. (London: Academic Press, 2008), 14.

26. Xing Xu et al., "A Gigantic Feathered Dinosaur from the Lower Cretaceous of China," *Nature* 484, no. 7392 (2012): 92–95. On *Beipiaosaurus* and *Microraptor,* see Xing Xu, Zhi-lu Tang, and Xiao-lin Wang, "A Therizinosauroid Dinosaur with Integumentary Structures from China," *Nature* 399, no. 6734 (1999): 350–354; and Xing Xu, Zhonghe Zhou, and Xiaolin Wang, "The Smallest Known Non-avian Theropod Dinosaur," *Nature* 408, no. 6813 (2000): 705–708.

27. Dong Zhiming, *Dinosaurs in Asia* (Kunming, China: Yunnan Science and Technology, 2009), vi, 2.

28. The word *jehol* is a transliteration of the Chinese word 热河 in the outdated Wade-Giles romanization system, which is rendered as *rehe* in the more modern Pinyin system. *Jehol* or *rehe* literally means "hot river," in reference to the numerous hot springs that still exist in the vicinity of a mountain resort to which members of the Qing court often escaped from the summer heat of Beijing during the eighteenth and nineteenth centuries. The name Jehol was first applied to fossil-bearing strata in the coal-bearing regions of Liaoning by the American paleontologist Arthur W. Grabau and was subsequently brought into much wider circulation by the Chinese scientist Gu Zhiwei. See Arthur W. Grabau, "Cretaceous Mollusca

from North China," *Bulletin of the Geological Survey of China* 5 (1923): 183–198; Arthur W. Grabau, *Stratigraphy of China,* vol. 2, *Mesozoic* (Beijing: Geological Survey of China, 1928); and Gu Zhiwei, *Jurassic and Cretaceous of China* (Beijing: Science Press, 1962).

29. See Xiao-lin Wang and Zhong-he Zhou, "Mesozoic Pompeii," in Chang et al., *The Jehol Fossils,* 19–36.

30. Jakob Vinther et al., "The Colour of Fossil Feathers," *Biology Letters* 4, no. 5 (2008): 522–525. As recently as 2007, a philosopher of science predicted that we would never know the color of dinosaurs. See Derek Turner, *Making Prehistory: Historical Science and the Scientific Realism Debate* (Cambridge: Cambridge University Press, 2007); and Derek D. Turner, "A Second Look at the Colors of the Dinosaurs," *Studies in History and Philosophy of Science* 55 (Feb. 2016): 60–68.

31. Fucheng Zhang et al., "Fossilized Melanosomes and the Colour of Cretaceous Dinosaurs and Birds," *Nature* 463, no. 7284 (2010): 1075–1078.

32. Quanguo Li et al., "Plumage Color Patterns of an Extinct Dinosaur," *Science* 327 (2010): 1369–1372. For a recent review essay, see Jakob Vinther, "A Guide to the Field of Palaeo Colour," *BioEssays* 37, no. 6 (2015): 643–656. On the study of organic microfossils, see Derek E. G. Briggs and Roger E. Summons, "Ancient Biomolecules: Their Origins, Fossilization, and Role in Revealing the History of Life," *BioEssays* 36, no. 5 (2014): 482–490.

33. For an overview, see Amber Dance, "Prehistoric Animals, in Living Color," *Proceedings of the National Academy of Sciences of the United States of America* 113, no. 31 (2016): 8552–8556. Critics argue that instead of melanosomes, Vinther might actually be measuring the size and shape of intracellular bacteria. On the controversy, see Alison E. Moyer et al., "Melanosomes or Microbes: Testing an Alternative Hypothesis for the Origin of Microbodies in Fossil Feathers," *Scientific Reports* 4 (Mar. 5, 2014): 4233; Johan Lindgren et al., "Interpreting Melanin-Based Coloration through Deep Time: A Critical Review," *Proceedings of the Royal Society B* 282, no. 1813 (2015): 20150614; Mary H. Schweitzer, Johan Lindgren, and Alison E. Moyer, "Melanosomes and Ancient Coloration Re-examined: A Response to Vinther 2015," *BioEssays* 37, no. 11 (2015): 1174–1183.

34. Mark A. Norell, "Fossilized Feathers," *Science* 333, no. 6049 (2011): 1590–1591.

35. Julia Clarke, "Feathers before Flight," *Science* 340, no. 6133 (2013): 690–692; Richard O. Prum and Alan H. Brush, "The Evolutionary Origin and Diversification of Feathers," *Quarterly Review of Biology* 77 (2002): 261–295; Richard O. Prum and Alan H. Brush, "Which Came First: The Feather or the Bird?," *Scientific American* 288, no. 3 (2003): 84–93; Xu Xing and Guo Yu, "The Origin and Early Evolution of Feathers: Insights from Recent Paleontological and Neontological Data," *Vertebrata PalAsiatica* 47 (2009): 311–329.

36. Dongyu Hu et al., "A Bony-Crested Jurassic Dinosaur with Evidence of Iridescent Plumage Highlights Complexity in Early Paravian Evolution," *Nature Communications* 9, no. 1 (2018): 217; Li et al., "Plumage Color Patterns"; Marie-Claire Koschowitz, Christian Fischer, and Martin Sander, "Beyond the Rainbow," *Science* 346, no. 6208 (2014): 416–418; Darla K. Zelenitsky et al., "Feathered Non-avian Dinosaurs from North America Provide Insight into Wing Origins," *Science* 338, no. 6106 (2012): 510–514. See also John M. Grady et al., "Evidence for Mesothermy in Dinosaurs," *Science* 344, no. 6189 (2014): 1268–1272; Johan Lindgren et al., "Skin Pigmentation Provides Evidence of Convergent Melanism in Extinct Marine Reptiles," *Nature* 506, no. 7489 (2014): 484–488; and Jakob Vinther et al., "3D Camouflage in an Ornithischian Dinosaur," *Current Biology* 26, no. 18 (2016): 2456–2462.

37. On model organisms, see Angela N. H. Creager, *The Life of a Virus: Tobacco Mosaic Virus as an Experimental Model, 1930–1965* (Chicago: University of Chicago Press, 2002); Robert Kohler, *Lords of the Fly: Drosophila Genetics and the Experimental Life* (Chicago: University of Chicago Press, 1994); Tania Munz, *The Dancing Bees: Karl von Frisch and the Discovery of the Honeybee Language* (Chicago: University

of Chicago Press, 2016); and Karen Rader, *Making Mice: Standardizing Animals for American Biomedical Research, 1900–1955* (Princeton, NJ: Princeton University Press, 2004).

38. For a powerful articulation of this claim, see James Secord, "Knowledge in Transit," *Isis* 95, no. 4 (2004): 654–672. See also Aileen Fyfe and Bernard Lightman, *Science in the Marketplace: Nineteenth-Century Sites and Experiences* (Chicago: University of Chicago Press, 2007); and James Secord, *Victorian Sensation: The Extraordinary Publication, Reception, and Secret Authorship of Vestiges of the Natural History of Creation* (Chicago: University of Chicago Press, 2000).

39. On the way so-called boundary objects promote collaboration in science, see Susan Leigh Star and James R. Griesemer, "Institutional Ecology, 'Translations' and Boundary Objects: Amateurs and Professionals in Berkeley's Museum of Vertebrate Zoology, 1907–1939," *Social Studies of Science* 19 (1989): 387–420. For a different view, see Thomas Gieryn, *Cultural Boundaries of Science: Credibility on the Line* (Chicago: University of Chicago Press, 1999); and Thomas Gieryn, "Boundary-Work and the Demarcation of Science from Non-science," *American Sociological Review* 48, no. 6 (1983): 781–795. See also Lukas Rieppel, "Hoaxes, Humbugs, and Fraud: Distinguishing Truth from Untruth in Early America," *Journal of the Early Republic* 38, no. 3 (2018): 501–529.

40. The canonical example is George Basalla, "The Spread of Western Science," *Science* 156, no. 3775 (1967): 611–622. Compare to the "centers of calculation" in Bruno Latour, *Science in Action* (Cambridge, MA: Harvard University Press, 1987). For a critical response, see Warwick Anderson, "Remembering the Spread of Western Science," *Historical Records of Australian Science* 29, no. 2 (2018): 73–81; and Marwa Elshakry, "When Science Became Western: Historiographical Reflections," *Isis* 101, no. 1 (2010): 98–109.

41. On science, see Harold J. Cook, *Matters of Exchange: Commerce, Medicine, and Science in the Dutch Golden Age* (New Haven, CT: Yale University Press, 2007); Simon Schaffer et al., eds., *The Brokered World: Go-Betweens and Global Intelligence, 1770–1820* (Sagamore Beach, MA: Science History, 2009); Kapil Raj, *Relocating Modern Science: Circulation and the Construction of Knowledge in South Asia and Europe, 1650–1900* (Houndmills, UK: Palgrave Macmillan, 2007); and Lissa Roberts, "Situating Science in Global History: Local Exchanges and Networks of Circulation," *Itinerario* 33, no. 1 (2009): 9–30. On capitalism, see Sven Beckert, *Empire of Cotton: A Global History* (New York: Alfred A. Knopf, 2014). See also Andre Gunder Frank, *ReOrient: Global Economy in the Asian Age* (Berkeley: University of California Press, 1998); Kenneth Pomeranz, *The Great Divergence: China, Europe, and the Making of the Modern World Economy* (Princeton, NJ: Princeton University Press, 2000); Roy Bin Wong, *China Transformed: Historical Change and the Limits of European Experience* (Ithaca: Cornell University Press, 1997).

42. For a related argument, see Warwick Anderson, "Making Global Health History: The Postcolonial Worldliness of Biomedicine," *Social History of Medicine* 27, no. 2 (2014): 93–113; and Fa-ti Fan, "The Global Turn in the History of Science," *East Asian Science, Technology and Society* 6, no. 2 (2012): 249–258. See also Anna L. Tsing, *Friction: An Ethnography of Global Connection* (Princeton, NJ: Princeton University Press, 2011).

43. Ren Naiqiang, *Hua yang guo zhi jiaobu tuzhu*, trans. Brian Lander (Shanghai: Shanghai guji chubanshe, 1987), 166. On the medicinal properties of dragon bones, see Carla Nappi, *The Monkey and the Inkpot: Natural History and Its Transformations in Early Modern China* (Cambridge, MA: Harvard University Press, 2009), 50–68.

44. Dong Zhiming, for example, states, "As we know nowadays, these 'dragon bones' actually belonge [*sic*] to dinosaurs." Dong, *Dinosaurs in Asia*, 84. Elsewhere, Dong writes, "Mesozoic rocks containing dinosaur bones are exposed today in that area of Sichuan province, so it is highly probable that Chang Qu's description is the earliest recorded occurrence of dinosaur bones." Dong Zhiming, *Dinosaurs from China*, trans. Angela Milner (Beijing: China Ocean, 1988), 9. On dinosaur fossils in ancient China, see Li Chung-chun, "Records of Vertebrate Fossils in Old Chinese Classics," *Vertebrata PalAsiatica* 12 (1974): 174–180; Lida Xing et al., "The Folklore of Dinosaur Trackways in China: Impact on Paleontology," *Ichnos*

18, no. 4 (2011): 213–220; and Zhen Shou-nan, "Records on Fossil Vertebrates in Ancient Chinese Literature," *Vertebrata PalAsiatica* 4 (1961): 370–373.

45. Richard Owen, *Report on British Fossil Reptiles,* pt. 2, Report of the British Association for the Advancement of Science (London: printed by R. and J. E. Taylor, 1842), 102–103. Note that while Qu could not have invoked the taxonomic category Dinosauria in 350 CE, he could nonetheless have been referring to dinosaur bones as material objects, just as two words in different languages can refer to the same thing. That is, while "dinosaur bone" and "dragon bone" may have partially overlapping referents, they do not have the same sense. On the distinction, see Gottlob Frege, "Über Sinn und Bedeutung," *Zeitschrift für Philosophie und philosophische Kritik* 100, no. 1 (1892): 25–50.

46. Here I draw on Dipesh Chakrabarty, *Provincializing Europe: Postcolonial Thought and Historical Difference* (Princeton, NJ: Princeton University Press, 2000).

47. Dong Zhiming, *Dinosaurian Faunas of China* (Beijing: China Ocean, 1992), 9–29, esp. 9–11.

48. See Fa-ti Fan, "Circulating Material Objects: The International Controversies over Antiquities and Fossils in Twentieth-Century China," in *The Circulation of Knowledge between Britain, India, and China: The Early-Modern World to the Twentieth Century,* ed. Bernard V. Lightman, Gordon McOuat, and Larry Stewart (Leiden: Brill, 2013), 209–236. In contrast, the Chinese geological community at the time was primarily concerned with natural resource extraction. See Grace Yen Shen, *Unearthing the Nation: Modern Geology and Nationalism in Republican China* (Chicago: University of Chicago Press, 2014); and Shellen Xiao Wu, *Empires of Coal: Fueling China's Entry into the Modern World Order, 1860–1920* (Stanford, CA: Stanford University Press, 2015).

49. "Beijing xueshu tuanti fandui wairen caiqu guwu zhi xuanyan zuori yeyi fabiao" [The manifesto of the Association of Learned Societies against foreigners collecting ancient relics was announced yesterday], *Chen bao,* Mar. 10, 1927, quoted in Hsiao-pei Yen, "Constructing the Chinese: Paleoanthropology and Anthropology in the Chinese Frontier, 1920–1950" (PhD diss., Harvard University, 2012), 62.

50. Quoted in Wang Chen, ed., *Gaoshang Zhe de Muzhiming* [The epitaph of the nobles] (Beijing: Zhongguo wenlian chubanshe, 2005), 10. See also Yen, "Constructing the Chinese," 63.

51. Andrews to Howard E. Cole, Nov. 9, 1921, and Andrews to W. B. Walker, Apr. 19, 1924, box 7, folder 17, Central Asiatic Expedition Papers, MSS C446, AMNH.

52. Doc. 512-206-2, filed June 11, 1920, and Parker G. Tenney, "Report of Travel with the Central Asiatic Expedition," Sept. 12, 1930, Folder 2055-682-1 through 2055-682-3, Record Group 165: Military Intelligence Division Numeric Files, NARA. Sometimes Andrews even hid sensitive information in mundane correspondence using invisible ink. See entry 78, doc. 21012-93, Record Group 38: Office of Naval Intelligence Files, NARA. See also Roy Chapman Andrews, "Explorations in the Gobi Desert," *National Geographic* 63, no. 6 (1933): 653–716; and Vincent L. Morgan and Spencer G. Lucas, *Walter Granger, 1872–1941, Paleontologist,* Bulletin of the New Mexico Museum of Natural History and Science 19 (Albuquerque: New Mexico Museum of Natural History and Science, 2002), 29–31.

53. Wang Ping-Chih, "The Detained Fossils," *China Press,* Sept. 18, 1928, 4.

54. Andrews, *New Conquest of Central Asia,* 418; Granger to Ma Hen, Feb. 2, 1929, box 3, folder 19, Central Asiatic Expedition Papers, AMNH.

55. Osborn to Wu Chao Chu, May 9, 1929, box 4, folder 7, Central Asiatic Expedition Papers, AMNH.

56. Andrews to Osborn, July 18, 1929, dispatch 2235, State Department Decimal File, Record Group 59, NARA; "The Squabble over the Mongolian Fossils," *China Weekly Review,* May 4, 1929, 405; "China's Fossils," *Seattle Times,* June 29, 1929; "Expedition Called Off by Andrews: Peking Society Wins 'Face' at Expense of China and Word," *Peking Leader,* July 19, 1929; Henry Fairfield Osborn, "Interruption of Central Asiatic Exploration by the American Museum of Natural History," *Science* 70, no. 1813 (1929): 291–293.

57. See the correspondence between Nelson T. Johnson to Andrews in the autumn and winter of 1932, documents 031.11, Record Group 59: State Department Decimal File, NARA. See also Andrews to Kosaburo Shimamura, Dec. 19, 1932, RC Andrews Administrative Papers, box 7, folder 3, Archives of the AMNH; "Exploration in China: Statement by Antiquities Commission and Mr. Roy Chapman Andrews," *The North-China Herald and Supreme Court & Consular Gazette,* Sep 14, 1932.

58. David Norman, "Yang Zhongjian," in *Complete Dictionary of Scientific Biography* (New York: Scribner's, 2008), 25:383–385. On the early history of paleontology in China, see Xiaobo Yu, "Chinese Paleontology and the Reception of Darwinism in Early Twentieth Century," *Studies in History and Philosophy of Science Part C: Studies in History and Philosophy of Biological and Biomedical Sciences* 66 (Dec. 1, 2017): 46–54. On biology more broadly, see Laurence A. Schneider, *Biology and Revolution in Twentieth-Century China* (Lanham, MD: Rowman and Littlefield, 2003).

59. Osborn, "Interruption of Central Asiatic Exploration by the American Museum of Natural History," *Science* 70, no. 1813 (1929): 293.

60. "Manifesto of the Association of Learned Societies," quoted in Hsiao-pei Yen, "Constructing the Chinese," 62. A similar sentiment was expressed in Wang Ping-Chih, "The Detained Fossils," *The China Press,* Sept. 18, 1928, 4.

61. Zhongjian Yang, *Yang Zhongjian huiyi lu* (Beijing: Dizhi chubanshe, 1983), 70, quoted in Yen, "Constructing the Chinese," 68.

62. Yang's student Dong Zhiming divides the history of Chinese paleontology into four distinct periods. After a time of Western exploration, Yang worked to build up a Chinese paleontological community. Then, between 1949 and 1979, a number of Yang's former students, including Chang Meeman and Dong, took on that leadership role. Finally, a period of international collaborative research followed, which led to the discovery of feathered dinosaurs in the late twentieth century. See Dong, *Dinosaurs in Asia,* 86–88.

63. Zhiming Dong, "The Field Activities of the Sino-Canadian Dinosaur Project in China, 1987–1990," *Canadian Journal of Earth Sciences* 30, no. 10 (1993): 1997–2001.

64. Thomas Piketty, *Capital in the Twenty-First Century*, trans. Arthur Goldhammer (Cambridge, MA: Belknap Press of Harvard University Press, 2014). Any discussion of accumulation alongside circulation in the political economy of global capitalism must also cite Rosa Luxemburg, *Die Akkumulation des Kapitals: ein Beitrag zur ökonomischen Erklärung des Imperialismus* (Berlin: Frankes Verlag, 1922) and Karl Marx, *Das Kapital: Kritik der politischen Oekonomie* (Hamburg: O. Meissner, 1872).

65. See Steve Fiffer, *Tyrannosaurus Sue: The Extraordinary Saga of the Largest, Most Fought Over* T. rex *Ever Found* (New York: Freeman, 2000); and Peter Larson and Kristin Donnan, *Rex Appeal: The Amazing Story of Sue, the Dinosaur That Changed Science, the Law, and My Life* (Montpelier, VT: Invisible Cities, 2002).

66. Jim Ritter, "Experts Fear Fossil Free for All," *Chicago Sun-Times,* July 21, 1996, 40.

67. "Guide to Manuscript Preparation for the *Journal of Vertebrate Paleontology,*" Society of Vertebrate Paleontology, last revised Feb. 2011, http://vertpaleo.org/Publications/Journal-of-Vertebrate-Paleontology/Information-for-Authors/Guide-to-manuscript-preparation.aspx.

68. On Prokopi, see Paige Williams, *The Dinosaur Artist: Obsession, Betrayal, and the Quest for Earth's Ultimate Trophy* (New York: Hachette Books, 2018). On Cage, see, for example, Katie Rogers, "Nicolas Cage Agrees to Return Stolen Dinosaur Skull to Mongolia," *New York Times,* Dec. 22, 2015.

69. See, for example, the Society of Vertebrate Paleontology's "Member Bylaw on Ethics Statement," which states that the "barter, sale or purchase of scientifically significant vertebrate fossils is not condoned, unless it brings them into, or keeps them within, a public trust. Any other trade or commerce in scientifically significant vertebrate fossils is inconsistent with the foregoing, in that it deprives both the public and professionals of important specimens, which are part of our natural heritage." "Member Bylaw on Ethics Statement," Society of Vertebrate Paleontology, accessed Jan. 28, 2017, http://vertpaleo.org/Membership

/Member-Ethics/Member-Bylaw-on-Ethics-Statement.aspx. In contrast, commercial collectors argue that all fossil hunting ought to be encouraged because specimens that go uncollected will eventually be lost to erosion. See, for example, Michael Triebold, Kirby Siber, and Paul R. Janke, "Commercial Collectors," in *Fossils and the Future: Paleontology in the Twenty-First Century,* ed. Richard H. Lane et al. (Frankfurt am Main: Senckenbergische Naturforschende Gesellschaft, 2000), 53–64; Neal Larson, Walter Stein, Michael Triebold, and George Winters, "What Commercial Fossil Dealers Contribute to the Science of Paleontology," *The Journal of Paleontological Sciences*, no. 7 (2014): 1–13; and Neal Larson, "Fossils for Sale: Is it Good for Science?" *The Geological Curator* 7, no. 6 (2001): 219–222.

70. See, for example, Kevin Padian, "Feathers, Fakes, and Fossil Dealers: How the Commercial Sale of Fossils Erodes Science and Education," *Palaeontologia Electronica,* no. 2 (2000), available online at http://palaeo- electronica.org.

71. The commercial (and often illegal) exchange of dinosaurs is surprisingly common on eBay. A search that was performed on January 29, 2017, for example, returned a mounted *Triceratops* skull from the Hell Creek Formation in South Dakota being sold for $830,000 plus $5,000 in shipping and handling costs. A search on May 30, 2014, returned a mounted *Tenontosaurus* specimen measuring about ten feet in length being sold for only $110,000 by the dealer "triassica." See Lukas Rieppel, "Prospecting for Dinosaurs on the Mining Frontier: The Value of Information in America's Gilded Age," *Social Studies of Science* 45, no. 2 (2015): 161–186.

72. See http://www.ebay.com/gds/Fake-Fossils-caveat-emptor-knowledge-is-power-/10000000006892670 /g.html, accessed on Jan. 29, 2017.

73. "Fake Chinese Fossils," Paleo Direct, accessed Jan. 29, 2017, http://www.paleodirect.com/fakechinesefossils1.htm.

74. Lucas Laursen, "Reunion of Fossil Halves Splits Scientists," *Nature News,* May 19, 2009. See also Jens L. Franzen et al., "Complete Primate Skeleton from the Middle Eocene of Messel in Germany: Morphology and Paleobiology," *PLOS ONE* 4, no. 5 (2009): e5723.

75. Octavio Mateus, Marvin Overbeeke, and F. Rita, "Dinosaur Frauds, Hoaxes and 'Frankensteins': How to Distinguish Fake and Genuine Vertebrate Fossils," *Journal of Paleontological Techniques,* no. 2 (2008): 1–5.

76. Timothy Rowe et al., "Forensic Palaeontology: The *Archaeoraptor* Forgery," *Nature* 410, no. 6828 (2001): 539–540. On the *Archaeoraptor* saga, see Lewis M. Simons, "*Archaeoraptor* Fossil Trail," *National Geographic,* vol. 198, no. 4 (2000): 128–132. For the original article, see Christopher P. Sloan, "Feathers for *T. rex*?," *National Geographic* 196, no. 5 (1999): 98–107. See also Rex Dalton, "Feathers Fly over Chinese Fossil Bird's Legality and Authenticity," *Nature* 403, no. 6771 (2000): 689–690; Rex Dalton, "Fake Bird Fossil Highlights the Problem of Illegal Trading," *Nature* 404, no. 6779 (2000): 696–696; and Xu Xing, "Response to 'Feathers for *T. rex*?,'" *National Geographic* 197, no. 3 (2000): Forum Section (pages unnumbered). See also Pickrell, *Flying Dinosaurs,* 66–82.

77. On the history of the "Chinese copycat," see Eugenia Lean, "Making the Chinese Copycat: Trademarks and Recipes in Early Twentieth-Century Global Science and Capitalism," *Osiris* 33, no. 1 (2018): 271–293.

78. Quoted in Pickrell, *Flying Dinosaurs,* 71.

79. Quoted in Richard Stone, "Altering the Past: China's Faked Fossils Problem," *Science* 330, no. 6012 (2010): 1740. The paleontologist Alan Feduccia even argues that concerns over fake fossils undermines recent theories about avian evolution. See Alan Feduccia, *Riddle of the Feathered Dragons: Hidden Birds of China* (New Haven, CT: Yale University Press, 2012).

80. Ivan Franceschini and Nicholas Loubere, eds., *Gilded Age,* Made in China Yearbook (Acton: ANU Press, 2017).

81. On the gene's eye view of evolution, see Richard Dawkins, *The Selfish Gene* (New York: Oxford University Press, 1976); George C. Williams, *Adaptation and Natural Selection: A Critique of Some*

Current Evolutionary Thought (Princeton, NJ: Princeton University Press, 1966). For a more nuanced perspective, see David Haig, "The Strategic Gene," *Biology and Philosophy* 27, no. 4 (2012): 461–479; David Haig, "The Social Gene," in *Behavioral Ecology,* ed. John R. Krebs and Nicholas B. Davies (Hoboken, NJ: Wiley-Blackwell, 1997), 284–304.

82. On biological altruism, see William D. Hamilton, "The Genetical Evolution of Social Behaviour I and II," *Journal of Theoretical Biology* 7, no. 1 (1964): 1–16 and 17–52. See also Oren Harman, *The Price of Altruism: George Price and the Search for the Origins of Kindness* (New York: W. W. Norton, 2010). On the history of neoliberal economic theory, see Angus Burgin, *The Great Persuasion: Reinventing Free Markets since the Depression* (Cambridge, MA: Harvard University Press, 2012); Philip Mirowski and Dieter Plehwe, eds., *The Road from Mont Pèlerin: The Making of the Neoliberal Thought Collective* (Cambridge, MA: Harvard University Press, 2009).

Acknowledgments

Much like the dinosaur, this book is a chimera: a complex assemblage made up of countless distinct elements, each with its own history and developmental trajectory. Often, the origin of each part can be traced back to a conversation with friends, colleagues, and interlocutors who lent me their ears, eyes, and time. I thank them all for their generosity and their insight.

This book has taken almost exactly a decade to fully take form, which makes the experience of sitting down to compose these acknowledgments especially sweet. I want to begin by thanking an extraordinary group of mentors, many of whom have since become friends. Janet Browne shepherded the project, as well as its author, with expert guidance and care. Mario Biagioli, Peter Galison, and Harriet Ritvo also deserve special thanks for teaching me how to write about the history of science. It was about halfway through graduate school that I realized this book required making a connection to the history of capitalism, and I thank Sven Beckert for offering inspiration and facilitating that conversation. Finally, Naomi Pierce has been unfailingly generous, welcoming me into her lab, giving me access to a thriving community of biologists, and sending me off on my own natural history expedition.

In the years since, I have been fortunate to find a new intellectual home, first in Chicago and, more recently, in Rhode Island. At Northwestern, Ken Alder was the coolest postdoctoral mentor that I could imagine. I especially thank him, Teri Chettiar, Daniel Immerwahr, Tania Munz, and Helen Tilley for making my time in Chicago so productive and stimulating. Following my move to Providence, Brown has proved to be an ideal community. I especially thank Hal Cook, Bathsheba Demuth, Anne Fausto-Sterling, Dan Hirschman, Nancy Jacobs, Jennifer Johnson, Jennifer Lambe, Brian Lander, Steve Lubar, Iris Montero, Tara Nummedal, Adi Ophir, Emily Owens, Ethan Pollock, Joan Richards, Seth Rockman, Daniel Rodriguez, Neil Safier, Naoko Shibusawa, Deborah Weinstein, and Sandy Zipp for their intellectual comradery. As department chairs, Cynthia Brokaw and Robert Self have been unfailingly supportive. Finally, Brown students are really the best! I especially thank Eliza Cohen, who spent part of a summer doing research for this book in New York, and Kate Duffy, who provided archival material on the mysterious "Bumpus Affair."

I am indebted to a large group of scholars who have engaged with my ideas and helped hone my arguments. Jonathan Levy, Lynn Nyhart, Adi Ophir, Joan Richards, Robert Self, Naoko Shibusawa, and Sandy Zipp all read an earlier version of this manuscript in its entirety, as did three anonymous reviewers for the press. I wrote a first draft of this book while on sabbatical leave in Cambridge, Massachusetts, where I benefited from sustained engagement with two extraordinary intellectual communities. At the Charles Warren Center for American History, I want to thank Nicolas Barreyre, Sven Beckert, Gabrielle Clark, Lily Geismer, Martin Giraudeau, Paul Kershaw, John Larson, Stefan Link, Noam Maggor, Rebecca Marchiel, Abby Spinak, Zsuzsanna Vargha, and Michael Zakim. At the American Academy of Arts and Sciences, I greatly benefited from feedback by Les Beldo, Lawrence Buell, Merve Emre, Rachel Guberman, Katherine Marino, Emily Owens, Joy Rankin, and Rachel Wise. While writing this book, I have also been fortunate to visit the Max Planck Institute for the History of Science on two occasions, and I thank Lorraine Daston and Dagmar Schäfer for making that possible. In the late stages of writing, I benefited enormously from a semester at Brown's Cogut Institute for the Humanities, especially conversations with Amanda Anderson and Tapati Guha Thakurta. I cannot possibly remember everyone else who has given helpful advice, comments, and criticisms over the years, but I want to offer particular thanks to Tal Arbel, Jeremy Blatter, Henry Cowles, Alex Csiszar, James Delbourgo, Will Deringer, Stephanie Dick, Daniela Helbig, Kuang-chi Hung, Sheila Jasanoff, Sally Gregory Kohlstedt, Eugenia Lean, Arnon Levy, Bernie Lightman, Laura Martin, Gregg

Mitman, Joanna Radin, Christian Reiss, Ruth Rogaski, Alistair Sponsel, Laura Stark, Hallam Stevens, Anke te Heesen, Wenfei Tong, Lee Vinsel, Hsiao-pei Yen, Xiaobo Yu, and Nasser Zakariya. Finally, it has been a pleasure to watch the history and philosophy of paleontology become a lively topic of scholarly research, and I have learned a great deal from Paul Brinkman, Adrian Currie, Chris Manias, Ilja Nieuwland, Lydia Pyne, David Sepkoski, Derek Turner, and Caitlin Wylie. Before I forget, a big shout-out, as always, to Sounding Board!

Various pieces of this book have appeared in print elsewhere, and I acknowledge their publishers for allowing them to reappear here. Portions of Chapter 1 were first published in "Prospecting for Dinosaurs on the Mining Frontier: The Value of Information in America's Gilded Age," *Social Studies of Science* 45, no. 2 (2015): 161–186, and are reprinted with permission of Sage Publications Ltd. Chapter 5 builds on ideas first presented in "Organizing the Marketplace," *Osiris* 33 (2018): 232–252, and Chapter 6 builds on "Bringing Dinosaurs Back to Life: Exhibiting Prehistory at the American Museum of Natural History," *Isis* 103, no. 3 (September 2012): 460–490.

Libraries and archives supply the lifeblood of historians, and I would be remiss if I did not acknowledge the kind assistance of Ronnie Broadfoot, Dana Fisher, and Mary Sears at Harvard's Museum of Comparative Zoology; Dan Brinkman and Barbara Narendra at the Peabody Museum of Natural History; Susan Bell, Barbara Mathe, Gregory Raml, and Mai Reitmeyer at the American Museum of Natural History; Armand Esai and Christine Giannoni at the Field Museum of Natural History; Amy Henrici, Elizabeth Hill, Xianghua Sun, and the late Bernadette Callery at the Carnegie Museum of Natural History; John R. Waggener at the American Heritage Center in Laramie, Wyoming; Clare Flemming and Robert McCracken Peck at the Philadelphia Academy of Natural Sciences; and Anita Hermannstädter and Hannelore Landsberg at the Berlin Natural History Museum.

This book was acquired for Harvard University Press by Brian Distelberg, and it was expertly stewarded through the review process by Janice Audet. It is a rare thing indeed for a first-time author to work with an editor who puts in the time to give incisive and detailed comments on structure, organization, and style as Janice has done not once but twice. I can only hope the final product has been worth the effort!

Finally, I want to thank my friends and my family. I was introduced to the science of paleontology and its history by my father, whose intellectual journey often seems to run parallel with my own. I thank him, and my mother, for making me who I am. Thanks to my big brother, Mike, for all the late-night philosophizing. But most of all, thanks to Christine for reminding me that a whole world exists beyond books. I dedicate this one to her, with all of my love.

Illustration Credits

Figure 1.1. Document VPAR.002790, Yale Peabody Museum of Natural History.

Figure 1.2. Photograph by Lukas Rieppel.

Figure 1.3. E. D. Cope, "The Fossil Reptiles of New Jersey," *American Naturalist* 3, no. 2 (1869): 84–91, plate 2. Courtesy of the Brown University Library.

Figure 1.4. Othniel C. Marsh, *Dinosaurs of North America* (Washington, DC: Government Printing Office, 1896), plate 17. Courtesy of the Brown University Library.

Figure 2.1. *Los Angeles Herald,* Sunday supplement, Apr. 2, 1905, 37. Courtesy of the US Library of Congress.

Figure 2.2. Lithograph from the *Thirteenth Annual Report of the Board of Commissioners of the Central Park, for the Year Ending December 31, 1869* (New York: Evening Post Steam Presses, 1870), Y-Bind Central, New-York Historical Society, 96514d.

Figure 2.3. Image 17837, American Museum of Natural History Library.

Figure 2.4. Image 38715, American Museum of Natural History Library.

Figure 2.5. From an original watercolor by Benjamin Waterhouse Hawkins. Courtesy of the Academy of Natural Sciences Philadelphia, Archives Collection 803.

Figure 3.1. Image PH/3/1/807, Photography Collection, Archives of the London Natural History Museum, © The Trustees of the Natural History Museum, London.

Figure 3.2. From the Samuel H. Knight Collection, American Heritage Center, University of Wyoming.

Figure 3.3. From the Samuel H. Knight Collection, American Heritage Center, University of Wyoming.

Figure 3.4. *New York Journal,* Dec. 11, 1898, Wisconsin Historical Society, document WHS-86522.

Figure 3.5. © Carnegie Institute, Carnegie Museum of Natural History.

Figure 3.6. © Carnegie Institute, Carnegie Museum of Natural History.

Figure 3.7. John Bell Hatcher, "Diplodocus (Marsh): Its Osteology, Taxonomy, and Probable Habits, with a Restoration of the Skeleton," in *Memoirs of the Carnegie Museum of Natural History,* vol. 1, *1901–1904*, ed. W. J. Holland and J. B. Hatcher (Pittsburgh: Carnegie Institute, 1904), plate 13. Courtesy of the Ernst Mayr Library, Museum of Comparative Zoology, Harvard University.

Figure 3.8. © Carnegie Institute, Carnegie Museum of Natural History.

Figure 3.9. *Morning Leader,* June 3, 1905, image 046938_H, Newspaper Clippings, vol. 2, Archives of the London Natural History Museum, © The Trustees of the Natural History Museum, London.

Figure 3.10. From Wilhelm Bölsche, *Tiere der Urwelt,* ser. 1a (Wandsbek, Germany: Kakao Compagnie Reichardt, 1910). Courtesy of the Andrew Carnegie Birthplace Museum, Dunfermline, Scotland.

Figure 3.11. From the *Augsburger Kreisblatt,* Jan. 1938, Signiatur II, Tendaguru Expedition, 10.3 Newspaper Clippings, Historische Sammlung, Museum für Naturkunde Berlin.

Figure 4.1. Document VPAR.000554, Yale Peabody Museum of Natural History.

Figure 4.2. Image no. 17811, American Museum of Natural History Library.

Figure 4.3. Photograph by Lukas Rieppel.

Figure 4.4. Photograph by Lukas Rieppel.

Figure 4.5. Table 1, "Cost of Principal Mounted Specimens," in the 1910 Annual Report of the Acting Curator, *Annual Reports of the Department of Vertebrate Paleontology,* vol. 4, *1904–1912,* Archives of the Department of Vertebrate Paleontology, American Museum of Natural History.

Figure 5.1. Image no. 35575, American Museum of Natural History Library.

Figure 5.2. Oil painting by Charles R. Knight commissioned by J. P. Morgan and executed in consultation with Edward Drinker Cope, 1897. Courtesy of the American Museum of Natural History.

Figure 5.3. Mural painting by Charles R. Knight at the Field Museum of Natural History in Chicago, 1931.

Figure 5.4. Mural painting by Charles R. Knight at the American Museum of Natural History in New York, 1916.

Figure 5.5. Image no. 35517, American Museum of Natural History Library.

Figure 6.1. Poster by Kaufmann & Fabry. Image no. COP_01-0321_10210_001, Century of Progress records, series 1: General Correspondence, Special Collections and University Archives, University of Illinois at Chicago Library.

Figure 6.2. *Strand,* Apr. 1912, 362. Document EP85.St810, Houghton Library, Harvard University.

Figure 6.3. *Science and Invention,* May 1925. Document SF-2314, Houghton Library, Harvard University.

Figure 6.4. *Science and Invention,* May 1925. Document SF-2314, Houghton Library, Harvard University.

Figure 6.5. Image no. 17506, American Museum of Natural History Library.

Figure 6.6. Image no. 45615, American Museum of Natural History Library.

Figure 6.7. Image no. 311978, American Museum of Natural History Library.

Figure 6.8. © Carnegie Institute, Carnegie Museum of Natural History.

Figure 6.9. Image no. 311977, American Museum of Natural History Library.

Figure 6.10. Image no. 326557, American Museum of Natural History Library.

Figure C.1. Image no. 311977, American Museum of Natural History Library.

Figure C.2. Courtesy of the Field Museum of Natural History in Chicago.

Figure C.3. Courtesy of Brian Choo.

Figure C.4. Photograph by Lukas Rieppel.

Figure C.5. Photograph by Lukas Rieppel.

Index

Note: The letter *f* following a page number denotes a figure.